Comprehensive Practical Organic Chemistry

PREPARATION AND QUANTITATIVE ANALYSIS

Comprehensive Practical Organic Chemistry

PREPARATION AND QUANTITATIVE ANALYSIS

V K Ahluwalia

Dr B R Ambedkar Centre for Biomedical Research
University of Delhi

Renu Aggarwal

Gargi College
University of Delhi

Universities Press

COMPREHENSIVE PRACTICAL ORGANIC CHEMISTRY – PREPARATION AND QUANTITATIVE ANALYSIS

UNIVERSITIES PRESS (INDIA) PRIVATE LIMITED

Registered Office
3-6-747/1/A & 3-6-754/1, Himayatnagar, Hyderabad 500 029, Telangana, India
info@universitiespress.com; www.universitiespress.com

Distributed by
Orient Blackswan Private limited

Registered Office
3-6-752 Himayatnagar, Hyderabad 500 029, Telangana, India

Other Offices
Bengaluru, Chennai, Guwahati, Hyderabad,
Kolkata, Mumbai, New Delhi, Noida, Patna

502322

First Published in India by
Universities Press (India) Private Limited 2000
Reprinted 2004, 2006, 2007, 2008, 2009, 2010, 2013, 2016, 2018, 2026

ISBN: 978-81-7371-475-7

Printed at
Yash Printographics, Greater Noida 201 310

Published by
Universities Press (India) Private Limited
3-6-747/1/A & 3-6-754/1 Himayatnagar, Hyderabad 500 029, Telangana, India

Preface

The publication of this book is undertaken with a view to make available the methods of preparation of organic compounds which are normally described in the syllabi of undergraduate and postgraduate classes. A special feature of this book is that the preparations have been described on the basis of the name reactions arranged in alphabetical order. Besides the name reactions, some special procedures for acetylation, methylation, esterification, halogenation, nitration, oxidation, reduction, sulphonation and also some special types of compounds like dyes and drugs have also been described. The mechanism of each of these reactions is given.

The book also describes quantitative analysis and estimations of most of the compounds and groups as prescribed in the syllabi of various undergraduate and postgraduate courses of Indian Universities. Finally the different techniques of chromatography are also described.

It is hoped that the book will serve as a model for undergraduate and postgraduate courses in chemistry.

The authors express their thanks to Prof. G. B. V. Subramanian, Prof. M. R. Parthasarthy, Prof. R. N. Khanna, Prof. J. M. Khurana, Prof. Ramesh Chandra, Pooja Bhagat, Suman Dudeja, Bindu Goel, Madhu Bala and Aruna Dahiya for the help rendered during the preparation of the manuscript.

V K Ahluwalia
Renu Aggarwal

Contents

Section A: Preparations

1. ACETYLATION

Phenols, amines and quinones can be conveniently acetylated. The acetylating reagent used depends on the nature of the compound to be acetylated. The following acetylating agents are normally used:

(i) Acetic anhydride-sulphuric acid (catalytic amount) (see preparation of aspirin, hydroquinone diacetate.)
(ii) Acetic anhydride-sodium hydroxide (aqueous solution) (see preparation of β-naphthyl acetate, *p*-acetylaminophenol.)
(iii) Acetic anhydride-sodium acetate (see preparation of glucose α-acetate, mannitol hexa acetate.)
(iv) Acetic anhydride-zinc chloride (see preparation of glucose β-acetate.)
(v) Acetic anhydride-acetic acid-zinc dust (see preparation of acetanilide.)
(vi) Acetic anhydride-conc. sulphuric acid (excess). This is known as Thieles acetylation of quinones (see preparation of hydroxy-hydroquinone triacetate.)
(vi) Acetic anhydride-zinc dust (reductive acetylation.)
(viii) Acetyl chloride-pyridine (see preparation of aspirin.)

The acetylation of alcohols, e.g., *n*-butanol to *n*-butylacetate is discussed under esters.

Mechanism:

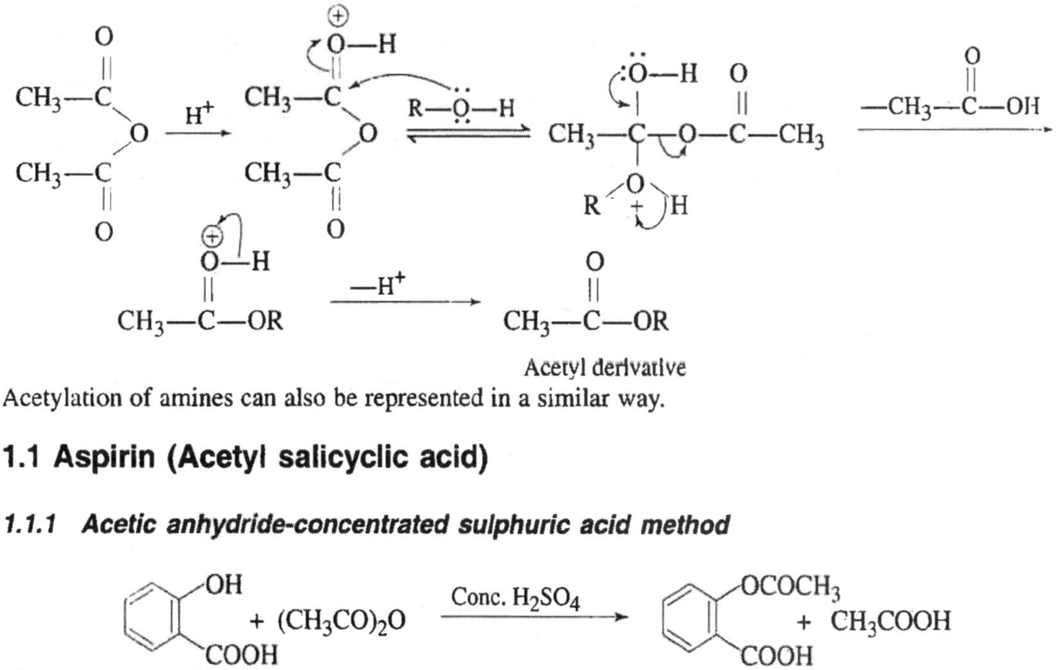

Acetyl derivative

Acetylation of amines can also be represented in a similar way.

1.1 Aspirin (Acetyl salicyclic acid)

1.1.1 Acetic anhydride-concentrated sulphuric acid method

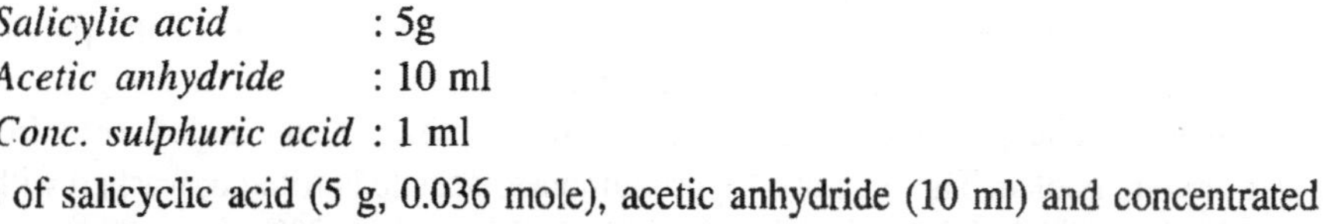

Salicylic acid   Aspirin

Salicylic acid : 5g
Acetic anhydride : 10 ml
Conc. sulphuric acid : 1 ml

A mixture of salicyclic acid (5 g, 0.036 mole), acetic anhydride (10 ml) and concentrated sulphuric acid (1 ml) is shaken thoroughly in a conical flask and heated on a water bath at 50–60°C for 10–15 min with occasional stirring (with a glass rod). The reaction mixture is cooled, diluted with water (75 ml) and the separated product filtered. It is crystallised from 50 per cent alcohol. The yield is 5 g (76.7%), m.p. 130–135°C (the m.p. varies due to partial decomposition).

1.1.2 Acetyl chloride-pyridine method

Alternatively a mixture of salicylic acid (5 g, 0.036 mole) and pyridine (4 ml) is cooled in an ice bath and acetyl chloride (3.5 ml) is added dropwise with continuous stirring. The reaction mixture is worked up as in the above procedure.

1.2 Hydroquinone Diacetate

Method 1 Acetic anhydride-concentrated sulphuric acid method

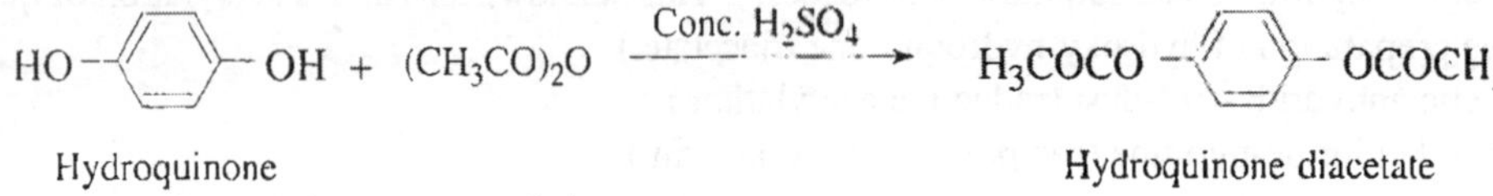

Hydroquinone : 5.6 g
Acetic anhydride : 10 ml
Conc. sulphuric acid : 1-2 drops

To a mixture of hydroquinone (5.6 g, 0.05 mole) and acetic anhydride (10 ml) in a conical flask is added concentrated sulphuric acid (1-2 drops). The reaction mixture is shaken gently (exothermic reaction). The mixture warms up and the hydroquinone dissolves. The reaction mixture is kept for 5 minutes and the clear solution poured on to crushed ice (40–50 g). The separated diacetate is filtered, washed with water and crystallised from 50 per cent alcohol. The yield is 9.2 g (94.8%); m.p. is 122°C.

Method 2 Acetic anhydride-zinc dust method

This method is used for the reductive acetylation of quinones. Thus *p*-benzoquinone on heating with acetic anhydride and zinc dust yields hydroquinone diacetate.

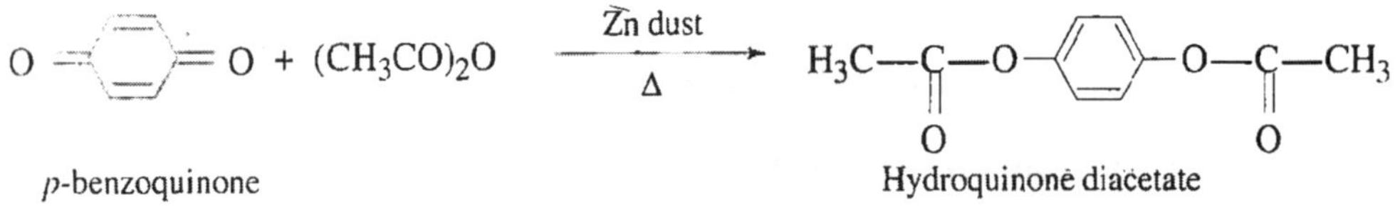

Hydroquinone : 1 g
Zinc dust : 1 g
Acetic anhydride : 5 ml
Anhydrous sodium acetate : 2 g

A mixture of hydroquinone (1 g, 0.009 mole), zinc dust (1 g), acetic anhydride (5 ml) and anhydrous sodium acetate (2 g) (preparation of anhydride sodium acetate is given on page 12) is gently warmed until the colour of the quinone has disappeared. The mixture is boiled for 2 minutes, glacial acetic acid (4 ml) added, and the mixture boiled again to dissolve the product and part of the precipitated zinc acetate. The hot solution is decanted and the residual zinc acetate is washed with hot glacial acetic acid (5 ml) by decantation. The combined solution is heated to boiling, water added to hydrolyse acetic anhydride and mixture cooled. The separated hydroquinone diacetate is filtered, washed with water and crystallised from dilute alcohol. The yield is 1.2 g; (67.2%); m.p. is 122°C.

1.3 β-Naphthyl Acetate

It is obtained by acetylation of β-naphthol with acetic anhydride in presence of aqueous sodium hydroxide.

To a cooled solution (ice bath, 0–5°C) of β-naphthol (8 g, 0.055 mole) in sodium hydroxide solution (10%, 40 ml) is added crushed ice (about 30 g) and then acetic anhydride (9.2 ml). The reaction mixture is shaken vigorously for 30 minutes. The separated product is filtered, washed with water and crystallised from dilute alcohol. The yield is 10 g; (96.8%); m.p. is 71°C.

1.4 *p*-Acetylaminophenol

It is obtained by acetylation of *p*-aminophenol with acetic anhydride in aqueous solution. Out of the two groups present, viz. hydroxyl and amino, only the amino group is acetylated in aqueous medium. However, the hydroxyl group can be acetylated in presence of aqueous alkali (see preparation of β-naphthyl acetate).

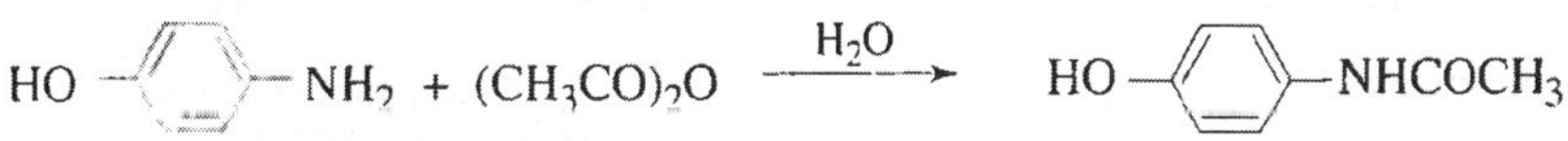

p-Aminophenol → *p*-Acetylaminophenol

p-Aminophenol : 4.4 g

Acetic anhydride : 4.8 ml

Water : 12 ml

To a suspension of *p*-aminophenol (4.4 g, 0.04 mole) in water (12 ml) in a conical flask is added acetic anhydride (4.8 ml). The mixture is stirred vigorously and warmed (water bath). The solid dissolves. After 10 minutes the solution is cooled and the separated *p*-acetylaminophenol is filtered and washed with cold water. It is crystallised from hot water (about 30 ml). The yield is 5 g (82.1%); mp. is 169°C.

If the product obtained is not pure (as indicated by its m.p.) it is purified by dissolving in dilute alkali in cold. The required product is reprecipitated by careful neutralization (caution).

1.5 α- and β-Glucose Pentaacetate

Glucose can be represented by the ring structures I and II. These differ only with regard to the disposition of the H and OH groups about the carbon atoms marked with an astrick. If the ring be regarded as being in the plane of the paper, α-glucose may have this hydrogen atom above and the OH group below the plane of the paper, while β-glucose will have this hydrogen atom below and the OH group above the plane. Both the cyclic forms are in equilibrium with a minute amount of the open chain molecule.

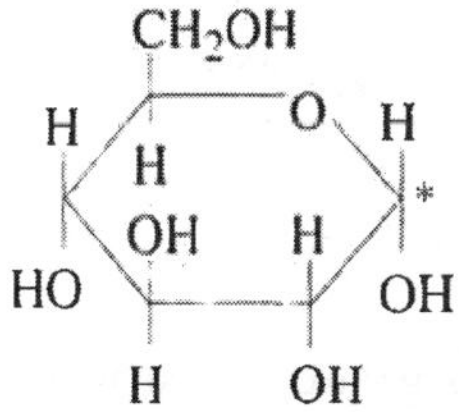

(I) α-*D*-Glucose

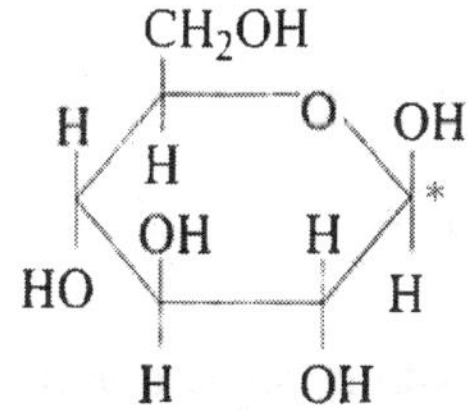

(II) β-*D*-Glucose

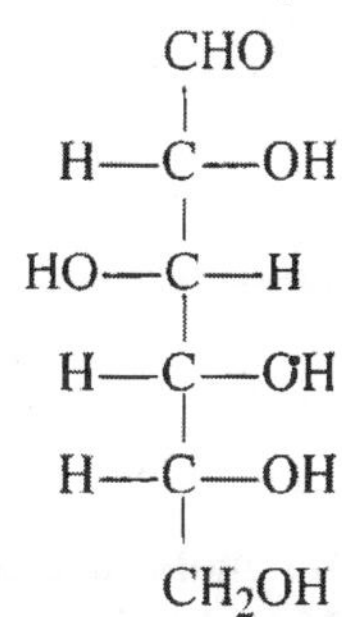

(III) *D*-Glucose
(Open chain aldehyde)

In the acetylation of *D*-glucose, the formation of the α- and β-pentaacetate depends on the nature of the catalyst used. The catalyst not only increases the speed of the reaction but also influences its stereochemical course. With zinc chloride as the catalyst, the α-pentaacetate is the main product, whereas with sodium acetate, β-anomer is obtained as the major product.

1.5.1 α-Glucose pentaacetate (1,2,3,4,6-penta-O-acetyl-α-D-glucopyranose)

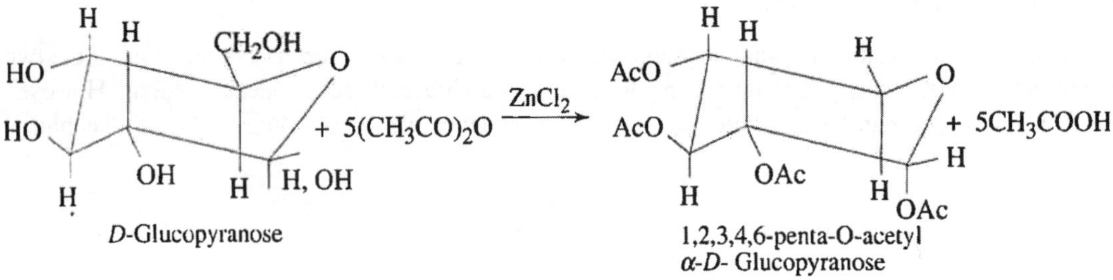

D-Glucose : 1 g
Acetic anhydride : 5 ml
Zinc chloride (anhyd) : 0.25 g

Powdered anhydrous zinc chloride (0.25 g) is heated in a R.B. flask with acetic anhydirde (5 ml) on a water bath (anhydrous condition). The mixture is refluxed (5–10 minutes) in a boiling water bath. Glucose (1 g, 0.0056 mole) is added to the reaction mixture and the contents are heated for one more minute on a water bath. The reaction mixture is poured into ice water (75 ml) while stirring. The mixture is stirred till the oily acetate solidifies. It is filtered, washed with cold water and crystallised from methanol. The yield is 1.5 g; (69.4%) m.p. is 112–114°C.

1.5.2 β-Glucose pentaacetate (1,2,3,4,6-penta-O-acetyl-β-D-glucopyranose)

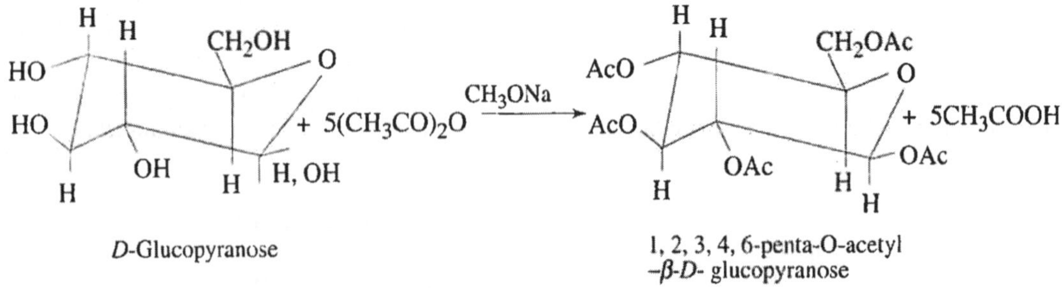

D-glucose : 2 g
Acetic anhydride : 10 ml
Sodium acetate (anhyd) : 1.5 g

Dry glucose (2 g, 0.011 mole) and anhydrous sodium acetate (1.5 g) are mixed with a pestle and transferred to a R.B. flask. Acetic anhydride (10 ml) is added and the mixture heated (water bath) for

2.5 hours under anhydrous conditions. The mixture is shaken occasionally during heating and finally poured with stirring into ice-water (200 ml). The separated product is filtered, washed with cold water and crystallised from alcohol or hot water. The yield is 2.8 g; (65%); m.p. is 131°C.

1.6 D-Mannitol hexaacetate

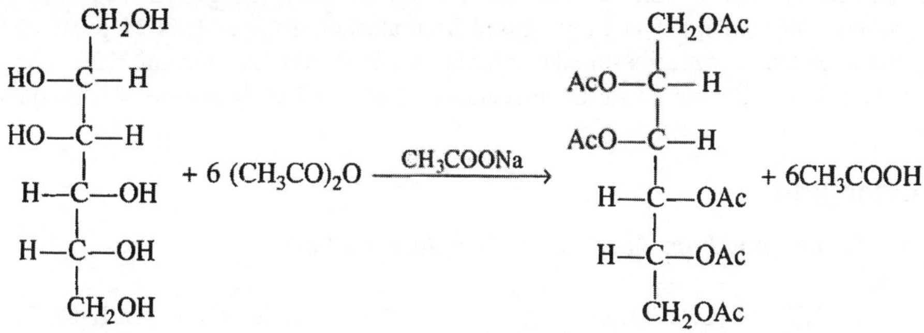

D Mannitol　　　　*D*-Mannitol hexaacetate

D-Mannitol : 3.8 g
Acetic anhydride : 22 ml
Sodium acetate (anhyd) : 3g

A mixture of *D*-Mannitol (3.8 g, 0.02 mole), acetic anhydride (22 ml) and anhydrous sodium acetate (3 g) is refluxed gently (sand bath) for 15–30 minutes. The reaction mixture is poured with vigorous stirring into ice water (200 ml). The mixture is continuously stirred for 10 minutes. The separated solid is filtered, washed with cold water and crystallised from alcohol. The yield is 8 g, (88.9%); m.p. is 121°C.

1.7 Hydroxyhydroquinone Triacetate

p-Benzoquinone on reaction with acetic anhydride in presence of concentrated sulphuric acid yields hydroxyhydroquinone triacetate. This reaction is known as Thiele's acetylation. The mechanism is shown below.

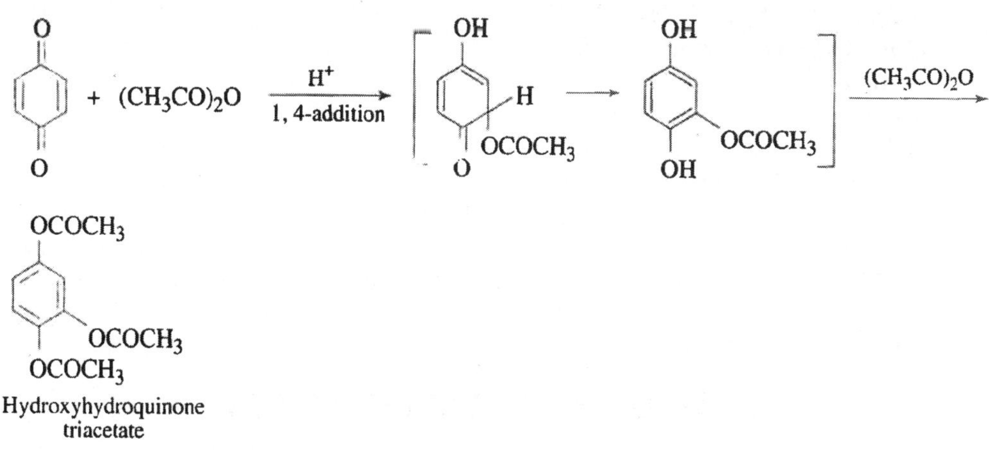

Hydroxyhydroquinone triacetate

p-Benzoquinone : 16 g
Acetic anhydride : 48 ml
Conc. sulphuric acid : 1.8 ml

Concentrated sulphuric acid (1.8 ml) is added slowly to acetic anhydride (48 ml) contained in a R.B. flask (250 ml capacity). Finely powdered *p*-benzoquinone (16 g, 0.148 mole) is then added in small lots at a time to the stirred mixture during 10–15 minutes. The temperature of the reaction is maintained between 40–45°C during the addition. After the addition is over the stirring is further continued for 5 minutes and the reaction mixture poured onto crushed ice (300–350 g). The separated product is filtered, washed with cold water and crystallised from alcohol. The yields is 34 g (91.1%); m.p. is 96-97°C. The triacetoxy benzene is stored in a desiccator over anhy. calcium chloride.

Note: *p*-Benzoquinone required for the preparation is obtained by oxidation of hydroquinone (see page 169).

1.8 Acetanilide

Method 1 Acetic anhydride-Acetic acid-Zinc dust method

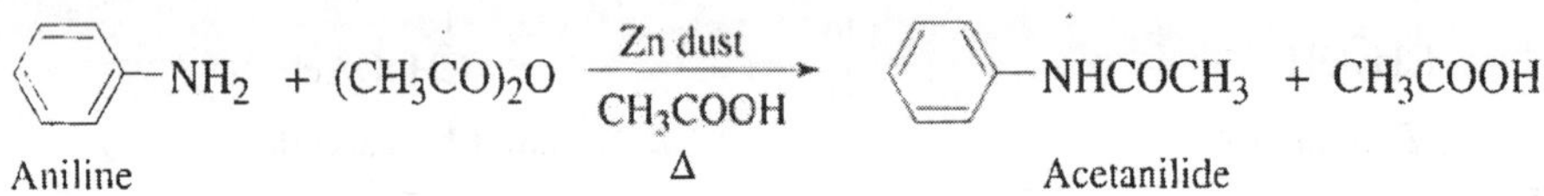

Aniline	: 12 ml
Glacial acetic acid	: 12 ml
Acetic anhydride	: 12 ml
Zinc dust	: 0.1 g

A mixture of aniline (12 ml, 0.0132 mole), acetic anhydride (12 ml), glacial acetic acid (12 ml) and zinc dust (0.1 g) is refluxed gently in a R.B. flask under anhydrous conditions for 10–15 minutes. The hot reaction mixture is poured with stirring into ice-cold water (200 ml). The separated product is filtered, washed with cold water and crystallised from boiling water containing 3-4 ml alcohol. The yield is 12 g (67.4%); m.p. is 114°C.

Note: Zinc helps to prevent the oxidation of aniline during the reaction.

Method 2 Acetic anhydride method

Aniline	: 4 ml
Conc. hydrochloric acid	: 3.7 ml
Acetic anhydride	: 5.3 ml
Sodium acetate	: 5.7 g

Aniline (4 ml, 0.045 mole) is added to dilute hydrochloric acid (3.7 ml conc. HCl and 100 ml water) contained in a beaker. The mixture is stirred until a clear solution is obtained. To the clear solution (filter if necessary) acetic anhydride (5.3 ml) is added and the mixture stirred. Then a solution of crystalline sodium acetate (5.7 g) in water (20 ml) is added. The mixture is stirred vigorously and cooled (ice). The separated acetanilide is filtered, washed with cold water and crystallised from boiling water containing 2–5 ml methylated spirit. The yield is 4 g (67.5%); m.p. is 114°C.

2. ACIDS (CARBOXYLIC ACIDS)

The carboxylic acids are obtained by the following methods:

(i) Oxidative method (see under oxidations).

(ii) The Arndt-Eistert method.
(iii) Carboxylation of aromatic ring system (see under Kolbe reaction).
(iv) Hydrolysis of nitriles.
(v) Carbonation of Grignard reagents.

In the present section only the last two methods are discussed. Carboxylic acids are conveniently obtained by the hydrolysis of readily available nitriles under acidic or basic conditions. Thus,

$$R{-}C{\equiv}N \xrightarrow{\bar{O}H/H_2O} R\,CONH_2 \longrightarrow RCO_2^- + NH_3$$

$$R{-}C{\equiv}N \xrightarrow{H_3\overset{+}{O}} R\,CONH_2 \longrightarrow RCOOH + NH_3$$

In this case, the experimental conditions can be selected to stop the reaction at the amide stage (this is especially possible for aromatic amides, eg, *o*-toluamide from *o*-tolunitrile).

The probable mechanism for hydrolysis under basic conditions is given below:

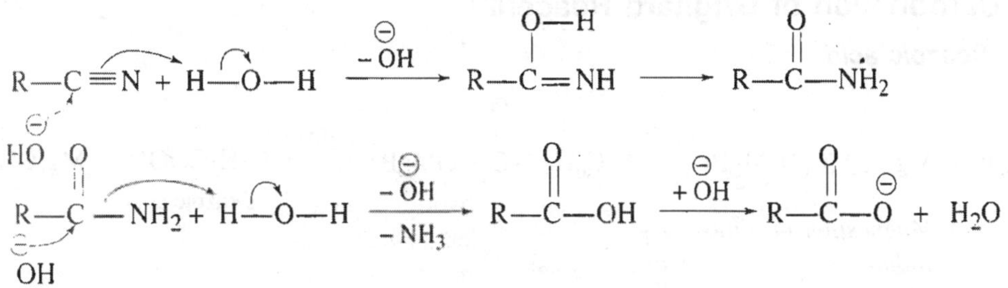

The carbonation of Grignard reagent is carried out by the addition of a Grignard reagent to carbon dioxide. Thus,

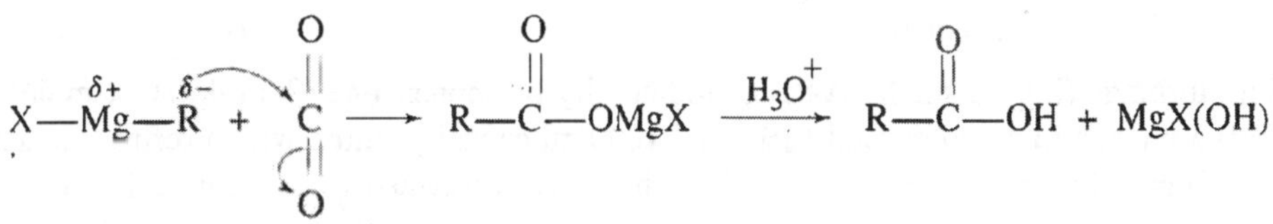

2.1 Hydrolysis of Nitriles

2.1.1 Phenylacetic acid

$$\underset{\text{Benzyl cyanide}}{C_6H_5CH_2CN} + H_2SO_4 \xrightarrow{H_2O} \underset{\text{Phenyl acetic acid}}{C_6H_5CH_2COOH} + NH_4SO_4$$

Benzyl cyanide : 10.8 ml
Conc. sulphuric acid : 10 ml
Glacial acetic acid : 10 ml

A mixture of benzyl cyanide (10.8 ml, 0.094 mole), water (10 ml) glacial acetic acid (10 ml) and concentrated sulphuric acid (10 ml) is refluxed for 45 minutes in a R.B. flask (100 ml capacity). The reaction mixture is poured into water (50 ml). The crude acid is filtered. It is purified by dissolving in sodium carbonate solution and acidifying the clear filtrate. Phenyl acetic acid is finally crystallised from hot water or petroleum ether. The yield is 4.8 g (37.7%); m.p. is 77°C.

2.1.2 p-Toluic acid

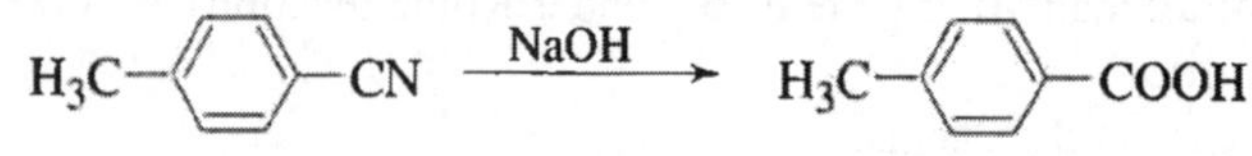

p-Tolunitrile *p*-Toluic acid

p-Tolunitrile : 10 g
Sodium hydroxide solution (10%) : 80 ml
Ethyl alcohol : 15 ml

A mixture of *p*-tolunitrile (10 g, 0.085 mole) aqueous sodium hydroxide solution (10%, 80 ml) and ethyl alcohol (15 ml) is refluxed for 1.5 hr in a R.B. flask (100 ml capacity). The clear solution is acidified with conc. hydrochloric acid. The separated *p*-toluic acid is filtered, washed with cold water and crystallised from dilute alcohol (1:1) or from benzene. The yield is 5.5 g (47.3%); m.p. is 178°C.

Note: *p*-Tolunitrile is prepared as described on page 205.

2.2 Carbonation of Grignard Reagent

2.2.1 Benzoic acid

$$C_6H_5Br + Mg \longrightarrow C_6H_5MgBr \xrightarrow{CO_2} C_6H_5-\overset{\overset{\Large O}{\|}}{C}-OMgBr \xrightarrow{H_2O} C_6H_5COOH + Mg(OH)Br$$

Benzoic acid

Magnesium turnings : 2 g
Iodine : small crystal
Dry bromobenzene : 8.9 ml
Sodium dried ether : 40 ml
Solid carbon dioxide : 20 g
(dry ice)

Magnesium turnings (2 g), a small crystal of iodine, dry bromobenzene (2.5 ml, 0.024 mole) and dry ether (5 ml) are taken in a two necked R.B. flask (200 ml capacity) fitted with a reflux condenser and a dropping funnel. In case there is no reaction, the flask is warmed (water bath). The formation of Grignard reagent is indicated by the disappearence of iodine colour and production of cloudiness. A solution of dry bromobenzene (6.4 ml, 0.06 mole) in dry ether (35 ml) is added dropwise (15 minutes) through the dropping funnel at such a rate that the ether boils gently by the heat of the reaction without the application of external heating. The mixture is refluxed for 15 minutes (water bath).

The above solution of the Grignard reagent is poured on to small lumps of dry ice (20 g) slowly with stirring (caution vigorous reaction). Continue stirring; crushed ice (100 g) and dilute hydrochloric acid (1:1, 15 ml) are added until most of the solid has decomposed. The mixture is extracted with ether (2 × 30 ml) and the combined ether solution extracted with sodium hydroxide solution (5%, 2 × 20 ml). The clear alkaline extract is acidified. The precipitated benzoic acid (obtained after cooling) is filtered. It is crystallised from hot water. The yield is 5 g (48.4%); m.p. is 121°C.

3. ACID ANHYDRIDES

Carboxylic acids can be converted into anhydrides by elimination of a molecule of water from two molecules of a carboxylic acid.

$$\underset{\text{Corboxylic acid}}{\text{RCOO}\boxed{\text{H + HO}}\text{OC R}} \xrightarrow{-H_2O} \underset{\text{Anhydride}}{\text{RCOOCOR}}$$

This can be achieved by treating a carboxylic acid in an inert solvent with thionyl chloride in presence of pyridine. In this case the acid is first converted into acid chloride which in turn reacts with carboxylic acid to form the anhydride. Thus,

$$\underset{\text{Carboxylic acid}}{\text{RCOOH} + \text{SOCl}_2} \xrightarrow{\text{Pyridine}} \underset{\text{Acid chloride}}{\text{RCOCl}}$$

$$\underset{\text{Acid chloride}}{\text{RCOCl} + \text{HOOCR}} \xrightarrow{-\text{HCl}} \underset{\text{Acid anhydride}}{\text{R—}\overset{\text{O}}{\overset{\|}{\text{C}}}\text{—O—}\overset{\text{O}}{\overset{\|}{\text{C}}}\text{—R}}$$

Alternatively, the acid chloride can be treated with the sodium salt of the same acid. Thus,

$$\text{RCOCl} + \overset{+}{\text{Na}}\,\overset{-}{\text{O}}\text{—CO—R} \xrightarrow{-\text{NaCl}} \text{R—}\overset{\text{O}}{\overset{\|}{\text{C}}}\text{—O—}\overset{\text{O}}{\overset{\|}{\text{C}}}\text{—R}$$

However, the anhydrides of 1,4-dioic acids can be prepared either by sublimation or by heating with acetic anhydride. Thus,

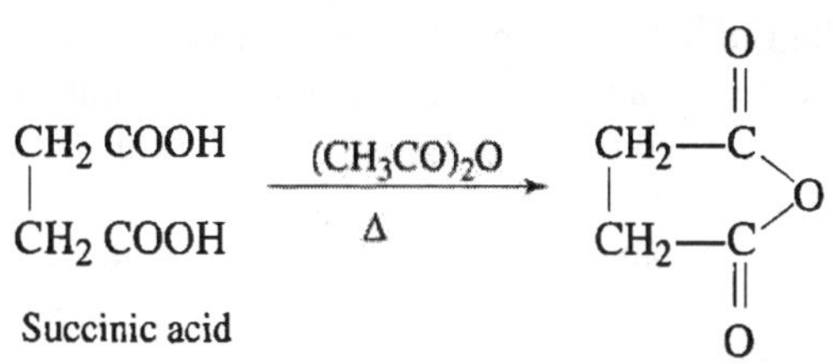

Succinic acid

Succinic anhydride

3.1 Acetic Anhydride

$$\underset{\text{Acetyl chloride}}{\text{CH}_3\text{COCl}} + \underset{\text{Sodium acetate}}{\text{CH}_3\text{COONa}} \xrightarrow{\Delta} \underset{\text{Acetic anhydride}}{\text{CH}_3\text{—}\overset{\text{O}}{\overset{\|}{\text{C}}}\text{—O—}\overset{\text{O}}{\overset{\|}{\text{C}}}\text{—CH}_3} + \text{NaCl}$$

Sodium acetate : 28 g
(freshly fused)
Acetyl chloride : 20 ml

Freshly fused sodium acetate (28 g, 0.34 mole) is taken in a distillation flask fitted with a dropping funnel and the apparatus for distillation assembled (anhydrous condition). The flask is cooled (ice-water) and acetyl chloride (20 ml, 0.28 mole) gradually added from a dropping funnel. The contents are stirred thoroughly with the help of a glass rod. After the addition is over and the exothermic reaction has subsided, the reaction flask is heated slowly with a luminous flame. Uniform heating is ensured by the flame. The heating is continued until no more liquid distills. Powdered anhydrous sodium acetate

(2–3 g) is added to the distillate and the acetic anhydride obtained by redistillation. The yield is 21 g (73.2%); b.p. is 139–40°C.

Note: Sodium acetate (anhydrous) is prepared as follows:

Crystalline sodium acetate (about 40 g) is heated in a porcelin basin over a free flame (wireguaze). A glass rod is used to stir the material. The salt first melts, steam is evolved and the mass solidifies. The solid is heated carefully with a larger flame until the solid just melts. Too much heating is avoided to prevent charring. The melted sodium acetate is allowed to solidify. It is removed from the dish and powdered in a pestle-mortar and stored in a dry bottle.

3.2 Phthalic Anhydride

It is obtained by sublimation of phthalic acid.

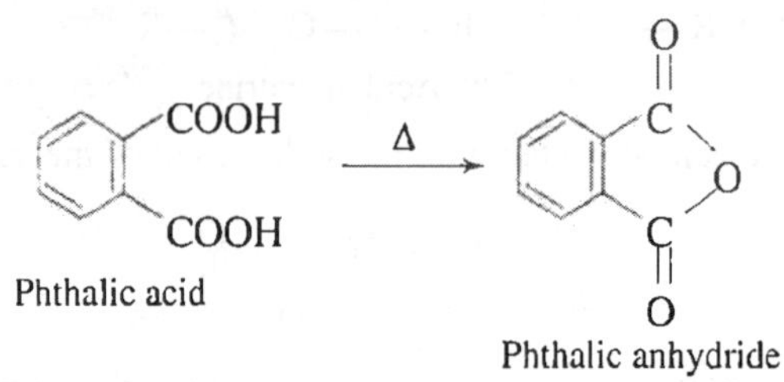

Phthalic acid (10 g, 0.06 mole) is placed in a porcelin dish (3 cm diameter), and a circular filter paper having circular holes is placed on the dish. The filter paper is covered by an inverted water-jacketed glass funnel and its stem is plugged loosely with cotton or filter paper. An ordinary inverted funnel can also be used. The dish is heated on a sand bath with the help of a flame. The anhydride formed rises through the holes in the filter paper and condenses in the hollow of the funnel and on the filter paper. The yield is 8 g (89.9%); m.p. is 132°C.

Phthalic anhydride can also be obtained by heating with acetic anhydride (see the preparation of succinic anhydride).

3.3 Succinic Anhydride

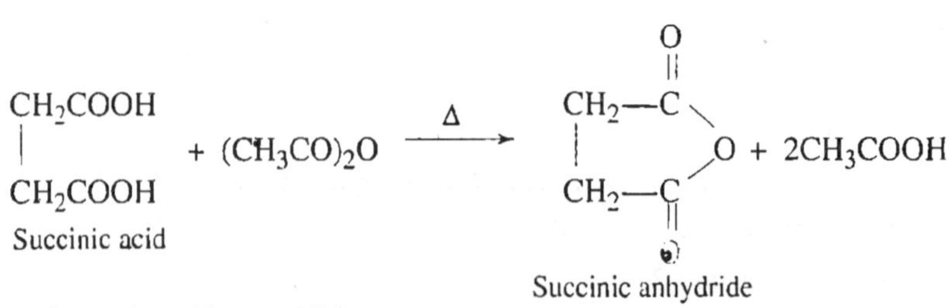

Succinic acid : 17.2 g
Acetic anhydride : 30 ml

A mixture of succinic acid (17.2 g, 0.146 mole) and redistilled acetic anhydride (30 ml) is taken in a R.B. flask (100 ml capacity) fitted with a reflux condenser (anhydrous conditions). The mixture is gently warmed over a small luminous flame with continuous shaking of the flask until a clear solution is obtained (3–4 min). The mixture is refluxed for 15 minutes. The flask is allowed to cool (first at room temperature and then in an ice bath). The separated succinic anhydride is filtered in a Buchner funnel. It is washed with ether (2 × 10 ml) and dried in a desiccator over anhydrous calcium chloride. The yield is 13 g (89.2%); m.p. is 119–20°C.

3.4 Maleic Anhydride

It is obtained by distilling maleic acid with an inert solvent having a high boiling point. The water passes over with the solvent leaving behind the anhydride.

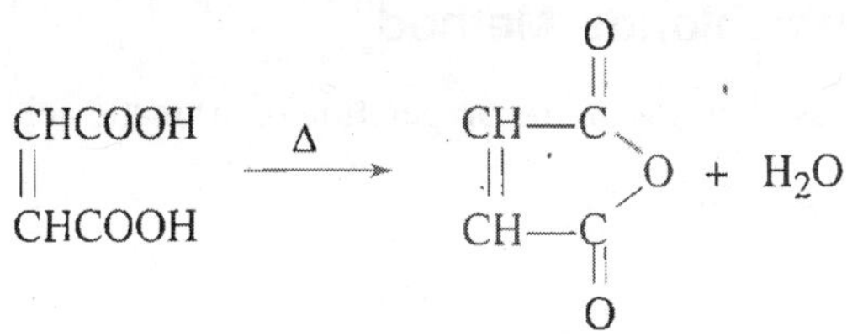

Maleic acid (25 g, 0.215 mole) and sym-tetrachloroethane (b.p. 146°C, 25 ml) are placed in a distillation flask (50 ml capacity). The distillation assembly is fitted up. The mixture is heated slowly and the distillate up to 150°C is collected. This consists of the solvent along with water. The distillation is continued (this time using an air condenser) using a new receiver. Maleic anhydride is collected at 185–200°C. It is crystallised from chloroform, m.p. 54°C, yield 18 g (85.2%).

3.5 Benzoic Anhydride (Thionyl chloride-pyridine method)

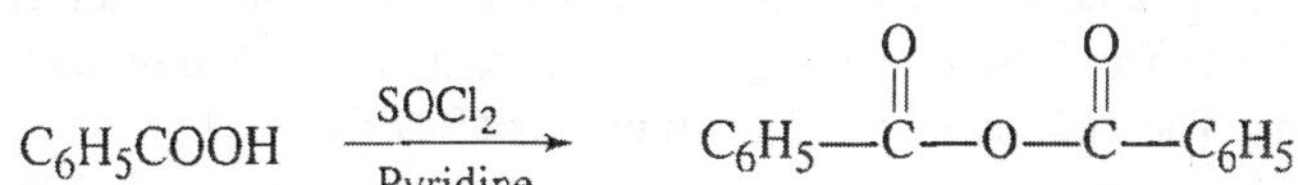

Anhydride of benzoic acid

Benzoic acid : 24.4 g
Thionyl chloride : 12 ml
Pyridine : 18 ml
Anhydrous ether : 45 ml

A solution of thionyl chloride (12 ml) in anhydrous ether (20 ml) is added gradually to a suspension of benzoic acid (24.4 g, 0.2 mole) in dry ether (25 ml) and pyridine (18 ml). The mixture is shaken during addition and the temperature kept at 5–8°C. The reaction mixture is stoppered and kept for 12 hours in a refrigerator. It is treated with crushed ice (30 g) and dilute hydrochloric acid (1:1, 15 ml). The separated anhydride is filtered, washed with water, dried and crystallised from benzene-petroleum ether. The yield is 20 g (96%); m.p. 42°C.

Note: This is a convenient and general method of converting aromatic carboxylic acids into their anhydrides.

4. ACID CHLORIDES (CARBOXYLIC ACID CHLORIDES) (ACYL HALIDES)

Acyl halides are obtained from carboxylic acids by heating with phosphorus trichloride, phosphorus pentachloride or thionyl chloride. Thus,

$$3RCOOH + PCl_3 \longrightarrow 3RCOCl + H_3PO_3$$

$$RCOOH + PCl_5 \longrightarrow RCOCl + HCl + POCl_3$$

$$RCOOH + SOCl_2 \longrightarrow RCOCl + HCl\uparrow + SO_2\uparrow$$

A very convenient method consists in treatment of anhydrous sodium salt of a carboxylic acid with phosphorus oxychloride—this method is used commercially and gives pure products.

$$2RCOONa + POCl_3 \longrightarrow 2RCOCl + NaCl + NaPO_3$$

4.1 Phosphorus Pentachloride Method

Phosphorus pentachloride is used mostly in the preparation of aromatic acid chlorides.

4.1.1 3,5-Dinitrobenzoyl chloride

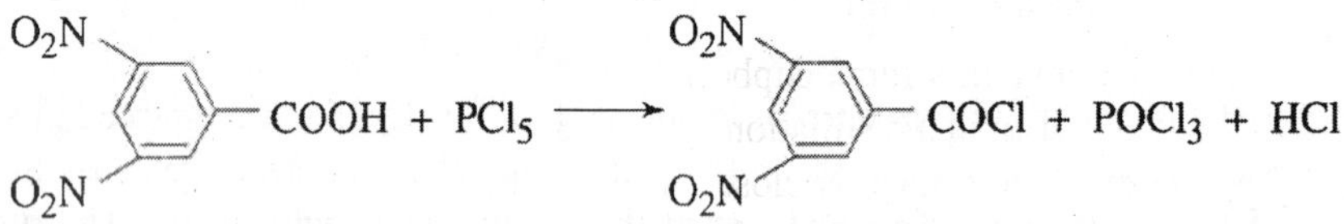

3,5-Dinitrobenzoic acid : 15 g
Phosphorus pentachloride : 16.5 g

A mixture of 3,5-dinitrobenzoic acid (15 g, 0.07 mole) and phosphorus pentachloride (16.5 g) is heated to 120–130°C (oil bath) for 60–75 minutes in a R.B. flask (100 ml capacity) fitted with a reflux condenser. The reaction mixture is cooled and the formed phosphorus oxychloride removed under reduced pressure (25°C/20 mm). 3, 5-Dinitrobenzoyl chloride is then obtained from the residual product by crystallisation from carbon tetrachloride. The yield is 12 g (73.6%); m.p. is 69–70°C.

4.2 Phosphorus Trichloride Method

It is used for the preparation of aliphatic or aromatic acyl halides.

4.2.1 Acetyl chloride

$$3CH_3COOH + PCl_3 \longrightarrow 3CH_3COCl + H_3PO_3$$

Acetic acid (glacial) : 23 ml
Phosphorus trichloride : 13 ml

Glacial acetic acid (23 ml, 0.4 mole) is taken in a distillation flask (100 ml capacity) fitted with a dropping funnel at the top and a liebigs condenser, the lower end of which is connected to another distillation flask fitted with a calcium chloride guard tube. It is necessary to have the apparatus dry. The reaction flask is cooled (cold water) and phosphorus trichloride (13 ml) added dropwise through the dropping funnel (15 minutes). The reaction flask is kept for 1 hr and heated to 40–50°C (water bath) till no more evolution of hydrogen chloride is observed. Acetyl chloride is finally obtained by distillation using a water bath. It is purified by redistillation. The yield is 10 g (31.7%); b.p. is 52°C.

4.3 Thionyl Chloride Method

Thionyl chloride is used for the preparation of aliphatic and aromatic acyl halides. Since the by-products of this reaction are gases this is the most commonly used method. In practice, an excess of thionyl chloride is used, this prevents the formation of anhydrides. Thionyl chloride is used for the preparation of those acyl halides whose boiling points are reasonably different (20–30°C) from the boiling point of thionyl chloride. This permits the separation of excess thionyl chloride by fractional distillation. In case the boiling point of acyl chloride is too near to that of thionyl chloride, the excess of the reagent is destroyed by addition of pure formic acid.

$$HCOOH + SOCl_2 \longrightarrow CO + SO_2 + 2HCl$$

4.3.1 Benzoyl chloride

$$C_6H_5COOH + SOCl_2 \longrightarrow C_6H_5\,COCl + SO_2 + HCl$$

Benzoic acid : 20 g
Thionyl chloride : 15 ml

This preparation should be done in a fume cupboard. Benzoic acid (20 g, 0.164 mole) and redistilled thionyl chloride (15 ml) are taken in a distillation flask fitted with a reflux condenser and a guard tube, and the side arm of the distillation flask is closed with a small cork. The reaction mixture is heated (water bath) with occasional shaking for 20 to 30 minutes till no more hydrogen chloride is evolved. The reaction mixture is cooled and the distillation assembly fitted up. The mixture is distilled under anhydrous conditions. A forerun of thionyl chloride is first obtained. The receiver flask is changed and the fraction between 194–198°C is collected (flame is used for distillation). The yield is 16 g (69.5%); b.p. is 197°C.

4.3.2 Butyryl chloride

$$CH_3CH_2CH_2COOH + SOCl_2 \longrightarrow CH_3CH_2CH_2COCl + SO_2 + HCl$$

Butyric acid : 15.5 ml
Thionyl chloride : 14.5 ml

Redistilled thionyl chloride (14.5 ml) is taken in a two necked flask (100 ml capacity) fitted with a dropping funnel and a reflux condenser (anhydrous conditions). The reaction flask is heated gently (water bath) and butyric acid (15.5 ml, 0.17 mole) is added drop by drop through a dropping funnel during 30 minutes. When all the acid has been added, the reaction mixture is heated (water bath) for 30 minutes. The apparatus is rearranged for distillation and the crude acid chloride is collected between 70–100°C. It is purified by distillation using a short fractionating column. The yield is 15 g, (82.9%); b.p is 100–101°C.

5. ALKYL HALIDES

Alkyl chlorides are generally obtained from the corresponding alcohols by the use of thionyl chloride. Zinc chloride-hydrochloric acid or hydrochloric acid only. On the other hand, alkyl bromides are obtained by the use of hydrobromic acid-sulphuric acid or sodium bromide-sulphuric acid. However, the alkyl iodides are obtained from the corresponding alcohol by the action of red phosphorus and iodine or with potassium iodide-phosphoric acid.

5.1 Alkyl Chlorides

Method 1 Thionyl chloride method

Thionyl chloride is useful for the preparation of alkyl chlorides from primary alcohols.
This is an internal substitution reaction and is designated as S_Ni (substitution nucleophilic internal).

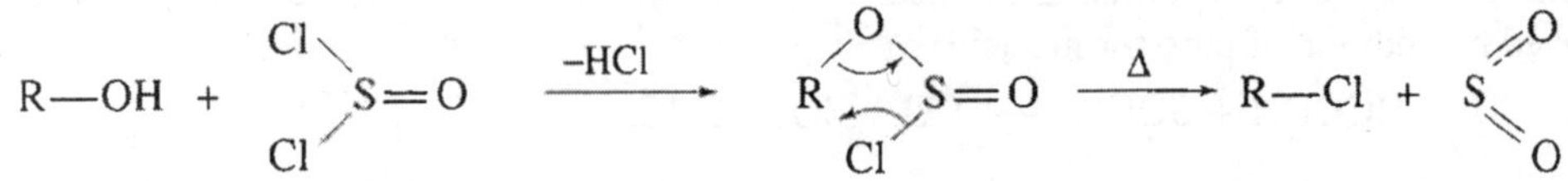

Primary alcohol Thionyl chloride Alkyl chlorosulphite Alkyl chloride

Sometimes pyridine is also used along with thionyl chloride for the replacement of the hydroxyl group of primary alcohols with chloro. In presence of pyridine, the chloride ion liberated in the first step (chlorosulphite formation) effects an S_N2 displacement. Thus,

$$HCl + C_5H_5N \longrightarrow C_5H_5\overset{+}{N}H + Cl^-$$

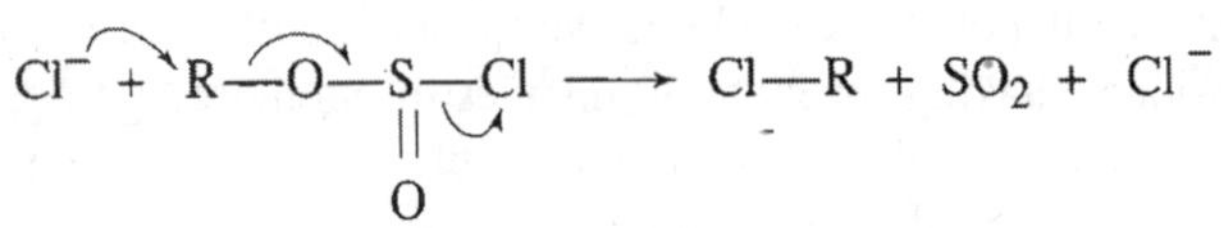

5.1.1 *n*-Hexyl chloride (1-Chlorohexane)

$$CH_3(CH_2)_4CH_2OH \xrightarrow{SOCl_2} CH_3(CH_2)_4CH_2Cl + SO_2 + HCl$$

n-Hexyl alcohol *n*-Hexyl chloride

n-Hexyl alcohol : 13.7 ml
Thionyl chloride : 24.2 ml

Redistilled thionyl chloride (24.2 ml) is taken in a two necked R.B. flask (100 ml capacity) fitted with a reflux condenser and a dropping funnel. The reaction is performed under anhydrous conditions. *n*-Hexyl alcohol (24.2 ml, 0.193 mole) is added dropwise during 45 minutes with stirring or shaking occasionally. After the addition of hexyl alcohol is complete, the mixture is refluxed for one hour. The apparatus is rearranged for distillation and the mixture is distilled slowly. Thionyl chloride passes over in the forerun (up to 120°C) and finally *n*-hexyl chloride is collected at 131–134°C. It is purified by washing with water, sodium carbonate solution (10%) and again with water. It is dried (anhydrous calcium chloride) and then distilled. Yield is 7.3 g (31.4%); b.p. is 132–133°C.

5.1.2 Isobutyl Chloride (1-Chloro-2-methylpropane)

$$CH_3\underset{\displaystyle CH_3}{\underset{|}{C}H}—CH_2OH + SOCl_2 \xrightarrow{C_5H_5N} CH_3—\underset{\displaystyle CH_3}{\underset{|}{C}H}—CH_2—Cl + SO_2 + HCl$$

Isobutanol Isobutyl chloride

Isobutanol (2-methyl propane-1-ol) : 18.5 ml
Thionyl chloride : 36.5 ml
Pyridine : 20.5 ml

Isobutanol (18.5 ml, 0.2 mole) and pyridine (20.5 ml) are taken in a three necked R.B. flask (250 ml capacity) which is fitted with a sealed stirrer, reflux condenser and a dropping funnel (anhydrous conditions) fitted with calcium chloride guard tube. Thionyl chloride (36.5 ml) is added dropwise through a dropping funnel during 3 hours. After the addition, the mixture is refluxed for 30 minutes. The reaction mixture is cooled and the upper layer decanted (the lower pyridine layer contained excess of thionyl chloride). The upper layer is washed with dilute sodium hydroxide solution (5%) and water (2 × 10 ml), and then dried (anhydrous calcium chloride). It is finally distilled using a short fractionating column. The yield is 13 g (70%); b.p. is 68–69°C.

Method 2 Zinc chloride-hydrochloric acid method

This method is used for the preparation of alkyl chlorides from secondary as well as primary alcohols. The zinc chloride-hydrochloric acid method is of S_N2 type displacement, especially in the case of primary alcohols. Thus,

$$R{-}OH + HCl + ZnCl_2 \longrightarrow Cl_2\overset{\ominus}{Zn}{-}Cl + R{-}\overset{\oplus}{O}H_2$$

$$Cl_2\overset{\ominus}{Zn}{-}Cl + R{-}\overset{\oplus}{O}H_2 \longrightarrow ZnCl_2 + RCl + H_2O$$

In this case, S_N1 mechanism is also possible.

$$ROH + ZnCl_2 \longrightarrow R{-}\overset{\oplus}{\underset{H}{O}}{-}\overset{\ominus}{Zn}Cl_2 \longrightarrow R^{\oplus} + HO\overset{\ominus}{Zn}Cl_2 \longrightarrow RCl + Zn(OH)Cl$$

Unsaturated alcohols like allyl alcohol give a very poor yield by the zinc chloride-hydrochloric acid method. However, the use of copper (I) chloride as a catalyst gives good yields of the corresponding chloride.

5.1.3 sec.Butyl chloride (2-Chlorobutane)

$$\underset{\textit{Sec.}\text{Butyl alcohol}}{CH_3CH_2\underset{OH}{CH}CH_3} + HCl \xrightarrow{ZnCl_2} \underset{\textit{Sec.}\text{Butyl chloride}}{CH_3CH_2\underset{Cl}{CH}{-}CH_3} + H_2O$$

Sec. Butyl alcohol : 5.8 ml
Zinc chloride (anhydrous) : 17 g
Conc. Hydrochloric acid : 17 ml

Concentrated hydrochloric acid (17 ml) and anhydrous Zinc chloride (17 g) are taken in a distillation flask fitted with a reflux condenser and the side arm closed with a small rubber stopper. *Sec.* butyl alcohol (5.8 ml, 0.063 mole) is added to the above mixture in the distillation flask and the contents gently refluxed (wire guaze) for one hour. The distillation flask is kept in an inclined position so that liquid is not collected in the side arm. The reaction mixture is cooled and the apparatus is fitted for distillation. The distillate upto 100°C is collected. The upper layer is separated (separatory funnel) from the distillate and washed with water (5 ml) and sodium hydroxide solution (5%), then dried (anhydrous calcium chloride) and distilled. The yield is 4 g (68.4%); b.p. 67–69°C.

5.1.4 *Allyl chloride (3-Chloropropene)*

$$\underset{\text{Allyl alcohol}}{CH_2{=}CHCH_2OH} + HCl \xrightarrow[CuCl]{H_2SO_4} \underset{\text{Allyl chloride}}{CH_2{=}CHCH_2Cl} + H_2O$$

Allyl alcohol : 25 ml
Conc. Hydrochloric acid : 40 ml
Conc. Sulphuric acid : 12 ml
Copper (I) chloride : 0.5 g

A mixture of allyl alcohol (25 ml, 0.368 mole), concentrated hydrochloric acid (40 ml) and freshly prepared copper (I) chloride (0.5 g) is taken in a R.B. flask (250 ml capacity) fitted with a reflux condenser. The reaction mixture is cooled (ice bath) and concentrated sulphuric acid (12 ml) is added dropwise through the condenser with occasional shaking of the flask. The turbid mixture is allowed to stand for 30 minutes for the reaction to be completed and the allyl chloride separation takes place. The upper layer and allyl chloride is separated (separatory funnel), washed with water (2 × 30 ml), dried (anhydrous calcium chloride) and distilled. The yield is 25 g (89%); b.p. is 46–47°C.

Method 3 Hydrochloric acid method

This method is used for the preparation of alkyl chlorides from tertiary alcohols, in which the hydroxyl group is most easily replaced (room temperature). The reaction is nucleophilic displacement of the S_N1 type involving formation of a comparatively stable carbonium ion intermediate. Thus,

$$R_3COH + H^{\oplus} \rightleftharpoons R_3C\overset{\oplus}{O}H_2 \rightleftharpoons R_3C^{\oplus} + H_2O$$

$$R_3C^{\oplus} + Cl^- \longrightarrow R_3\ C{-}Cl$$

5.1.5 *Tert. Butyl chloride (2-Chloro-2-methylpropane)*

$$\underset{\textit{tert.}\text{ Butanol}}{(CH_3)_3\ C{-}OH} + HCl \longrightarrow \underset{\textit{tert.}\text{ Butyl chloride}}{(CH_3)_3\ C{-}Cl} + H_2O$$

tert. Butyl alcohol : 30 g
Conc. hydrochloric acid : 100 ml

A mixture of *tert.* butyl alcohol (30 g, 0.4 mole) and concentrated hydrochloric acid (100 ml) is shaken in a separatory funnel during 20–25 minutes (occasional shaking). The mixture is allowed to stand and the lower layer discarded. The remaining halide is washed with sodium bicarbonate solution (5%, 20 ml) and then with water (20 ml). It is dried (anhydrous calcium chloride) and distilled. The yield is 33 g (88.2%); b.p. is 49–50°C.

5.2 Alkyl Bromides

Alkyl bromides are more easily obtained than alkyl chlorides. Secondary and tertiary alcohols on heating with constant boiling hydrobromic acid give the corresponding bromides. Thus,

$$R{-}OH + HBr \longrightarrow R{-}Br + H_2O$$

However, in the case of a primary alcohol the reaction is carried out in presence of concentrated sulphuric acid. Alternatively a mixture of sodium bromide and concentrated sulphuric acid can be used.

This method can also be used for the preparation of dibromides from diols. Alternatively, the dibromides can also be obtained from the corresponding cyclic ethers like tetrahydrofuran and tetrahydropyran.

Alkyl bromides can also be obtained by the treatment of appropriate alcohols with red phosphrous and bromine. Thus,

$$10RCH_2OH + 2P + 5Br_2 \longrightarrow 10RCH_2Br + 2H_3PO_4 + 2H_2O.$$

5.2.1 Isopropyl bromide (2-Bromopropane)

$$\underset{\text{Isopropyl alcohol}}{CH_3\underset{|}{\overset{}{C}}H(OH)\text{—}CH_3} + HBr \longrightarrow \underset{\text{Isopropyl bromide}}{CH_3CH(Br)\text{—}CH_3} + H_2O$$

Isopropyl alcohol : 40 ml
Hydrobromic acid (48%) : 250 ml

A mixture of isopropyl alcohol (40 ml, 0.522 mole) and constant boiling hydrobromic acid (b.p. 126°C, 48%, 250 ml) is taken in a distillation flask (500 ml capacity). The distillation assembly is fitted and a few boiling chips added and mixture is distilled slowly (1–2 drops per second) until half the liquid has distilled over. The lower layer of crude isopropyl bromide is separated (separatory funnel). It is washed successively with concentrated hydrochloric acid (50 ml), water, sodium bicarbonate solution (5%) and water and then dried (anhydrous calcium chloride). Finally, isopropylbromide is obtained by distillation. The yield is 52 g (80.9%); b.p. is 59°C.

5.2.2 n-Butyl bromide

Method 1 Hydrobromic acid-sulphuric acid method

$$\underset{n\text{-Butyl alcohol}}{CH_3CH_2CH_2CH_2OH} + HBr \xrightarrow{H_2SO_4} \underset{n\text{-Butyl bromide}}{CH_3CH_2CH_2CH_2Br} + H_2O$$

n-Butyl alcohol : 20 ml
Hydrobromic acid (48%) : 36 ml
Conc. sulphuric acid : 12 ml

Concentrated sulphuric acid (6 ml) is added in small portions (1 ml at a time) with shaking to hydrobromic acid (48%, 36 ml) contained in a R.B. flask (100 ml capacity). *n*-Butyl alcohol (20 ml, 0.22 mole) is added to the flask followed by addition of concentrated sulphuric acid (6 ml) in 3-4 lots. The mixture is refluxed gently (few boiling chips are added) for 1.5 hours on a wire guage. This experiment should be carried out in a fume cupboard. The reaction mixture is allowed to cool and the distillation assembly fitted up. The mixture is distilled until no more oily drops pass over. The distillate is washed successively with water (10 ml), concentrated hydrochloric acid (25 ml), water (10 ml), sodium bicarbonate solution (5%) and water (10 ml). The remaining crude *n*-butyl bromide is dried (anhydrous magnesium sulphate) and distilled. The yield is 25 g (83.5%); b.p. is 101–102°C.

Method 2 Sodium bromide-Sulphuric acid method

n-Butyl alcohol : 12 ml
Sodium bromide : 16.2 g
Conc. sulphuric acid : 13.7 ml

n-Butyl alcohol (12 ml, 0.13 mole) is added to a solution of sodium bromide (16.2 g) in water (18 ml) contained in a R.B. flask (100 ml capacity). Sulphuric acid (13.7 ml) is added slowly with constant shaking. The temperature of the reaction mixture is kept below 40°C using a cold water bath if necessary. The mixture is refluxed gently for one hour. Working up of the reaction mixture as in method (1) gives *n*-butyl bromide. The yield is 12 g (66.8%); b.p. is 101–102°C.

5.2.3 1,4-Dibromobutane

Method 1 From butane 1,4-diol

$$\underset{\text{Butane-1,4-diol}}{HOCH_2CH_2CH_2CH_2OH} + 2HBr \xrightarrow{\text{Conc. } H_2SO_4} \underset{\text{1,4-dibromobutane}}{BrCH_2CH_2CH_2CH_2Br} + 2H_2O$$

Butane-1,4-diol : 20 ml
Hydrobromic acid (48%) : 70 ml
Conc. sulphuric acid : 48 ml

Constant boiling hydrobromic acid (70 ml) is taken in a three necked flask (500 ml capacity) fitted with a reflux condenser, stirrer and a dropping funnel. The flask is cooled (ice bath) and concentrated sulphuric acid (48 ml) is added slowly (through the dropping funnel) with stirring. The resulting ice-cooled solution is treated with butane-1,4-diol (20 ml, 0.226 mole) (added dropwise through a dropping funnel). The reaction mixture is left for 24 hours and then heated for 3 hours (steam bath). The lower layer of dibromobutane is separated (separatory funnel), washed successively with water, sodium carbonate solution (10%), water and then dried (magnesium sulphate). Finally it is distilled. The yield is 36 g (73.9%); b.p. is 83–84°C/12 mm.

Method 2 From tetrahydrofuran

$$\underset{\text{Tetrahydrofurane}}{\underbrace{CH_2CH_2CH_2CH_2}_{O}} + 2HBr \xrightarrow{H_2SO_4} \underset{\text{1, 4-dibromobutane}}{BrCH_2CH_2CH_2CH_2Br} + H_2O$$

Tetrahydrofuran : 16.1 ml
Hydrobromic acid 48% : 136 ml
Conc. Sulphuric acid : 33 ml

A mixture of tetrahydrofuran (16.1 ml, 0.2 mole), hydrobromic acid (136 ml) and concentrated sulphuric acid (33 ml) is refluxed for 3 hours in a R.B. flask (500 ml capacity). The lower layer of dibromide is separated (separatory funnel) and worked up as described in method 1. The yield is 32 g (75%); b.p. is 83–84°C/12 mm.

5.2.4 Isobutyl bromide (phosphorus-bromine method)

$$\underset{\text{Isobutyl alcohol}}{10CH_3{-}\underset{\displaystyle CH_3}{\underset{|}{CH}}{-}CH_2OH} + 2P + 5Br_2 \longrightarrow \underset{\text{Isobutyl bromide}}{10CH_3{-}\underset{\displaystyle CH_3}{\underset{|}{CH}}{-}CH_2Br} + 2H_3PO_4 + 2H_2O$$

Iso butyl alcohol : 23 ml
Red phosphorus : 1.7 g
Bromine : 6.4 ml

Isobutyl alcohol (23 ml, 0.25 mole) and purified red phosphorus (1.7 g) are taken in a three necked R.B. flask (100 ml capacity) fitted with a sealed stirrer, a reflux condenser and a dropping funnel. Bromine (6.4 ml, 0.124 mole) is added dropwise through the dropping funnel to the stirred solution. When all the bromine has been added, the mixture is refluxed gently for 15–30 minutes. The assembly for distillation is fixed up. The mixture is distilled gently. When no more liquid distills over, water (10 ml) is added and the distillation continued. The crude isobutyl bromide is separated (separatory funnel) from the distillate, washed successively with water, concentrated hydrochloric acid (20 ml), sodium carbonate solution (10%) and water. It is dried (anhydrous calcium chloride) and distilled. The yield is 30 g (87.9%); b.p. is 92–93°C.

5.3 Alkyl Iodides

Alkyl iodides are conveniently obtained from the corresponding alcohol by slow distillation with constant boiling hydroiodic acid. Thus,

$$ROH + HI \longrightarrow RI + H_2O$$

An alternative method (especially useful for the conversion of diols into diiodo compounds) is to allow the alcohol to react with a mixture of potassium iodide and 95 per cent orthophosphoric acid. This reagent is also used for the conversion of tetrahydro furan and tetrahydropyran to the corresponding α, ω-diiodo compounds.

Alkyl iodides can also be obtained by the treatment of primary or secondary alcohol with red phosphorus and iodine.

A most widely used method for the preparation of alkyl iodides is by the halogen exchange reaction. Thus, alkyl chlorides or bromides on treatment with sodium iodide in acetone give alkyl iodides.

$$R—Cl + NaI \longrightarrow RI + NaCl$$

This method is particularly used for the generation of alkyl iodides in situ. As an illustration, benzylation of a hydroxy compound is very conveniently carried out by refluxing the hydroxy compound with benzyl chloride in acetone is presence of anhydrous potassium carbonate and sodium or potassium iodide. The iodides are more reactive than the chlorides or bromides.

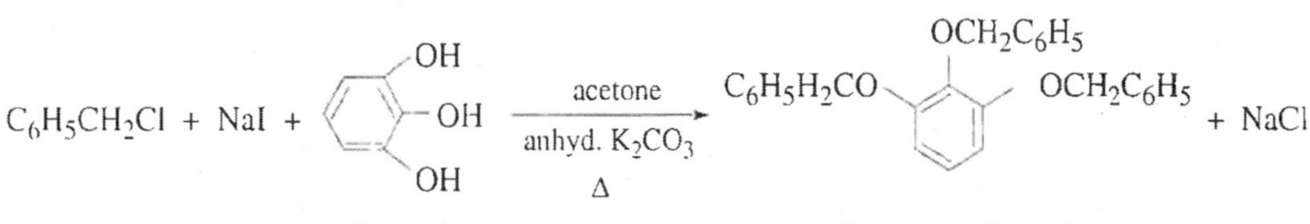

Pyrogallol   Pyragallol tribenzylether

5.3.1 Isopropyl iodide (2-Iodopropane)

$$\underset{\text{Isopropyl alcohol}}{CH_3\underset{\displaystyle OH}{\underset{|}{C}H}—CH_3} + HI \longrightarrow \underset{\text{Isopropyl iodide}}{CH_3—\underset{\displaystyle I}{\underset{|}{C}H}—CH_3} + H_2O$$

Isopropyl alcohol : 19 ml
Hydroiodic acid (57%) : 130 ml

A mixture of iso propyl alcohol (19 ml, 0.248 mole) and constant boiling hydroiodic acid (57%, 130 ml) is distilled in a R.B. flask (500 ml capacity) using an oil bath. A slow rate of distillation is maintained (1–2 drops per second). When approximately half the liquid has distilled, the lower layer of crude iodide is separated (separatory funnel). It is washed with concentrated hydrochloric acid (25 ml) and then successively with water, sodium carbonate solution (5%) and water. It is dried (anhydrous calcium chloride) and distilled. The yield is 30 g (71.1%); b.p. is 84°C.

5.3.2 1,4-Diiodobutane

Method 1 From butane-1,4-diol

$$\underset{\text{Butane-1,4-diol}}{HOCH_2CH_2CH_2CH_2OH} + 2KI + 2H_3PO_4 \longrightarrow \underset{\text{1,4-diiodobutane}}{ICH_2CH_2CH_2CH_2I} + 2\,KH_2PO_4 + 2H_2O$$

Butane-1,4-diol	: 11.25 g
Phosphorus pentoxide	: 16.25 g
Orthophosphoric acid 85%	: 33.75 g
Potassium iodide	: 83 g

A mixture of phosphorus pentoxide (16.25 g) and orthophosphoric acid (85%, 33.75 g) is stirred in a three necked flask (250 ml capacity) fitted with a sealed stirrer, reflux condenser and a thermometer. When the stirred mixture has attained room temperature, potassium iodide (83 g) and redistilled butane 1, 4-diol (11.25 g, 0.125 mole) are added. The mixture is stirred and heated [100–120°C, for 4 hours (oil bath)]. The reaction mixture is cooled to room temperature and water (40 ml) added. The resultant mixture is extracted with ether (100 ml). The ether extract is shaken with sodium thiosulphate solution (10%, 15 ml), washed with cold saturated sodium chloride solution (70 ml), dried (magnesium sulphate) and ether removed by distillation. The residue is distilled under reduced pressure. The yield is 33 g (85.2%); b.p. is 110°C/6 mm.

Method 2 From tetrahydrofuran

$$\underbrace{CH_2—CH_2—CH_2—CH_2}_{O} + 2KI + 2H_3PO_4 \xrightarrow[100-120^\circ]{P_2O_5} ICH_2CH_2CH_2CH_2I + 2KH_2PO_4 + H_2O$$

Tetrahydrofuran	: 10 ml
Phosphorus pentoxide	: 16.25 g
Orthophosphoric acid (85%)	: 33.75 g
Potassium iodide	: 83 g

A mixture of phosphorus pentoxide (16.25 g) and orthophosphoric acid (33.75 g) is stirred (as above) and treated with potassium iodide (83 g). To the mixture is added tetrahydrofuran (purified, 10 ml, 0.123 mole) and the mixture heated for 3-4 hours. The 1,4-diiodobutane is isolated as in method 1. The yield is 30 g (78.8%); b.p. is 110°C/6 mm.

5.3.3 n-Butyl iodide

Method 1 Red phosphorus-iodine method

$$\underset{n\text{-Butyl alcohol}}{10CH_3CH_2CH_2CH_2OH} + 5I_2 + 2P \longrightarrow \underset{n\text{-Butyl iodide}}{10\,CH_3CH_2CH_2CH_2I} + 2H_2O + 2H_3PO_4$$

n-Butyl alcohol : 19.2 ml
Red phosphorus : 2.5 g
Iodine : 25.4 g

A mixture of *n*-butyl alcohol (19.2 ml, 0.21 mole) and purified red phosphorus (2.5 g) is refluxed gently in a R.B. flask (100 ml capacity) fitted with a reflux condenser (anhydrous conditions) over a wire guage. The flame is removed the iodine (25.4 g) is added in small lots (1.5–2 g) through the condenser (20–30 minutes). The exothermic reaction is allowed to subside before each addition of iodine. After the addition of iodine has been completed, the mixture is refluxed for 40–50 minutes till no iodine is visible. The apparatus for distillation is assembled and the liquid distilled at a slow rate. When half of the liquid has distilled over, water (10 ml) is added and the distillation continued until no more oily drops pass over into the receiver. The crude iodide is separated from the distillate (separatory funnel) and washed successively with water (20 ml), concentrated hydrochloric acid (20 ml), water, sodium carbonate solution (10%) and water. It is dried (anhydrous calcium chloride) and distilled. The yield is 32 g (82.9%); b.p. is 129–130°C.

6. ALGAR-FLYNN-OYAMADA OXIDATION (AFO)

The AFO oxidation reaction involves the alkaline hydrogen peroxide oxidation of 2′-hydroxy chalcones into flavanoids and then into flavonols. If $-OCH_3$ or $-CH_3$ groups are present in the 6′-position of the chalcone, aurones rather than flavonols are obtained provided the 2- or 4-positions do not carry a hydroxyl group and the reaction is effected at room temperature. When the reaction is carried out in methanol or ethanol, 6′-methoxychalcones gives flavonols as the main products.

Mechanism The proposed mechanism of AFO oxidation involves the intermediate formation of epoxide.

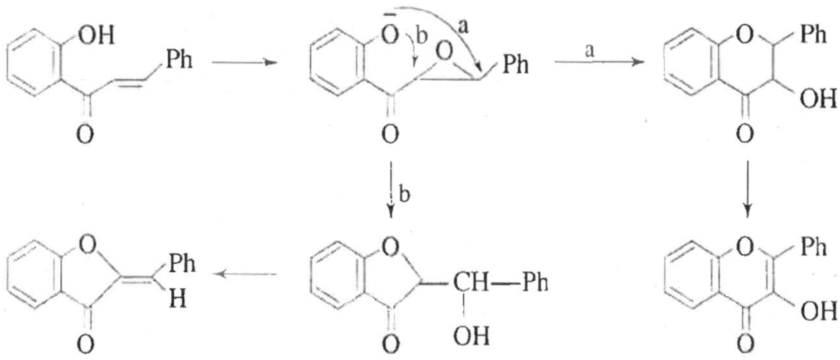

Aurone Flavonol

Dean and Podimuang suggested another mechanism for formation of flavonols as shown below:

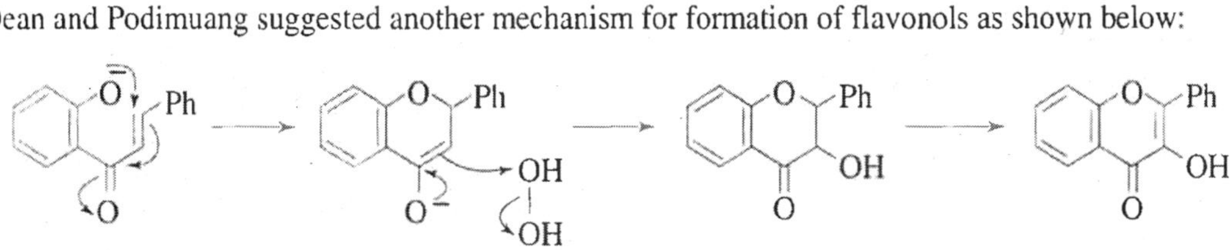

6.1 3-Hydroxyflavone

2′-Hydroxychalcone	: 1.42 g
Pyridine	: 10 ml
Aqueous potassium hydroxide (10%)	: 24 ml
Hydrogen peroxide (10%)	: 12 ml

2′-Hydroxychalcone (1.42 g, 0.006 mole) is dissolved in a mixture of pyridine (10 ml) and aqueous KOH (10%, 24 ml) and the resulting clear solution is warmed to 60–70°C in a water bath. Hydrogen peroxide (10%, 12 ml) is then added with stirring during a period of 5 minutes and the insoluble fraction separated is extracted with ether. The alkaline solution on acidification with hydrochloric acid gives 3-hydroxyflavone which is then filtered, washed with water and dried. It is finally recrystallized from ethanol. The yield is 0.75 g (50%); m.p. is 170°C.

7. ARNDT-EISTERT SYNTHESIS

It is a simple method for converting an acid into its next higher homologue or to a derivative of the homologous acid, such as an amide or ester. It involves three operations.

$$\underset{\text{Carboxylic acid}}{RCOOH} \xrightarrow[PCl_5]{SOCl_2 \text{ or}} \underset{\text{Acid chloride}}{RCOCl}$$

$$\underset{\text{Acid chloride}}{RCOCl} + \underset{\text{Diazomethane}}{2CH_2N_2} \longrightarrow \underset{\text{Diazoketone}}{RCOCHN_2}$$

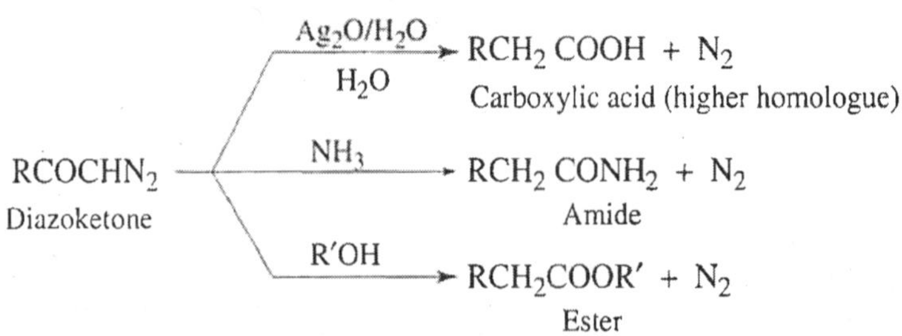

Mechanism

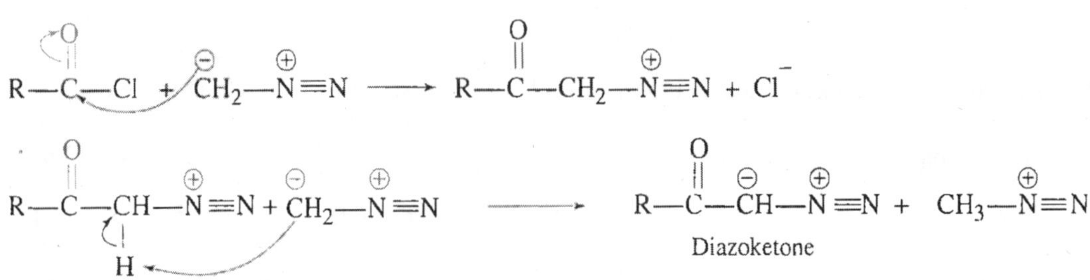

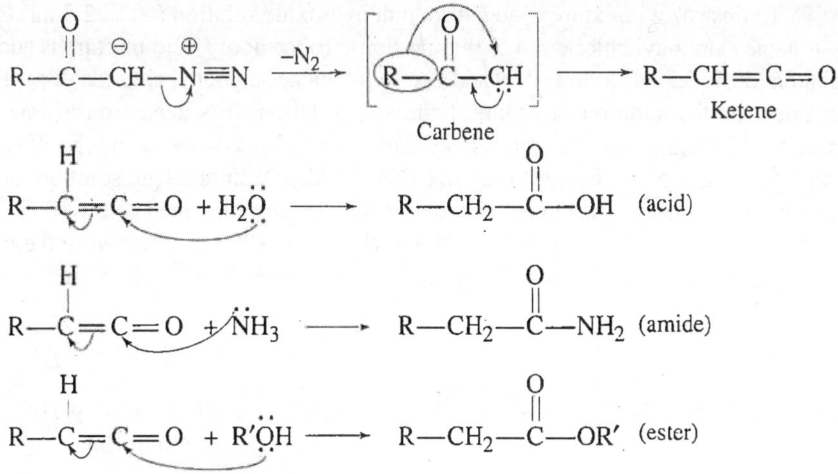

The final step involving the conversion of the diazoketone into an acid or its simple derivative is known as Wolff rearrangement.

7.1 α-Naphthyl Acetic Acid[1, 2]

α-Naphthoic acid	: 5 g
Diazomethane	: (prepared from 9 g p-tosyl sulfonyl methyl nitrosoamide)
Phosphorus pentachloride	: 6 g
Silver oxide	: 2 g
Sodium thiosulphate	: 3 g

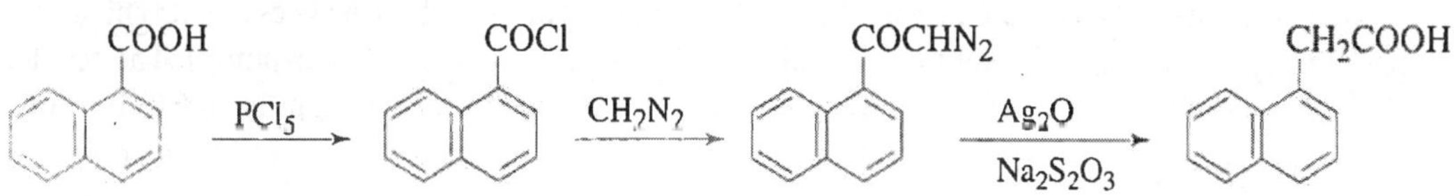

α-Naphthoic acid α-Naphthoyl chloride Diazoketone α-Naphthyl acetic acid

Step 1 Diazomethane

(a) p-Tosylsulfonylmethyl nitrosamide

p-Toluene sulphonyl chloride	: 16 g
Methylamine solution	: 40%, 8.7 ml
Sodium hydroxide solution	: 50%, 5 ml

p-Toluenesulphonyl chloride (9.5 g) is added in small lots at a time to a stirred solution of methylamine (8.7 ml, 40% solution) in a R.B. flask. The mixture becomes warm; the temperature is not allowed to

1. F. Arndt. B. Eistert, *Ber.*, 1935, **68**, 200.
2. A.B. Smith, *Chem. Commun.*, 1974, 695.

rise above 90°C. To this stirred mixture is added sodium hydroxide solution (50%, 2.5 ml) followed by addition of *p*-toluene sulphonyl chloride (4.5 g) in portions. To the above acid mixture is added sodium hydroxide solution (50%, 1.5 ml) followed by addition of *p*-toluene sulphonyl chloride (2 g). The mixture is rendered alkaline by the addition of sodium hydroxide solution. It is heated on a water bath with vigorous stirring for 15 minutes. The hot reaction mixture is poured into acetic acid (75 ml). The solution is cooled in an ice bath and to the cooled solution (5°C) is added sodium nitrite solution (6.3 g in 12.5 ml water) dropwise with stirring. The temperature during addition is maintained below 10°C. The nitroso derivative separates out as a yellow crystalline solid. Cold water (100 ml) is added to the mixture, the product is filtered, and washed with water. The yield is 16 g; m.p. is 58–60°C. It is used as such for the preparation of diazomethane.

$$p\text{-}CH_3C_6H_4SO_2Cl + CH_3NH_2 + NaOH \longrightarrow p\text{-}CH_3C_6H_4SO_2NHCH_3 + NaCl + H_2O$$

$$p\text{-}CH_3C_6H_4SO_2NHCH_3 + HNO_2 \xrightarrow{NaNO_2/HCl} \underset{p\text{-Tosylsulfonylmethyl-nitrosamide}}{p\text{-}CH_3C_6H_4SO_2N(NO)CH_3} + H_2O$$

(b) Diazomethane

A solution of the above nitroso derivative (4.3 g) in ether (60 ml) is added using a separatory funnel, to a solution of potassium hydroxide (1.0 g in 20 ml ethanol) in a distillation flask. The addition is made dropwise; the formed diazomethane is collected in ether in the receiver flask which is cooled externally by an ice-salt mixture (*precaution:* Only rubber corks should be used for the final assembly of the apparatus. Diazomethane explodes by the use of ground glass apparatus or by coming in contact with rough surface). The ethereal solution (distillate) contains 0.6–0.7 g diazomethane.

$$p\text{-}CH_3C_6H_4SO_2N(NO)CH_3 + KOH \longrightarrow p\text{-}CH_3C_6H_4SO_3H + CH_2N_2 + H_2O$$

Step 2 α-Naphthoyl chloride

α-Naphthoic acid (5 g, 0.029 mole) and phosphorus pentachloride (6 g) are placed in a distillation flask (50 ml capacity). The flask is heated (water bath) till no more hydrogen chloride is evolved. The formed phosphorus oxychloride is removed by distillation under reduced pressure (water pump). The residual product is then distilled to give *α*-naphthoyl chloride. The yield is 5 g (91%); b.p. is 168°C/10 mm.

Step 3 α-Naphthylacetic acid

α-Naphthoyl chloride (5 g, 0.026 mole) in dry ether (15 ml) is added dropwise to an ethereal solution of diazomethane (prepared from 9 g of *p*-tosylsulfonyl methylnitrosamide at 10°C. The reaction mixture is kept for 6 hours at 20–25°C and the ether removed in vacue at 30°C. The residual *α*-naphthyl diazoketone is crystallised from bezene. The yield is 5 g (98.1%) m.p. is 54–55°C.

A solution of the above diazoketone (3 g, 0.015 mole) in dry dioxane (25 ml) is added to a suspension of silver oxide (2 g), sodium thiosulphate (3 g) in water (100 ml) at 50–60°C with stirring. The stirring is continued for one hour till evolution of nitrogen ceases. The solution is then cooled, rendered acidic (dilute nitric acid) and the precipitated *α*-naphthylacetic acid is filtered. It is crystallised from water using animal charcoal. The yield is 1.4 g (50%); m.p. is 130°C.

8. BAEYER-VILLIGER OXIDATION

It involves the oxidation of aldehydes or ketones with peracids to give esters, which on hydrolysis give the hydroxy compounds. Thus,

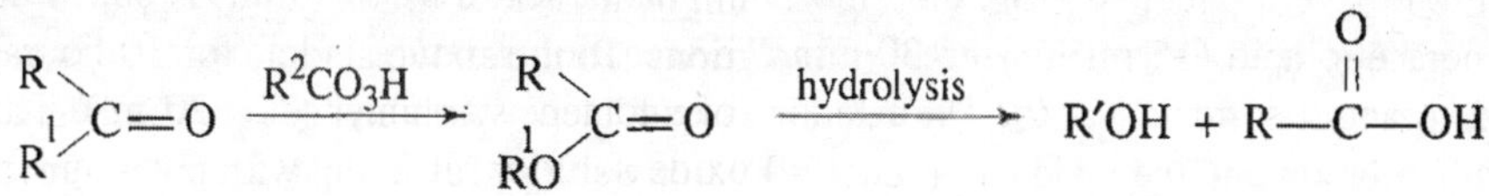

Typical peracids which are used include peroxybenzoic acid, $C_6H_5CO_3H$, performic acid, HCO_3H, *m*-chloroperoxybenzoic acid, *m*-$ClC_6H_4CO_3H$, peroxy-acetic acid, CH_3CO_3H, and trifluoroperoxy acetic acid, CF_3CO_3H.

Mechanism

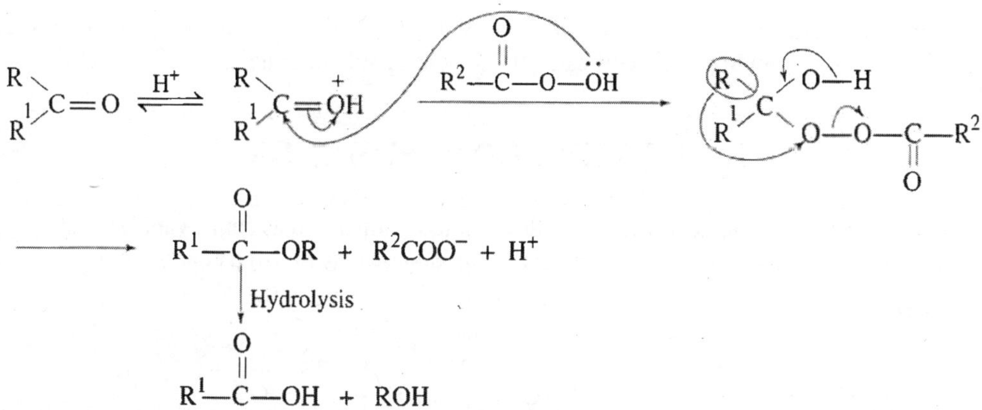

If $R' = H$, the aldehyde, RCHO, is converted into the hydroxy compound, R—OH.

8.1 3,4-Dimethoxyphenol

It is obtained by oxidation of veratraldehyde with peracetic acid.

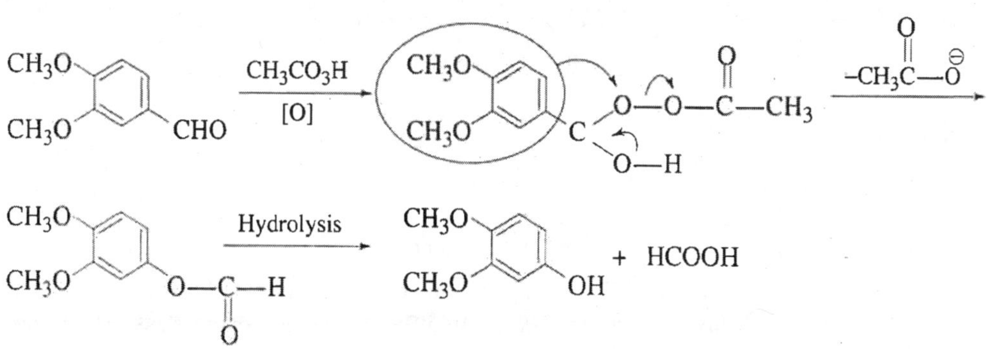

Veratraldehyde	: 5 g
Acetic acid (glacial)	: 30 ml
Peracetic acid solution	: 15 ml

To a solution of veratraldehyde (5g, 0.03 mole) in glacial acetic acid (30 ml) is added dropwise a solution of peracetic acid (15 ml) during 30 min. The reaction mixture is left for 10 hours and then concentrated to about 5 ml in vacuo. The residue is extracted with ether (2 × 20 ml). The ether is removed by distillation and the residual product is hydroxylsed by refluxing with potassium hydroxide (10 g) in aqueous alcohol (1:4, 100 ml) for 1 hour. The reaction mixture is concentrated in vacuo almost to dryness. It is dissolved in water (20 ml), the solution rendered acidic with dilute sulphuric acid and extracted with ether. The ether extract is then dried (sodium sulphate) and distilled. The oily residue thus obtained is purified by column chromatography over silica gel (2 × 30 cm). Elution with benzene gave 3, 4-dimethoxy phenol as yellow solid. The yield is 3 g (65.2%). It is crystallised from benzene, m.p. is 78–80°C.

Note: Peracetic acid solution is prepared in situ by adding 1 part of hydrogen peroxide (30%) to glacial acetic acid (3 parts) in presence of catalytic amount of sulphuric acid.

9. BENZOIN CONDENSATION

Aromatic aldehydes on treatment with cyanide ions (sodium or potassium cyanide) undergo self condensation to form α-hydroxy ketone (benzoin). This is known as Benzoin condensation.

Mechanism

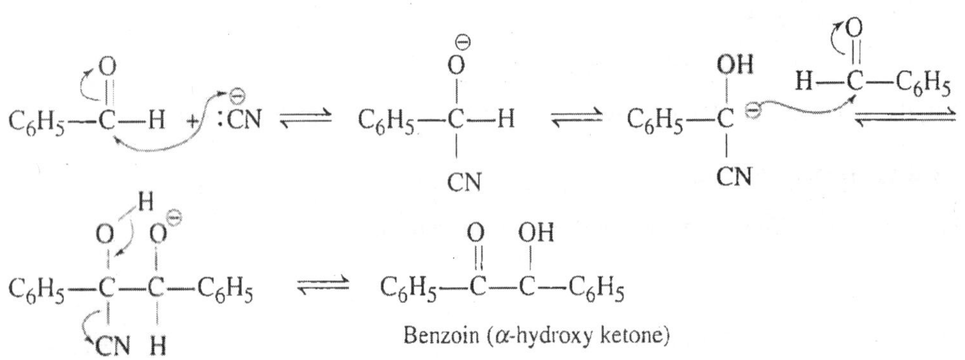

9.1 Benzoin

Benzaldehyde : 18 ml
Rectified spirit : 40 ml
Potassium cyanide : 2.5 g (in 10 ml water)

A mixture of benzaldehyde (18 ml, 0.177 mole), potassium cyanide solution (2.5 g in 10 ml water) and rectified spirit (40 ml) is refluxed gently for 30 min in a 250 ml round bottom flask. The solution is cooled in an ice-bath. The separated crude product is filtered at the pump, and washed with water and a little cold rectified spirit. It is crystallised from rectified spirit in colourless prisms. The yield is 16 g (42.7%); m.p. is 137°C.

Caution: Potassium cyanide is deadly poisonous. It should be handled with extreme care using hand gloves. The filtrate should be disposed off in the sink and washed with a large quantity of water.

10. BAKER-VENKATARAMAN REARRANGEMENT[1, 2]

The base catalysed rearrangement of *o*-acyloxyketones to β-diketones is known as Baker-Venkataraman rearrangement. Thus,

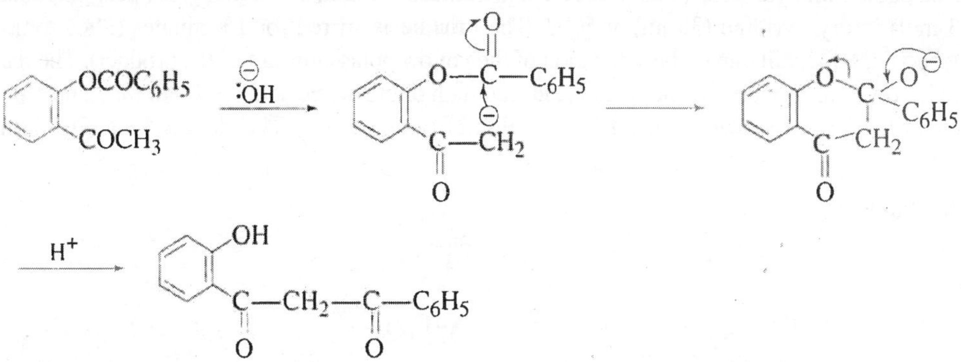

The β-diketones are important intermediates for the synthesis of flavones and chromones.

10.1 Flavone

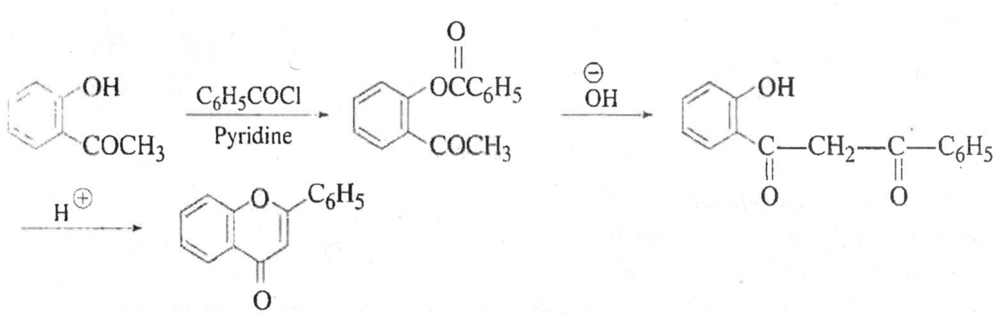

Step 1 o-Benzoyloxyacetophenone

o-Hydroxyacetophanone	: 6.8 g
Benzoyl chloride	: 8 ml
Pyridine (dry)	: 10 ml

A mixture of *o*-hydroxyacetophenone (6.8 g, 0.05 mole), benzoyl chloride (8 ml, 0.07 mole) and dry pyridine (10 ml) is shaken for 20 min. The reaction mixture which becomes warm is poured over crushed ice (about 100 g) and hydrochloric acid (1 M, 250 ml). The separated product is filtered, washed with ice-cold methanol (10 ml) and then with water (25 ml).

It is crystallised from methanol. The yield is 9.2g (80%); m.p. is 87–88°C.

1. W. BAKER, *J. Chem. Soc.*, **1933**, 1381.
2. H.S. MAHAL, K. VENKATARAMAN, *J. Chem. Soc.* **1934**, 1767.

Step 2 o-Hydroxydibenzoylmethane

o-Benzoyloxy'acetophenone : 7.2 g
Potassium hydroxide : 2.55 g
Pyridine : 30 ml

Powdered potassium hydroxide (2.55 g) is added to a stirred solution of *o*-benzoyloxy acetophonone (7.2 g, 0.03 mol) in dry pyridine (30 ml) at 50°C. The mixture is stirred for 15 minutes (after some time stirring becomes difficult due to the separation of the yellow potassium salt of the product). The reaction mixture is cooled to room temperature and acidified with dilute acetic acid (10%, 40 ml). The separated product is filtered, washed with water and crystallised from methanol. The yield is 5.8 g (80%); m.p. is 121°C.

Step 3 Flavone

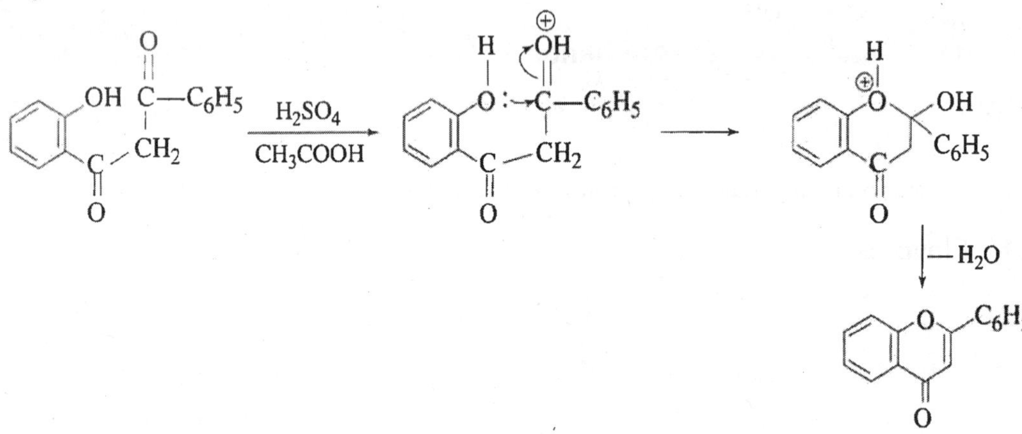

Flavone

o-Hydroxydibenzoylmethane : 3.6 g
Acetic acid (glacial) : 20 ml
Conc. sulphuric acid : 1 ml

Concentrated sulphuric acid (1 ml) is added with stirring to a solution of *o*-hydroxydibenzoylmethane (3.6 g, 0.015 mole) in glacial acetic acid (20 ml). The mixture is heated in a water bath for one hour with occasional stirring and poured over crushed ice (about 100 g). The separated flavone is filtered, washed with water and dried at 50°C. It is crystallised from petroleum ether. The yield is 3.2 g (95%); m.p. is 98°C.

11. BECKMANN REARRANGEMENT[1, 2]

Ketoximes on treatment with an acid catalyst (e.g., Phosphorus pentachloride, phosphorus pentoxide, phosphorus oxychloride, phosphoric acid, sulphuric acid) undergo a rearrangement to form *N*-substituted amides. The reaction is stereospecific in that the migrating group is *anti* to the oxime hydroxyl group; the migrating group approaches the nitrogen atom from the opposite to the departing oxygen.

1. S. Fujita, K. Koyona and Y. Inagaki, *Synthesis*, 1982, 68.
2. K. Hattori, Y. Matsumura T. Miyazaki, K. Marouka and Y. Yamamoto, *J. Am. Chem. Soc.*, 1981, **103**, 7368.

Mechanism:

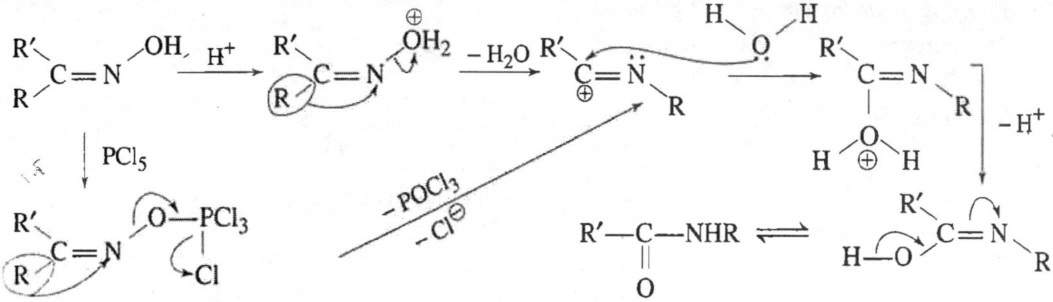

N-Substituted amide

11.1 Benzanilide from Benzophenone Oxime

Mechanism of oxime formation: Nucleophilic addition of hydroxylamine to the carbonyl group of an aldehyde or ketone takes place.

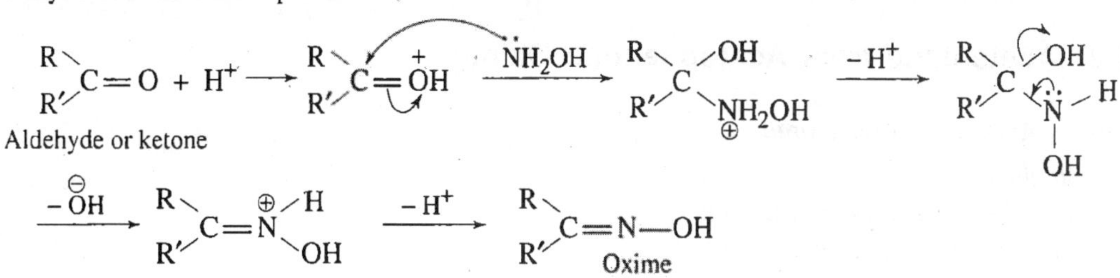

Step 1 Benzophenone Oxime

Benzophenone	: 4 g
Hydroxylamine hydrochloride	: 2.4 g
Sodium hydroxide	: 4.8 g
Ethanol	: 15 ml
Conc. hydrochloric acid	: 12 ml

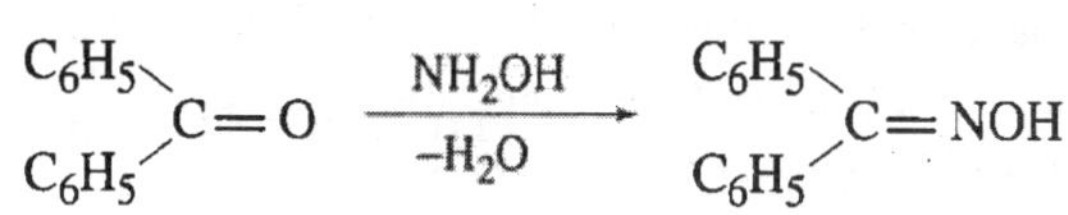

Benzophenone Benzophenone oxime

Benzophenone (4 g, 0.022 mole) and hydroxylamine hydrochloride (2.4 g, 0.035 mole) are dissolved in a mixture of ethanol (15 ml) and water (3 ml). The solution is continuously stirred and a solution of sodium hydroxide (4.8 g in 5 ml water) is added in portions. In case the reaction becomes vigorous, the flask is cooled under tap water. After the addition is over, the reaction mixture is refluxed for 20 minutes. It is cooled and diluted with water (40 ml). The separated unreacted benzophenone is filtered and the filtrate cooled and poured while stirring into dilute hydrochloric acid (12 ml concentrated hydrochloric acid in 75 ml water). The precipitated oxime is filtered and washed with cold water. It is crystallised from methanol. The yield is 4 g (93%); m.p. is 142°C. The oxime is stored in vacuum desiccator.

Step 2 Benzanilide

Benzophenone oxime : 3 g
Dry ether : 30 ml
Phosphorus pentachloride : 4.5 g
(or thionyl chloride, 4.5 ml)

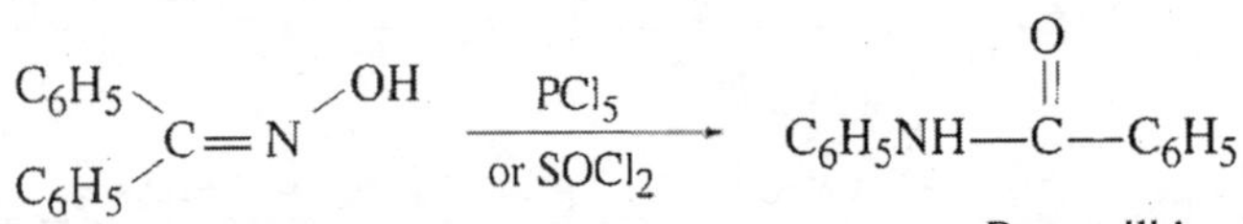

Benzanilide

Benzophenone oxime (3g, 0.015 mole) is dissolved in anhydrous ether (30 ml) in a conical flask. The solution is cooled (ice bath) and finely powdered phosphorus pentachloride (4.5 g) or thionyl chloride (4.5 ml) is added while stirring the ethereal solution. The reaction mixture is stirred thoroughly with a glass rod and the solvent removed (water bath). The residue is poured into ice-water and the mixture stirred. The separated product is filtered, washed with water and crystallised from alcohol. The yield is 2.5 g (83.3%); m.p. is 163°C.

11.2 Acetanilide from Acetophenone Oxime

Step 1 Acetophenone oxime

Acetophenone : 3 ml
Hydroxylamine hydrochloride : 3 g
Sodium acetate (dihydrate) : 3 g
Ethanol : 8 ml

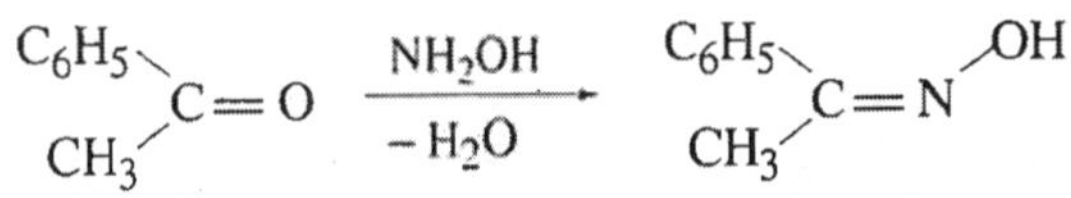

Acetophenone — Acetophenone oxime

Acetophenone (3 ml, 0.026 mole) is dissolved in aqueous alcohol (8 ml alcohol and 28 ml water) and then sodium acetate dihydrate (3g) and hydroxylamine hydrochloride (3g) are added. The mixture is heated in a boiling water bath for 10 minutes and then cooled (ice-salt bath). The separated oxime is filtered and crystallised from water (the crystallisation is made fast by cooling the solution and scratching the sides with a glass rod), yield 3 g (86.4%), m.p. 59°C.

Step 2 Acetanilide

Acetophenone oxime : 3 g
Conc. sulphuric acid : 6 ml
(or thionyl chloride 3 ml)

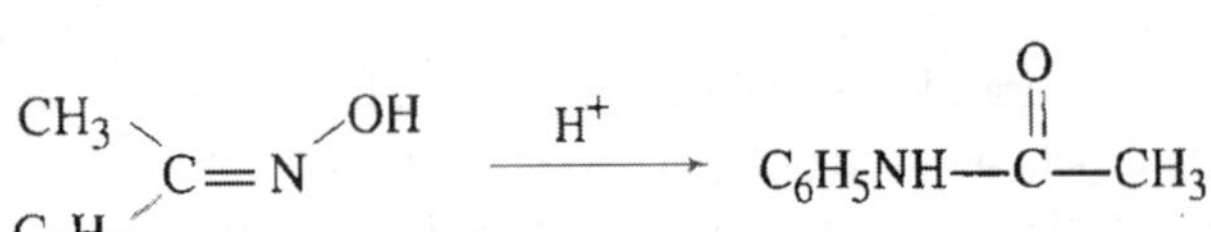

Acetophenone oxime — Acetanilide

Concentrated sulphuric acid (6 ml) is heated in a conical flask in a boiling water bath. In case thionyl chloride is used the flask is heated only upto 70–80°C. Acelophenone oxime (3 g, 0.022 mole) is added slowly with stirring to the warm acid. The reaction mixture is kept stirring for 10–15 minutes. It is cooled and poured slowly with stirring over crushed ice. The separated product is filtered and washed with water. It is recrystallised from boiling water. The yield is 2.7 g (90%); m.p. is 114°C.

12. BLANC REACTION[1] (CHLOROMETHYLATION)

This is a simple method for the replacement of a hydrogen atom in an aromatic compound by a chloromethyl (CH_2Cl) group in a single step. In this method, the aromatic compound is treated with formaldehyde and hydrogen chloride in presence of a catalyst such as zinc chloride or aluminium chloride. Thus,

$$H_2C{=}O + HCl \rightleftharpoons Cl^- + \left[H_2C{=}\overset{\oplus}{O}H \longleftrightarrow H_2\overset{\oplus}{C}{-}OH\right]$$

$$ArH + \overset{\oplus}{C}H_2OH \longrightarrow ArCH_2OH + H^+$$

$$ArCH_2OH + HCl \longrightarrow ArCH_2Cl + H_2O$$

12.1 1,2-Bis(chloromethyl)-4,5-dimethylbenzene

o-Xylene	: 17.5 g
Paraformaldehyde	: 15 g
Conc. hydrochloric acid	: 100 ml

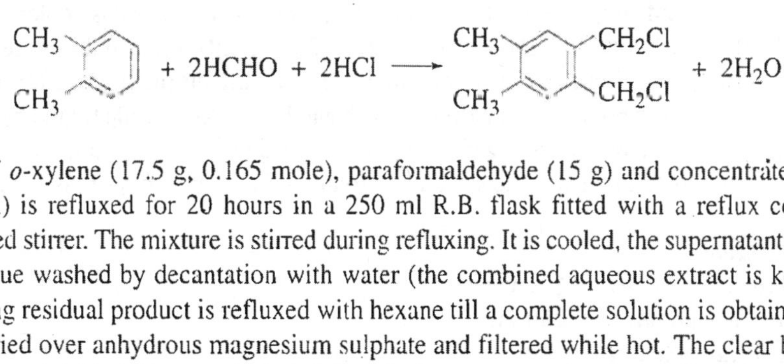

A mixture of *o*-xylene (17.5 g, 0.165 mole), paraformaldehyde (15 g) and concentrated hydrochloric acid (100 ml) is refluxed for 20 hours in a 250 ml R.B. flask fitted with a reflux condenser and a mercury sealed stirrer. The mixture is stirred during refluxing. It is cooled, the supernatant liquid decanted and the residue washed by decantation with water (the combined aqueous extract is kept separately). The remaining residual product is refluxed with hexane till a complete solution is obtained. The hexane solution is dried over anhydrous magnesium sulphate and filtered while hot. The clear hexane solution on cooling gave 1,2-bis (chloromethyl)-4,5-dimethylbenzene as colourless solid. It is crystallised from hexane. The yield is 10 g (30%); m.p. is 100°C.

The aqueous extract obtained above is stirred with 40 per cent aqueous sodium hydroxide solution and kept for 24 hours before it is discarded. This treatment is essential to destroy the side product, bis (chloromethyl ether). $ClCH_2O\ CH_2Cl$ and is known to cause lung cancer in experimental animals.

1. G. Blanc, *Bull. Soc. Chim.* France, 1923, **33**, 313.

13. BENZIDINE REARRANGEMENT[1]

The acid catalysed rearrangement of hydrazobenzene to 4,4′-diaminobiphenyl is known as benzidine rearrangement.

Mechanism

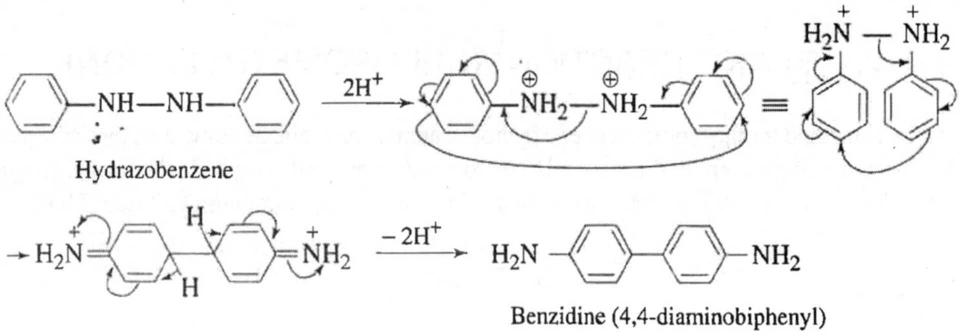

13.1 Benzidine

Step 1 Hydrazobenzene

(a) *From Azobenzene*

Azobenzene	: 5 g
Rectified spirit	: 90 ml
Hydrazine (95%)	: 10 ml
Aqueous copper sulphate solution (1%)	: 2 ml
Hydrogen peroxide (30%)	: 5 ml

To a solution of azobenzene (5g, 0.027 mole) in rectified spirit (90 ml) is added at room temperature hydrazine (95%, 10 ml) and copper sulphate solution (1%, 2 ml). The mixture is cooled in an ice bath, while shaking the flask, to about 0°C. Hydrogen peroxide (30%, 5 ml) is added dropwise to the cooled reaction mixture during 10–15 minutes. The mixture is kept stirring all the time. After the addition is complete, the reaction flask is removed from the cold bath and heated for a while till the hydrazobenzene dissolves. The solution is filtered and to the clear filtrate is added water (15–20 ml) until crystallisation starts. The mixture is cooled in ice and the separated product is filtered. The yield is 5.0 g; (99%) m.p. is 125–26°C.

(b) *From Nitrobenzene*

Nitrobenzene	: 10 ml
Absolute methanol	: 250 ml
Magnesium turnings	: 20 g
Iodine	: Few crystals

$$2C_6H_5NO_2 + 5Mg + 10CH_3OH \longrightarrow C_6H_5NHNHC_6H_5 + 5Mg(OCH_3)_2 + 4H_2O$$

A mixture of nitrobenzene (10 ml, 0.097 mole) and absolute methanol (150 ml) is taken in a dry round bottom flask fitted with a reflux condenser. A small crystal of iodine is added and then magnesium turnings (12 g) added. In case the reaction does not start within 2–3 minutes, the reaction mixture is

1. T. SHERADSKY and S. AURAMORK'S-GRISARV, *J. Heterocyclic Chemistry*, 1980, **17**, 189.

warmed on a water bath until the reaction starts. If the reaction becomes too vigorous, it is cooled momentarily in a cold water bath. More absolute methanol (100 ml) is added and the flask is shaken. When the reaction has subsided, more magnesium turning (8g) is added and the reaction is allowed to proceed as above. The reaction mixture is heated on a water bath (80–85°C) until mixture is colourless or light pale yellow. During heating, the reaction mixture is shaken frequently. The hot reaction mixture is poured into a beaker and water (200 ml) is added to it. The mixture is acidified by the dropwise addition of acetic acid till the mixture is acidic to litmus, the mixture is stirred during acidification. The separated magnesium hydroxide dissolves and a pale yellow residue of hydrazobenzene floats on the surface. It is filtered at the pump (the magnesium granules are allowed to stay behind in the beaker) and washed with water. It is recrystallised from 90 per cent ethanol containing a little dissolved sulphur dioxide (with minimum exposure to air to avoid oxidation of the product). The yield is 5.5g (30.7%); m.p. 125–26°C. The product is dried in a vacuum desiccator.

Step 2 Benzidine

Hydrazobenzene	: 10 g
Methanol	: 75 ml
Conc. hydrochloric acid	: 25 ml
Stannous chloride solution (obtained by dissolving 3 g stannous chloride dihydrate in 3 ml Conc. HCl and 10 ml water)	: 3 ml

A mixture of methanol (75 ml), water (40 ml), concentrated hydrochloric acid (25 ml) and stannous chloride solution (3 ml) is cooled to 0–2°C (ice bath) and then poured on the hydrazobenzene (10 g, 0.054 mole) placed in another flask. The mixture is shaken in an ice bath for 15 minutes and then heated gently to dissolve the precipitated benzidine dihydrochloride. In case the solution is yellow, a few drops of stannous chloride solution are added during heating. When the reaction mixture is being heated, concentrated hydrochloric acid (35 ml) is added dropwise, while constantly shaking the flask. The reaction mixture is cooled in an ice-salt mixture. The separated benzidine dihydrochloride is filtered, washed once with dilute hydrochloric acid and then with methanol. The yield is 14 g. The free base is regenerated from the salt by dissolving in water (100 ml) and then by heating and shaking it. A solution of sodium bicarbonate (7.2 g) in water (60 ml) (solution obtained by warming) is added dropwise to the above aqueous solution of the dihydrochloride till it is alkaline to litmus. The mixture is cooled (ice water) and the separated product filtered and washed with water. It is recrystallised from dilute alcohol and dried in a vacuum dessicator. The yield is 5.5 g (55%); m.p. is 128°C (anhydrous sample on rapid heating).

Note: (1)* Sodium bicarbonate is preferred for neutralisation. Sodium hydroxide is not to be used since it promotes air oxidation of benzidine. (2)* The yield is low, since 2, 4-diaminobiphenyl is also formed in 30 per cent yield as a byproduct. It remains dissolved in the methanolic solution after benzidine dihydrochloride has been isolated.

14. BENZIL-BENZILIC ACID REARRANGEMENT[1, 2]

α-Diketones on treatment with a base, rearrange to give the salts of α-hydroxy acids.

---

1. J. Liebig. *Ann.* 1838, **25**, 27.
2. N. Zinin. *Ann.* 1939, **31**, 329.

Mechanism:

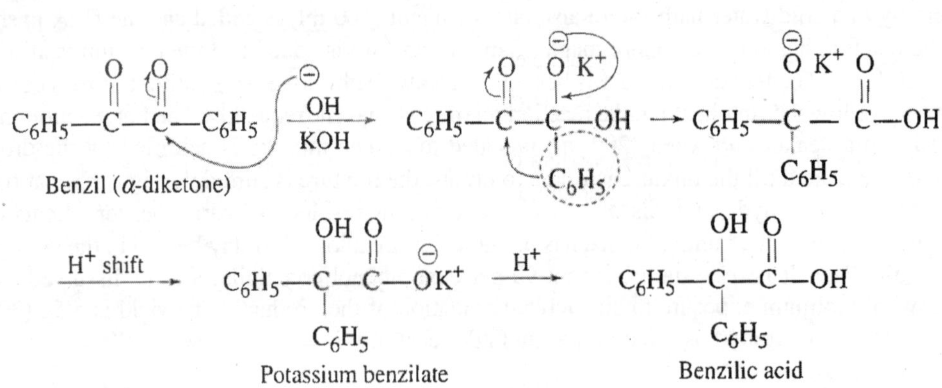

14.1 Benzilic Acid

Step 1 Benzil

Benzoin	: 10 g
Acetic acid	: 50 ml
Conc. nitric acid	: 25 ml

A mixture of benzoin (10 g, 0.047 mole), acetic acid (50 ml) and concentrated nitric acid (25 ml) is heated in a round bottom flask (boiling water bath) for 2 hours. The flask is connected to a trap containing sodium hydroxide solution to absorb oxides of nitrogen. Alternatively, the reaction is carried out in a fume-cupboard. The reaction mixture is cooled and poured with stirring into an ice-water mixture. The mixture is stirred well and the separated product filtered and washed with water. It is crystallised from rectified spirit as yellow prisms. The yield is 9 g (91%); m.p. is 95°C.

Step 2 Benzilic Acid

Benzil	: 8 g
Potassium hydroxide pellets	: 10 g (in 20 ml H_2O)
Rectified spirit	: 25 ml
Conc. hydrochloric acid	: 35 ml

A mixture of benzil (8 g, 0.038 mole), potassium hydroxide solution (10 g dissolved in 20 ml water) and rectified spirit (25 ml) is refluxed in a water bath for 12 minutes. The hot reaction mixture is poured into water (150 ml) contained in a beaker. In case the reaction does not go to completion, a colloidal solution is obtained due to separation of unreacted benzil. The mixture is treated with animal charcoal (2 g), stirred well and filtered. The clear filtrate is added with stirring to a mixture of crushed ice (150 g) and concentrated hydrochloric acid (35 ml). The separated benzilic acid is filtered and washed with cold water (until free from chloride). It is crystallised from benzene or hot water as colourless needles. The yield is 6 g (69%); m.p. is 150°C.

Note: Banzilic acid is also obtained by the oxidation of benzoin with alkaline potassium bromate (See page 169).

15. BOUVEAULT-BLANC REDUCTION[1]

Esters on reduction with sodium and absolute ethanol give primary alcohols.

$$RCOOC_2H_5 + 4[H] \xrightarrow{Na/C_2H_5OH} RCH_2OH + C_2H_5OH$$

The method can also be applied to the esters of dicarboxylic acids for the preparation of α, ω-diols. Esters can also be reduced to primary alcohol by lithium aluminium hydride (see page 190).

15.1 Pentan-1-ol (amyl alcohol)

It is obtained by the reduction of ethyl valerate with sodium and absolute ethanol.

$$\underset{\text{Ethyl Valerate}}{CH_3(CH_2)_2CH_2COOC_2H_5} \xrightarrow[Na/C_2H_5OH]{4[H]} \underset{\text{Amyl alcohol}}{CH_3(CH_2)_2CH_2CH_2OH} + C_2H_5OH$$

Ethyl valerate : 39 g (44.7 ml)
Absolute alcohol : 600 ml
Sodium : 71.3 g

Clean sodium (71.3 g, 3.1 mole) cut into small pieces is added in small lots to a mixture of ethyl valerate (39 g, 44.7 ml. 0.3 mole) and super dry ethyl alcohol (600 ml) in a two necked R.B. flask fitted with a reflux condenser. The addition of sodium is done at such a rate that the vigorous refluxing is continuous (25–30 minutes). The mixture is refluxed in an oil bath for one hour so that all the sodium dissolves. The mixture is subjected to fractional distillation so that approximately 200 ml of absolute alcohol is recovered. To the residue is added water (300 ml) and the distillation continued (oil bath at 110–120°C). Approximately 400–450 ml of approximately 90 per cent ethanol is recovered. The mixture is steam distilled till about 175 ml of the distillate is collected. The crude amyl alcohol is separated from the steam distillate, dried and distilled through a short column. The fraction boiling at 137°C–139°C is collected. The yield is 19 g (71 %).

16. BUCHERER HYDANTOIN SYNTHESIS[2-4]

It consists of heating cyanohydrins of aldehydes and ketones with ammonium carbonate

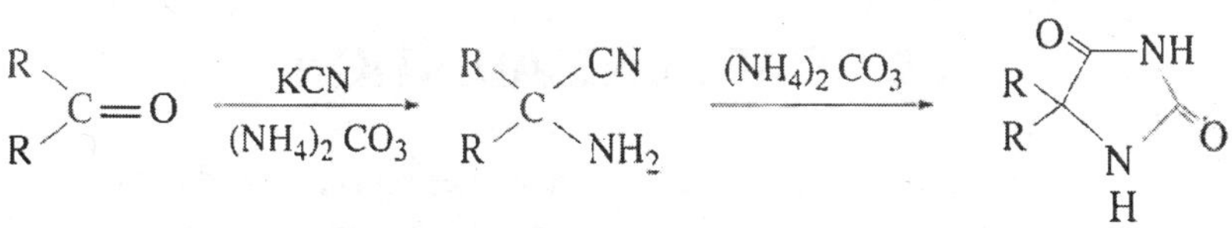

The hydantoins on hydrolysis yield α-amino acids (see strecker synthesis).

1. BOUVEAULT and G. BLANE *Compt Rect.*, 1903, *136*, 1676; Bull. Soc. Chim. France., 1904, **31**, 666.
2. H.T. BUCHERER, H.T. FISCHBECK, *J. Prakt. Chem.*, 1934, **140**, 19.
3. H.T. BUCHERER, W. STEINER, *J. Prakt. Chem.*, 1934, **140**, 291.
4. H.R. HENZA, R.J. SPEER, *J. Am. Chem. Soc.*, 1942, **64**, 522.

16.1 5,5-Dimethylhydantoin

It is obtained from acetone cyanohydrin by reaction with ammonium carbonate

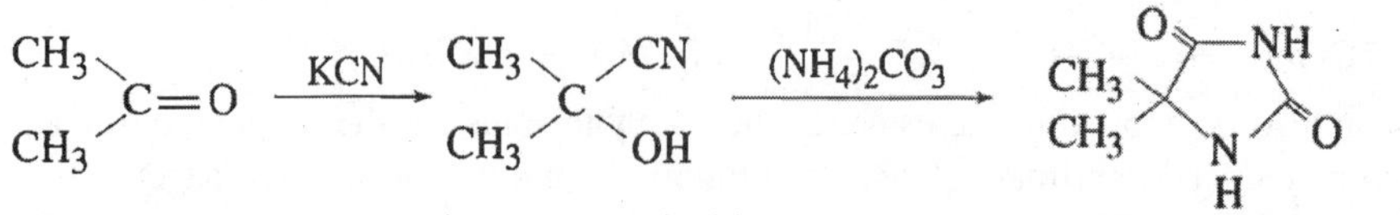

Acetone   Acetone cyanohydrin   5, 5-Dimethylhydantoin

Step 1 Acetone cyanohydrin

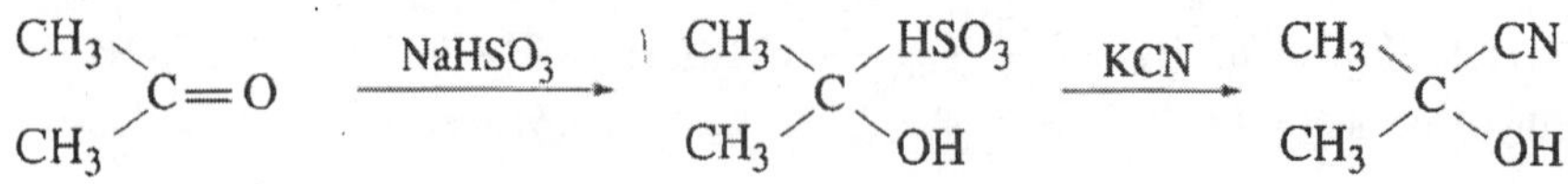

Acetone : 11.5 g
Sodium metabisulphite : 22 g
Potassium cyanide : 12 g

Acetone (11.5 g, 14.5 ml, 0.2 mole) is added with shaking to a solution of sodium metabisulphite (22 g) in water (40 ml) in a 250 ml R.B. flask. A solution of potassium cyanide (12 g) in cold water (40 ml) is added slowly with stirring. The upper layer of acetone cyanohydrin is separated (separatory funnel). It is then dried (sodium sulphate). The solvent is removed by distillation. The product can be used as such for the next step. The yield is 15 g (89%); b.p. is 80–82°C/15 mm.

Step 2 5,5-Dimethylhydantoin

Acetone cyanohydrin : 14.2 g

Ammonium carbonate : 25 g

A mixture of acetone cyanohydrin (14.2 g, 0.167 mole) and powdered ammonium carbonate (25 g) is warmed on a water bath (fume cupboard). The mixture is stirred and heated at 70–80°C for 3 hours. Subsequently, the reaction mixture is heated for 30 minutes at 90°C and cooled. The separated solid product is crystallised from hot water. The yield is 9.5 g (44.6%); m.p. is 178°C.

17. CARBENE ADDITION

Carbenes are neutral divalent carbon compounds which are very reactive and exhibit a high degree of stereospecificity in their reactions. The carbenes are generated (simplest method) by the treatment of chloroform or bromoform with bases. Thus,

$$\underset{\text{Bromoform}}{CHBr_3} \xrightarrow{\bar{OH}} :\bar{C}Br_3 \xrightarrow{-Br^-} \underset{\text{Dibromocarbene}}{:CBr_2}$$

The carbenes add on to the double bond of an alkene to form cyclopropane.

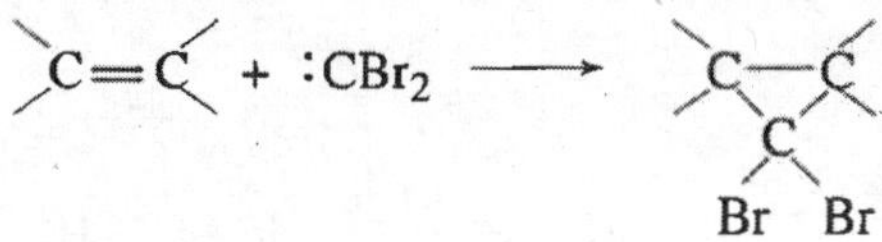

17.1 7,7-Dibromobicyclo(4.1.0) Heptane

Cyclohexene : 20 ml
Bromoform : 2.25 ml
Potassium metal : 8.85 g
tertiary Butanol : 25 ml

$$CHBr_3 + {}^{-}OC_4H_9 \longrightarrow C_4H_9OH + :\bar{C}Br_3 \xrightarrow{-Br^-} :CBr_2$$

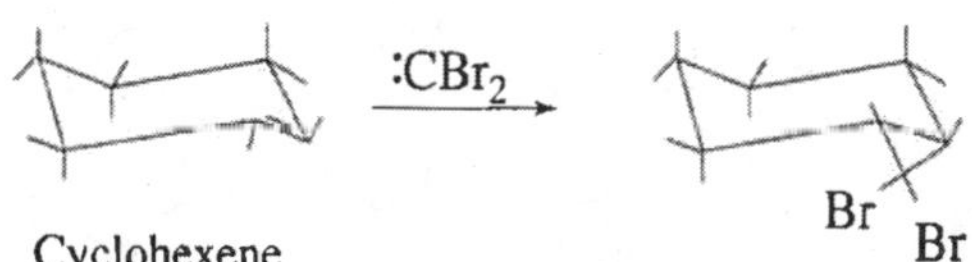

Cyclohexene — 7, 7-Dibromo [4.1.0] heptane

To *tert.* butanol (25 ml) in a dry conical flask is added clean potassium metal (8.85 g) in small lots. When all the potassium has reacted, the mixture is cooled in ice and dry cyclohexene (20 ml, 0.197 mole) is added with constant shaking. Subsequently bromoform (2.25 ml) is added slowly and the mixture shaken for 15 minutes more and poured into water. The aqueous mixture is extracted (separatory funnel) with petroleum ether (4 × 25 ml). The combined petroleum ether extract is dried over anhydrous sodium sulphate and distilled. The residual product of the 7, 7-dibromo bicyclo [4.1.0] heptane is distilled under reduced pressure. The yield is 10 g (20%); b.p. is 100°C/8 mm.

Note: For other reactions of carbenes and mechanism see Reimer-Tiemann reaction—preparation of salicylaldehyde and β-hydroxynaphthaldehyde.

18. CANNIZZARO REACTION[1-3]

Aldehydes (which do not have α-hydrogen) on treatment with concentrated alkali undergo self oxidation and reduction to produce alcohol and the salt of the corresponding acid. Thus,

$$\underset{\text{Benzaldehyde}}{C_6H_5CHO + C_6H_5CHO} \xrightarrow{KOH} \underset{\text{Benzyl alcohol}}{C_6H_5CH_2OH} + \underset{\text{Potassium benzoate}}{C_6H_5CO\bar{O}\overset{+}{K}}$$

This disproportionation reaction is known as cannizzaro reaction.

Mechanism: The mechanism most likely involves the production (by interaction of the aldehyde with hydroxide ion) of the reducing anions (A) and (B). Either of these anions may transfer a hydride ion to a carbonyl carbon atom in another aldehyde molecule.

1. S. Cannizzaro, *Ann.*, 1853, **88**, 129.
2. K. List and H. Limpricht, *Ann.*, 1854, **90**, 180.
3. C.G. Swain, A.L. Powell, W.A. Shippard and C.R. Morgan, *J. Am. Chem. Soc.*, 1979, **101**, 3576.

Thus,

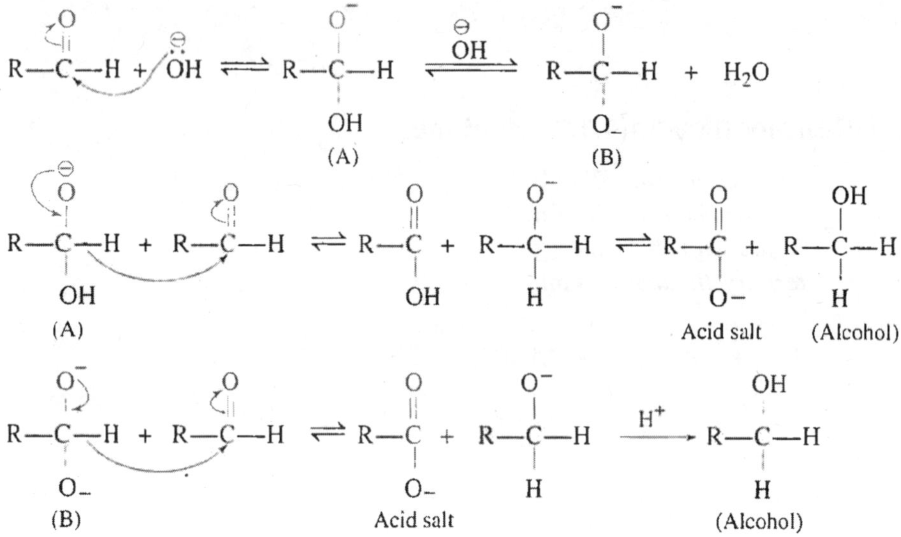

18.1 Benzyl Alcohol and Benzoic Acid from Benzaldehyde

Benzaldehyde : 20 ml
Potassium hydroxide : 12.5 g
(pellets)

$$2C_6H_5CHO \xrightarrow{KOH} C_6H_5COOK + C_6H_5CH_2OH$$

Potassium hydroxide pellets (12.5 g) are dissolved in water (60 ml) in a round bottom flask and freshly distilled benzaldehyde (20 ml, 0.197 mole) is added. The mixture is cooled and water added just to dissolve any precipitated potassium benzoate. The mixture is extracted with ether (3 × 30 ml). The clear alkaline solution is cooled and acidified with cold hydrochloric acid (1:1, 60 ml). The acidified solution is cooled and the separated benzoic acid is filtered and crystallised from hot water as colourless needles. The yield is 8.5 g (70.8%); m.p. is 121°C.

The combined ether extract (obtained above) is dried over anhydrous magnesium sulphate and distilled. The residue containing benzyl alcohol is distilled using a flame. The yield is 5.5 g (52.4%) b.p. is 205–206°C.

19. CHLOROSULPHONATION

An appropriately substituted organic compound (eg. acetanilide) can be chlorosulphonated with chlorosulphonic acid. The SO_2Cl group enters the reactive position. Thus,

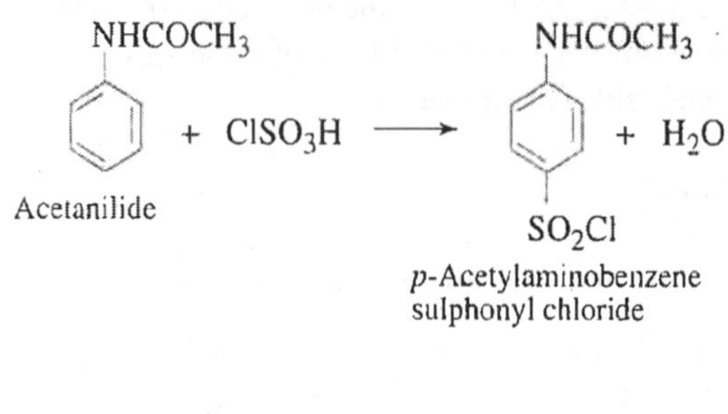

The SO_2Cl group is helpful for makikg molecules containing this moeity.
An example of chlorosulphonation is given in the preparation of sulphanilamide (see drugs).

20. CLAISEN CONDENSATION[1, 2]

Base catalysed condensation of an ester containing an α-hydrogen atom with a molecule of the same ester or a different one to give β-keto esters is known as Claisen condensation. Thus, the reaction of ethyl acetate with sodium ethoxide gives ethyl acetoacetate.

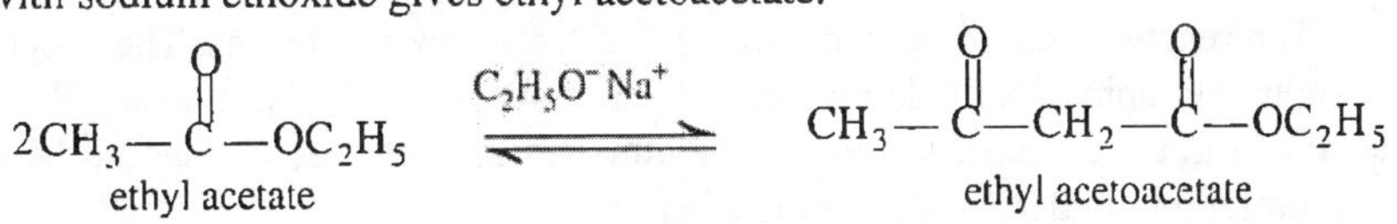

Mechanism:

The generally accepted mechanism for the Claisen condensation (shown here with ethyl acetate) is:

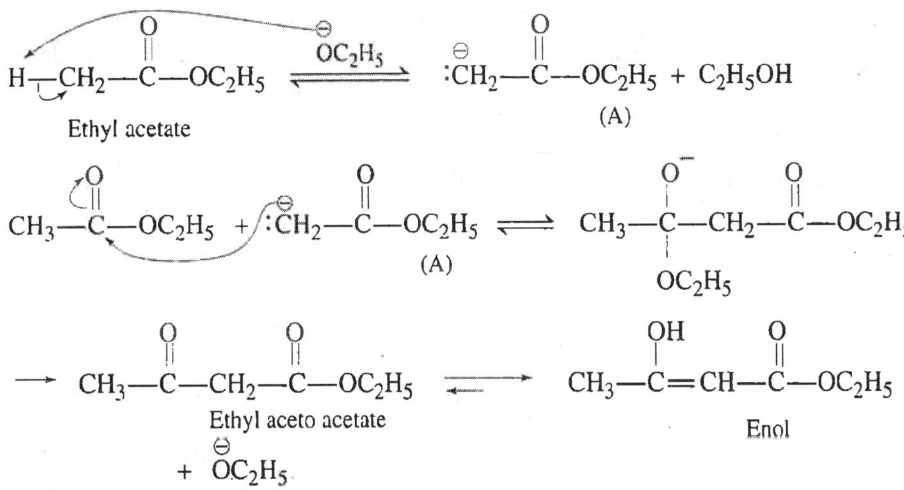

In practice, sodium metal is used instead of sodium ethoxide. Sodium reacts with minute traces of ethanol present in ethyl acetate and gives sufficient ethoxide ion to initiate the reaction.

20.1 Ethyl Acetoacetate

Ethyl acetate (dry) : 50 ml
Sodium metal : 4.5 g
Dil. acetic acid : (1:1)

In a round bottom flask fitted with a reflux condenser having a calcium chloride drying tube at the top, are taken ethyl acetate (50 ml, 0.5 mole) and small pieces of finely cut sodium metal (or preferably sodium wire) (4.5 g). The flask is warmed (oil bath) to start the reaction. Once the reaction starts, it

1. L. CLAISEN and O. LOWMAN, *Ber.* 1887, **20,** 651.
2. C.R. HAUSER and B.L. HUDGON, *Organic Reactions*, 1942, **1,** 266.

becomes vigorous and has to be cooled in cold water if necessary. When the vigorous reaction has subsided, the reaction mixture is gently heated for 1.5 hours or until all the sodium has reacted. Cool the reaction mixture. In case there is still a small amount of sodium left in the reaction mixture, add 2–3 ml of methanol to consume it. Acetic acid (dilute 1:1) is added slowly to make it slightly acidic to litmus paper. The resulting acidic solution is poured into cold saturated sodium chloride solution (50 ml). The total solution is transferred to a separatory funnel, shaken well and the upper layer of the ester separated. It is dried with anhydrous magnesium sulphate and distilled under reduced pressure. The initial fraction contains ethyl acetate. The ethyl acetoacetate is collected at 72–76°C (14 mm). The yield is 10 g (15%); b.p. is 181°C.

Precautions: The reaction should be performed in perfectly dry conditions. The ethyl acetate should be dried overnight over anhydrous calcium chloride. Sodium should be cut into small pieces under dry ether in order to avoid exposure of sodium to the atmosphere. The reaction should be performed in a fume cupboard preferably under nitrogen atmosphere.

21. CLAISEN REARRANGEMENT[1-3]

The allyl phenyl ether on heating undergoes thermal rearrangement to give 2-allylphenol. This is known as claisen rearrangement.

Mechanism:

The mechanism of the intramolecular rearrangement involves a cyclic transition state as shown below.

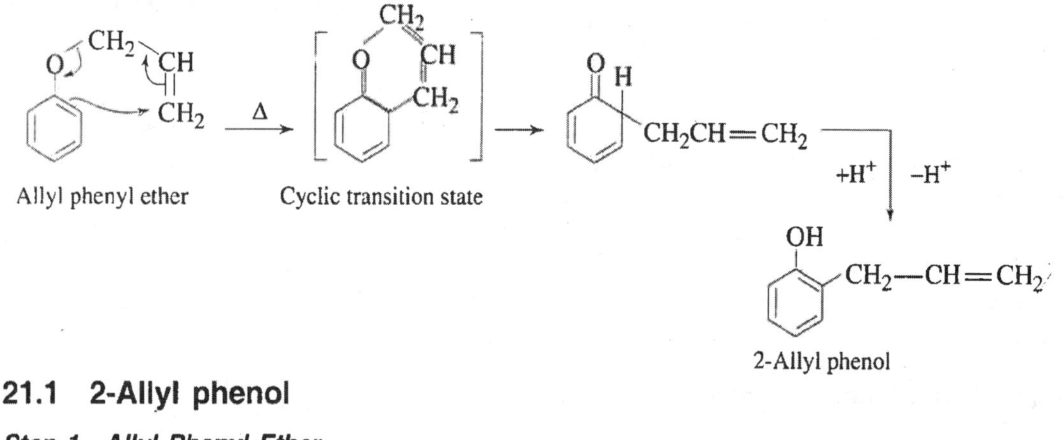

21.1 2-Allyl phenol

Step 1 Allyl Phenyl Ether

Phenol	: 18.8 g
Allyl bromide	: 24.2 g
Potassium carbonate (freshly ignited)	: 30 g
Acetone (dry)	: 50 ml

1. L. CLAISEN, *Ber.*, 1912, **45**, 3157.
2. L. CLAISEN and E. TIELZE, *Ber.*, 1925, **58**, 275.
3. L CLAISEN and E. TIEIZE, *Ber.*, 1926, **59**, 2344.

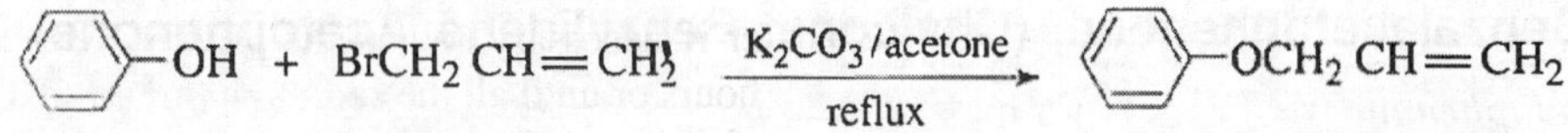

Phenol Allyl bromide Allyl phenyl ether

A mixture of phenol (18.8 g, 0.2 mole), allyl bromide (24.2 g, 0.2 mole), anhydrous potassium carbonate (30 g) and dry acetone (50 ml) is refluxed for 8 hours in a round bottom flask using a reflux condenser, and a calcium chloride drying tube in a water bath. Acetone is distilled away (as much as possible) on the boiling water bath and water (150 ml) is then added to the residue. The mixture is extracted with ether (separatory funnel) (2 × 50 ml). The ether extract is washed with sodium hydroxide solution (10%, 2 × 20 ml) and finally with water (2 × 30 ml). It is dried over anhydrous potassium carbonate.

The ether solution is filtered and distilled. After removal of ether, the residue consisting of allyl phenyl ether is distilled under reduced pressure. The yield is 21 g (78.4%); b.p. is 65°C/19 mm.

Step 2 2-Allyl phenol

The allyl phenyl ether (obtained above) is heated in a round bottom flask using a sand bath and an air condenser. The liquid is kept on gentle refluxing for 5–6 hours. The completion of the reaction is indicated by the refractive index (measured using a Abbe refractometer) rising to 1.5433 from its initial value of 0.9831. After the reaction is complete, the reaction mixture is allowed to cool and is then dissolved in sodium hydroxide solution (20%, 50 ml). The alkaline solution is extracted with ether (30 ml) to remove any starting material. The clear alkaline solution is acidified with dilute hydrochloric acid and extracted with ether (3 × 30 ml). The combined ether extract is dried over anhydrous calcium chloride and the ether is distilled away. The residue of 2-allyl phenol is distilled under reduced pressure. The yield is 14 g (66.7%); b.p. is 103–105°C/19 mm.

22. CLAISEN–SCHMIDT CONDENSATION[1, 2]

The condensation of an aromatic aldehyde with an aliphatic aldehyde or ketone (having α-hydrogen) in presence of a relatively strong base (hydroxide or alkoxide) to form an α, β-unsaturated aldehyde or a ketone is known as Claisen-Schmidt condensation.

$$C_6H_5CHO + CH_3{-}\overset{\overset{\displaystyle O}{\|}}{C}{-}R \xrightarrow{\text{NaOH}} C_6H_5CH{=}CH{-}\overset{\overset{\displaystyle O}{\|}}{C}{-}R$$

Mechanism:

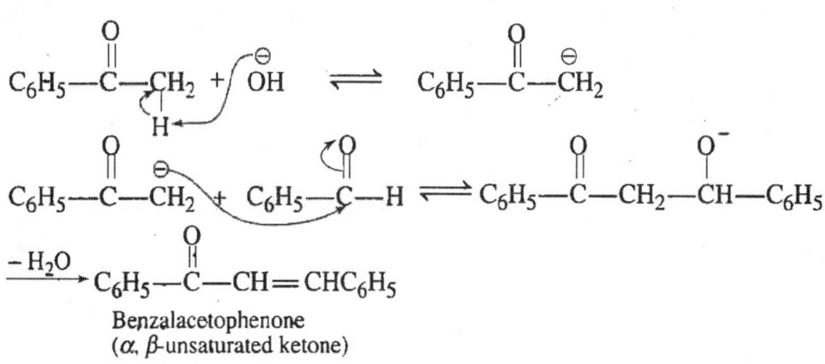

Benzalacetophenone
(α, β-unsaturated ketone)

1. L. Claisen and A. Claparede, *Ber.*, 1981, **14**, 2460.
2. J.B. Schmidt, *Ber.*, 1881, **14**, 1459.

22.1 Benzalacetophenone, (Chalcone, Benzylidene Acetophenone)

Acetophenone	: 5 ml
Benzaldehyde	: 4.4 ml
Rectified spirit	: 15 ml
Sodium hydroxide solution	: 2.2 g in 20 ml water

A solution of sodium hydroxide (2.2 g) in water (20 ml) and rectified spirit (15 ml) is cooled in a conical flask kept in an ice bath. To the cooled solution, acetophenone (5 ml, 0.043 mole) is added followed by the addition of benzaldehyde (4.4 ml, 0.043 mole). The reaction mixture is stirred all the time with a mechanical stirrer (or vigorous hand shaking) and the temperature of the reaction mixture kept at about 25°C. The mixture is stirred for 2–3 hours till the mixture becomes viscous and no more stirring is effective. The mixture is kept overnight in a refrigerator. The separated product is filtered under suction and washed well with cold water. It is recrystallised from rectified spirit. The yield is 7.7 g (86.5%); m.p. 56–57°C.

Precaution: The product acts as a skin irritant and should be handled with great care.

22.2 Benzalacetone, (Benzylideneacetone)

Acetone	: 15.9 g(20 ml),
Benzaldehyde	: 8.6 g (8.1 ml)
Sodium hydroxide solution	: 10%, 2 ml

$$C_6H_5CHO + CH_3COCH_3 \xrightarrow{-H_2O} C_6H_5CH{=}CH{-}\overset{\overset{\displaystyle O}{\|}}{C}{-}CH_3.$$

A solution of benzaldehyde (8.1 ml, 0.08 mole) and acetone (20 ml, 0.027 mole) is taken in a round bottom flask equipped with a mechanical stirrer (alternatively the flask can be shaken with hand during the experiment). The mixture is cooled in a bath of cold water. A solution of aqueous sodium hydroxide (2 ml, 10%) is added dropwise with a dropper during 20 minutes and the reaction mixture stirred (or shaken with hand). The temperature should be maintained between 25–30°C. The mixture is stirred at room temperature for 2 hours more. The reaction mixture is acidified (litmus paper) by the addition of dilute hydrochloric acid and extracted (separatory funnel) with ether (2 × 20 ml). The ether extract is washed with water, dried (magnesium sulphate) and the solvent removed by distillation. Benzalacetone is finally obtained by distillation under reduced pressure. The yield is 10 g (85.4%); b.p. 143–144°C/20 mm. It can be recrystallised from petroleum ether; m.p. is 42°C.

22.3 Dibenzalacetone (Dibenzylideneacetone)

$$2C_6H_5CHO + CH_3COCH_3 \xrightarrow{-2H_2O} C_6H_5CH{=}CH{-}\overset{\overset{\displaystyle O}{\|}}{C}{-}CH{=}CHC_6H_5$$

Acetone	: 7.3 g (9.3 ml),
Benzaldehyde	: 26.5 g (25.5 ml)
Ethanol	: 200 ml
Sodium hydroxide	: 25 g in 250 ml water

A solution of alcohol (200 ml) and sodium hydroxide solution (25 g in 250 ml water) is taken in a 500 ml R.B. flask fitted with a mechanical stirrer. The flask is cooled in cold water so as to maintain the temperature between 20–25°C. The solution is vigorously stirred and one half of the previously prepared mixture of benzaldehyde (25.5 ml, 0.25 mole) and acetone (9.3 ml, 0.126 mole) is added. A fluffy white precipitate is formed within 1–2 minutes. After 15 minutes, the remaining of the benzaldehyde-acetone mixture is added. The reaction mixture is stirred for 30 minutes more. The separated product is filtered, washed with water (to remove as much alkali as possible) and dried. It is crystallised from ethyl acetate or from alcohol. The yield is 25g (84.7%); m.p. is 112°C.

23. CLEMMENSEN REDUCTION[1, 2]

Carbonyl compounds on reduction with zinc amalgam and hydrochloric acid gives the corresponding reduced product, i.e. >C=O is converted into CH_2. This is known as clemmensen reduction.

Mechanism:

The mechanism of clemmensen reduction is uncertain. The reaction pathway involving transfer of electrons from the metal surface to the carbonyl carbon atom has been proposed:

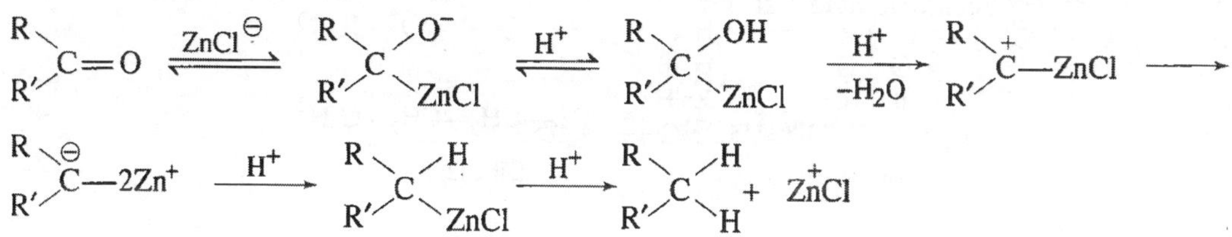

23.1 Ethyl Resorcinol from Resacetophenone

Resacetophenone	: 5g
Zinc amalgam	: from 20g Zn
Conc. hydrochloric acid	: 15 ml

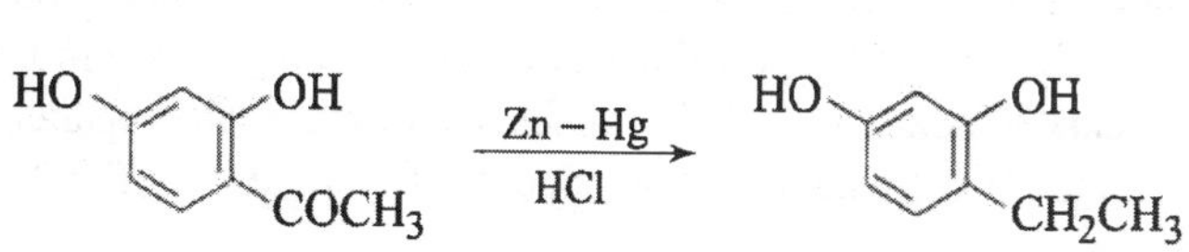

Resacetophenone — Ethyl Resorcinol

Preparation of Zinc amalgam Zinc (20 g) (either as strips from foils, zinc turnings or small pieces of granulated zinc) is placed in a 250 ml round bottom flask and washed thrice with warm dilute hydrochloric acid (1:1). To the washed zinc is added mercuric chloride (1.5 g), concentrated hydrochloric acid (1 ml) and water (20 ml). The mixture is vigorously stirred for 10 minutes. The aqueous solution is decanted, the amalgamated zinc is washed once with distilled water and covered with dilute hydrochloric acid (30 ml, 1:1). It is used as such for reduction.

---

1. W.T. BORDEN and T. RAVINDRANATHAN, *J. Org. Chem.*, 1971, **36**, 4125.
2. T. NABAYASKI, *J. Am. Chem Soc.*, 1960, **82**, 3900, 3906, 3909.

Step 1 β-Resacetophenone

It is obtained as described on page 151 (Nencki reaction).

Step 2 Ethyl Resorcinol

A mixture of amalgamated zinc (obtained from 20 g zinc as described above) and resacetophenone (5g, 0.033 mole) is refluxed in a R. B. flask for 3–4 hours, and concentrated hydrochloric acid (2.5 ml) is added to it every hour. At the end of each hour a drop of the liquid from the reaction mixture is taken in ethyl alcohol (0.5–1 ml) and tested with 2, 4-DNP. If no colour change is seen it means that the reaction is over. When the reaction is complete, the supernatant liquid is decanted from the unreacted amalgam. The solution is then saturated with solid sodium chloride (table salt), cooled and extracted with ether (5 × 20 ml). The combined ether extract is dried over anhydrous sodium sulphate and the solvent removed. A light yellow solid is obtained. It is crystallised either from ether-petroleum ether mixture or chloroform. Ethyl resorcinol crystallises in white glistening needles. The yield is 4g (88%); m.p. is 97°C.

23.2 Dibenzyl from Benzil

Benzil	: 7 g
Amalgamated zinc	: 30 g
Hydrochloric acid (1:1)	: 100 ml
Conc. Hydrochloric acid	: 20 ml

$$\underset{\text{Benzil}}{C_6H_5\overset{O}{\overset{\|}{C}}-\overset{O}{\overset{\|}{C}}-C_6H_5} \xrightarrow[\text{HCl}]{\text{Zn|Hg}} \underset{\text{Dibenzyl}}{C_6H_5-CH_2-CH_2-C_6H_5}$$

Step 1 Benzil

It is obtained as described on page 36.

Step 2 Dibenzyl

A mixture of amalgamated zinc (30 g), benzil (7g 0.033 mole) and dilute hydrochloric acid (1:1, 100 ml; addition of dilute hydrochloric acid is not necessary) is refluxed for 4–5 hours in a R.B. flask. Concentrated hydrochloric acid (2.5 ml) is added after every hour. The completion of the reaction is seen by the negative 2, 4 DNP reaction (test done as in case of ethyl resorcinol). The clear liquid is decanted from the amalgamated zinc and cooled in an ice-salt mixture. The separated dibenzyl is filtered and washed thrice with water. It is crystallised from alcohol. The yield is 5g (83.3%); m.p. is 52°C; b.p. is 280°C.

Note: Also see Wolff-Kishner reduction (page 220).

24. COMBES QUINOLINE SYNTHESIS[1]

It consists in the formation of quinolines by the condensation of β-diketones with primary arylamines followed by the acid catalysed ring closure of the intermediate Schiff's base.

---

1. A. Combes, *Bull. Soc. Chim.* France, 1888, **49**, 89.

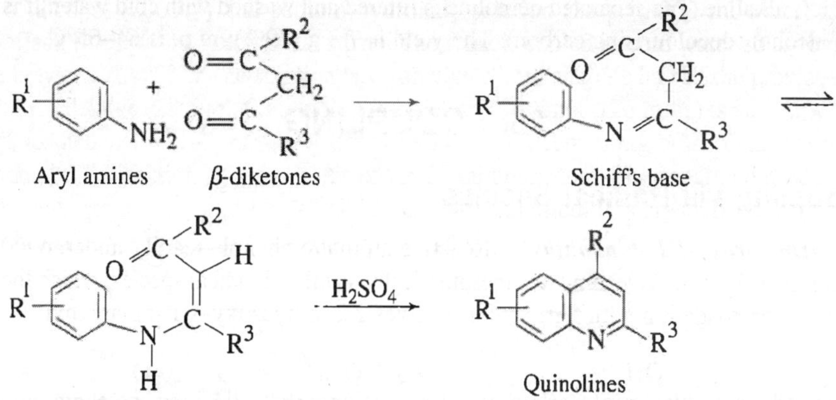

Aryl amines β-diketones Schiff's base Quinolines

24.1 2,4,6-Trimethyl quinoline

It is obtained by the condensation of *p*-toluidine with pentane-2,4-dione to form 4-(*p*-tolylamino) pent-3-ene-2-one, which is subsequently cyclised with concentrated sulphuric acid.

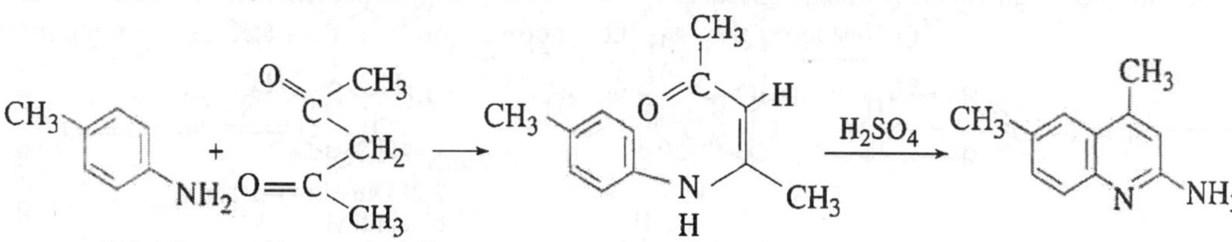

p-Toluidine Pentane-2, 4-dione 4-(*p*-tolyl amino) pent 3-ene-2-one 2, 4, 6- Trimethyl-quinoline

Step 1 4-(p-Tolylamino)pent-3-ene-2-one.

Pentane-2,4-dione : 7.65 g

Anhydrous calcium sulphate : 15 g

p-Toluidine : 7.98 g

To a mixture of pentane-2,4-dione (p.00) (7.65 g, 0.075 mol) and *p*-toluidine (7.98 g, 0.075 mol) is added granular anhydrous calcium sulphate (15 g). The mixture is heated for 1 hour on a steam bath with occasional shaking (anhydrous conditions). It is cooled, ether (60 ml) added and the mixture then filtered. The residual calcium sulphate is washed with ether (2 × 30 ml). The combined ether extract is filtered and distilled. The residual product is crystallised from hexane as a straw coloured product. The yield is 10.7 g (76%); m.p. is 68–69°C.

Step 2 2,4,6-Trimethylquinoline

4-(p-Tolylamino)pent-3-ene-2-one : 9 g

Conc. sulphuric acid : 40 ml

4-(*p*-Tolylamino) pent-3-ene-2-one (9 g, 0.48 mol) is added to concentrated sulphuric acid (d; 1.84, 40 ml) contained in a R.B. flask (250 ml capacity). The mixture is gently heated to 60–70°C (water bath) with occasional shaking till a homogeneous solution is obtained. It is heated for 30 minutes on a steam bath, cooled and cold water (350 ml) then added. Solid sodium carbonate is added in small lots to the solution till it is alkaline. The separated quinoline is filtered and washed with cold water. It is crystallised from dilute alcohol (decolourising carbon). The yield is 7.8 g (79%); m.p. is 64–65°C.

25. COUPLING

25.1 Coupling via Radical Species

25.1.1 2,2′-Dihydroxy-1,1′-binaphthyl. Reactive aromatic phenols readily undergo oxidation by a single electron transfer process leading to a mesomerically stabilised radical species, which then dimerises. Thus 2-naphthol on oxidation with ferric chloride gives 2,2′-dihydroxy-1,1′-binaphthyl

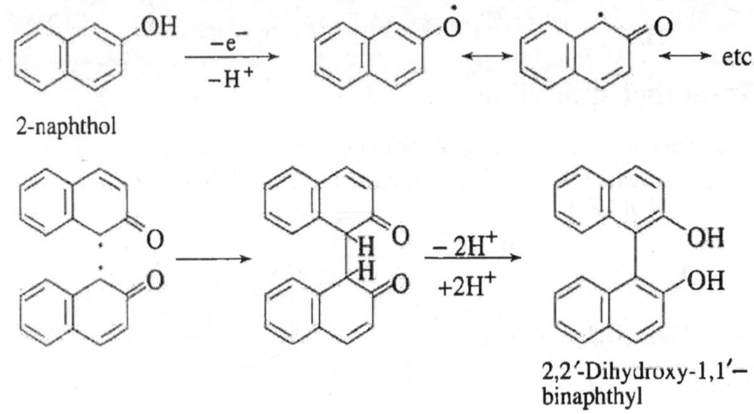

2-Naphthol : 7.2 g
Ferric chloride : 14 g

A mixture of 2-naphthol (7.2 g, 0.05 mole) and water (300 ml) is boiled in a three necked flask (500 ml capacity) fitted with a dropping funnel and a reflux condenser. To the boiling liquid containing 2-naphthol in suspension is slowly added (dropping funnel) with vigorous stirring, a solution of ferric chloride (14 g) in water (30 ml). The oily drops of 2-naphthol disappear and the bis-2-naphthol separates out. The mixture is boiled for 10–15 minutes more and the separated product filtered while hot and washed with hot water and dried. It is crystallised from toluene. The yield is 4 g (28%); m.p. is 218°C.

25.1.2 1,1′-Binaphthyl

α-Bromo naphthalene : 10.4 g
Magnesium turnings : 1.2 g
Cupric chloride : 6.8 g

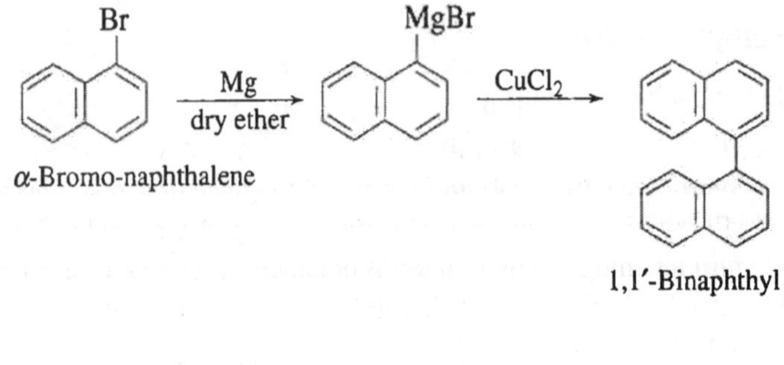

1,1′-Binaphthyl

A mixture of α-bromonaphthalene (10.4 g, 0.05 mole), dry ether (20 ml) and magnesium turnings (1.2 g) is warmed in a R.B. flask fitted with a reflux condenser, which is protected by a calcium chloride drying tube. A small crystal of iodine is added through the condenser to initiate the reaction. A precipitate is formed. It is dissolved by adding anhydrous benzene (15 ml). The heating is continued till all the magnesium is consumed. The suspension of α-naphthyl magnesium bromide thus obtained, is poured with cooling and stirring to a suspension of cupric chloride (6.8 g) in dry ether (25 ml). The mixture is refluxed for 1 hour (anhydrous conditions) and treated with water and dilute hydrochloric acid until cupric chloride dissolves. The organic layer is separated (separatory funnel). It is washed with dilute hydrochloric acid, dilute ammonia solution and water and dried (calcium chloride). 1,1′-Binaphthyl is obtained after evaporation of ether. The yield is 5 g (39.4%); m.p. is 154°C.

25.1.3 Biphenyl-2,2′-dicarboxylic acid (diphenic acid) It is obtained via the dimerisation of an aryl radical obtained by treatment of the appropriate diazonium salt with copper-su' ..ie in aqueous ammonia and hydroxylamine hydrochloride.

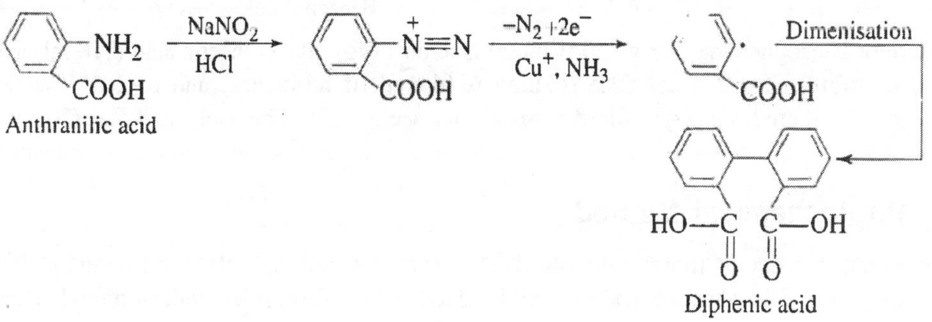

Anthranilic acid : 10 g
Sodium nitrite : 5.7 g
Conc. hydrochloric acid : 83 ml
Reducing agent (prepared as given below)
(a solution of cupra-ammonium ions)

The reducing agent (a solution of cupra-ammonium ions) is first prepared as follows: To a solution of copper sulphate (31.5 g) in water (125 ml) is added concentrated ammonia solution (sp. gr. 0.88, 50 ml). The resulting solution is cooled to 10°C and a cold solution (10°C) of hydroxylamine hydrochloride (8.9 g) in water (30 ml) and sodium hydroxide solution (6M, 21.8 ml) (in case the solution is not clear, it is filtered) is added with stirring. Reduction occurs, a gas is evolved and the solution assumes blue colour. The reducing agent is protected from air if not used immediately.

The anthranilic acid (10 g, 0.073 mole) is diazotised with sodium nitrite (5.7 g) and hydrochloric acid as has been done in the case of the preparation of methyl red.

To the cold reducing solution (10°C, obtained above) is added dropwise the diazotised anthranilic acid solution (as obtained above). The diazotised solution is added with the help of a dropping funnel, the stem of which dips below the surface of the liquid and is bent upwards at the end. The reducing solution is stirred all the time and the diazonium salt solution is added at the rate of about 10 ml/min. After the addition is complete, the mixture is stirred for 10 minutes. The solution is heated carefully to the boiling point and concentrated hydrochloric acid (63 ml) added. The diphenic acid precipitates as pale brown crystals. It is purified by dissolving in sodium bicarbonate solution and acidifying the clear hot solution with concentrated hydrochloric acid. The yield is 9 g (51%); m.p. is 227–228°C.

25.2 Reductive Coupling

25.2.1 Benzopinacol

Benzophenone : 7.5 g
Zinc dust : 4 g
Acetic acid (glacial) : 110 ml

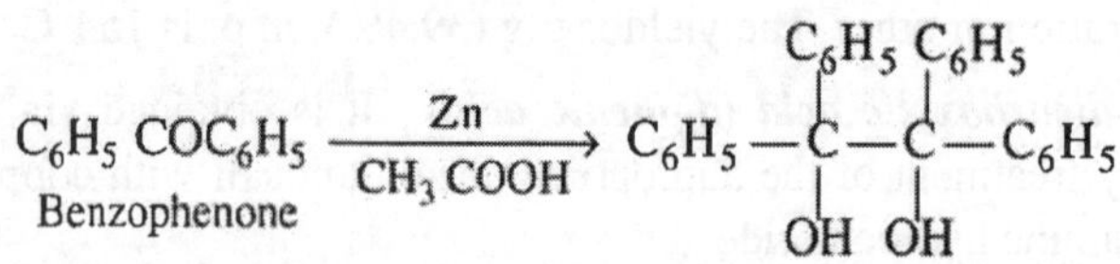

Benzopinacol

A mixture of benzophenone (7.5 g, 0.041 mole) zinc dust (4g), glacial acetic acid (110 ml) and water (22 ml) is refluxed for 2 hours. The solution is filtered (if necessary) and cooled. The separated benzpinacol is filtered and crystallised from glacial acetic acid. The yield is 4.5 g (30%); m.p. is 188–189°C.

25.3 Photochemical Method

Benzophenone (7.5 g, 0.04 mole) is dissolved in isopropyl alcohol (5.5 ml) by warming in a R.B. flask (100 ml capacity). Glacial acetic acid (0.5 ml) is added and the flask filled with isopropyl alcohol. The flask is stoppered tightly with a cork and placed in bright sunlight. The reaction is completed by the separation of a solid product. The reaction takes 5–10 days. The separated product is filtered and crystallised from glacial acetic acid.

Note: In case, benzophenone starts crystallising (as indicated by the separation of large thick prisms), it is redissolved by warming on a steam bath.

25.4 Ullmann's Coupling

See ullmann condensation (page 214). Using this method, the following compounds can be prepared: 2, 2′-dinitrobiphenyl (p. 215); N-phenyl anthranilic acid (p. 127).

26. DAKIN REACTION[1]

The replacement of the aldehyde or acetyl group, in phenolic aldehydes or ketones, by a hydroxyl group by means of hydrogen peroxide in presence of dilute sodium hydroxide is known as Dakin reaction. Thus,

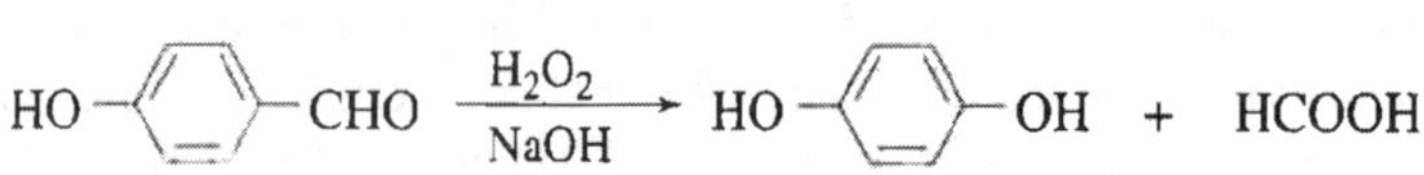

1. H.D. Dakin, *Am, Chem, J.* 1909, **42**, 477.

The, phenolic group must be ortho or para to the carbonyl substituent.

Mechanism:

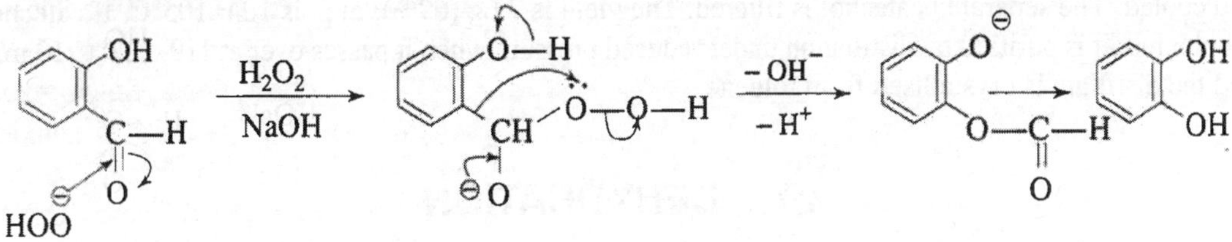

26.1 2,3-Dihydroxy Anisole (Pyrogallol monomethyl ether)

It is obtained by the oxidation of 2-hydroxy-3-methoxybenzaldehyde (o-vanillin) with hydrogen peroxide in presence of aqueous sodium hydroxide

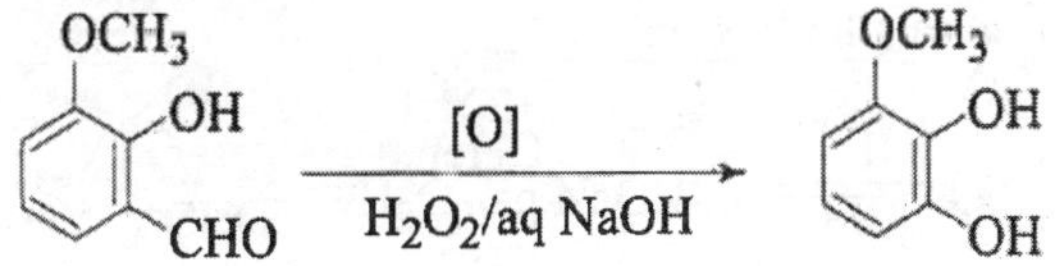

2-hydroxy-3-methoxybenzaldehyde (o-vanillin) : 20.3 g
Aqueous sodium hydroxide (2 N) : 66.6 ml
Hydrogen peroxide (6%) : 95 ml

To a solution of 2-hydroxy-3-methoxybenzaldehyde (o-vanillin) (20.3 g, 0.113 mol) in aqueous sodium hydroxide (2N, 66.6 ml, 0.133 mol) is dropwise added hydrogen peroxide (6%, 95 ml, 0.166 mol) during one hour at 40–50°C (the solution is cooled if necessary). After the reaction is over, the reaction mixture is cooled to room temperature, saturated with sodium chloride and extracted with ether (3 × 70 ml). The combined ether extract is dried (sodium sulphate) and distilled. The residual product is distilled and the fraction boiling at 136–138°C, 22 mm is collected. The yield is 11 g (68%).

26.2 Catechol

It is obtained by Dakin's oxidation of salicylaldehyde with hydrogen peroxide and sodium hydroxide solution,

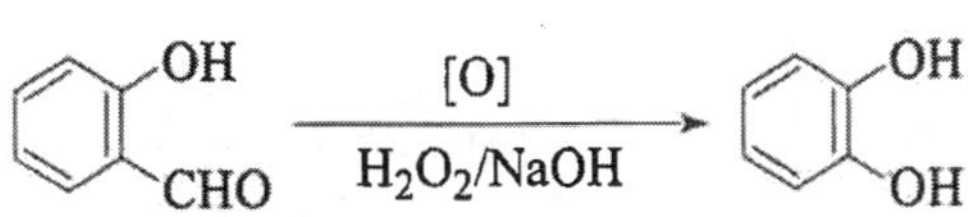

Salicylaldehyde : 12.2 g
Sodium hydroxide solution : 1N, 100 ml
Hydrogen peroxide (3%) : 14.2 g

Hydrogen peroxide (3%, 14.2g, 0.12 mole) is added at room temperature to a solution of salicylaldehyde (12.2 g, 0.1 mol) in sodium hydroxide solution (1N, 100 ml). The mixture darkens in colour and its temperature rises to 45–50°C. The solution is allowed to stand for 15–20 hours. Excess alkali is neutralised by adding a few drops of acetic acid. The solution is evaporated to dryness on a water bath under reduced pressure. The dried solid residue is powdered and repeatedly extracted with boiling toluene (10 × 20 ml). Alternatively, it can be Soxhlet extracted. The combined toluene solution is concentrated and cooled. The separated catechol is filtered. The yield is 7.6g (69%); m.p. is 104–105°C. If catechol is coloured, it is purified by distillation under reduced pressure when it passes over at 119–121°C/10 mm and the distillate is crystallised from toluene.

27. DEHYDRATION

27.1 Dehydration of Alcohols

27.1.1 Cyclohexene

Cyclohexanol : 20 ml

Phosphoric acid : 85%, 6 ml

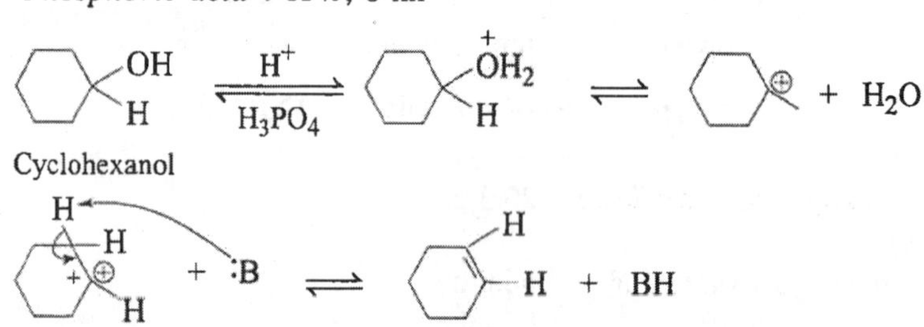

(proton transfer to base B, which may be, HSO_4^-, $H_2PO_4^-$, H_2O or cyclohexanol)

Cyclohexanol (20 ml, 0.2 mole) is taken in a distillation flask (250 ml capacity). Phosphoric acid (85%, 6 ml) is added slowly with constant shaking. The distillation assembly is fitted up and the distillation flask is gently heated with a small flame. Cyclohexene starts distilling along with some water. The temperature is kept below 105°C. The distillation is stopped when no more liquid distils. The distillate is shaken with saturated sodium chloride solution and the upper layer of cyclohexene is separated (separatory funnel).

It is dried (anhydrous calcium chloride, which also removes unchanged cyclohexanol) and distilled. The yield is 10 ml (59%); b.p. is 83°C.

27.1.2 Maleic Acid It is obtained by dehydration of maleic acid by heating with acetyl chloride followed by hydrolysis of the formed maleic anhydride. Thus,

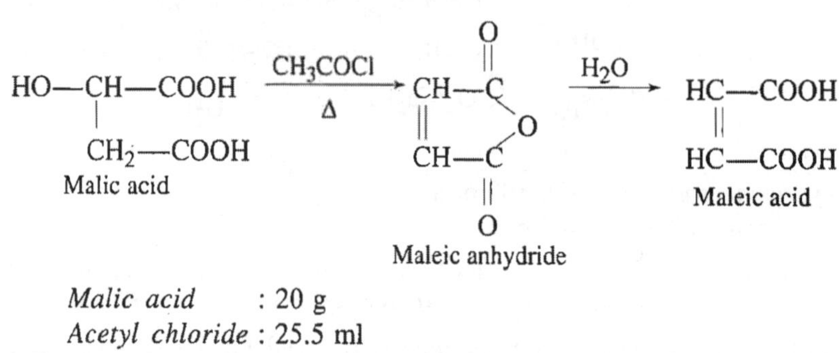

Malic acid : 20 g

Acetyl chloride : 25.5 ml

To malic acid (20 g, 0.15 mole) placed in a dry distillation flask (100 ml capacity) is cautiously added acetyl chloride (25.5 ml). The flask is warmed gently on a water bath to start the reaction, which then proceeds exothermically. Maleic acid goes into solution and hydrogen chloride is evolved. After the evolution of hydrogen chloride gas subsides, the reaction mixture is heated on a water bath for 1 hour. The distillation assembly is fitted and distillation carried out. Initially, a low boiling fraction passes over when the temperature rises rapidly to 185–190°C (the water from the water condenser is run out). The distillation is continued. Maleic anhydride is collected at 195–200°C. It is crystallised from chloroform. The yield is 8 g (54.4%); m.p. is 54°C.

In order to obtain maleic acid, the anhydride obtained above is heated with water (5 ml) on a water bath. The water is removed by leaving the reaction mixture in a desiccator over concentrated sulphuric acid. The resulting maleic acid has a m.p. of 143°C. It can be crystallised from acetone-petroleum ether, m.p. is 144°C.

27.2 Dehydration of Amides

27.2.1 ***Benzonitrile.*** It is obtained by the dehydration of benzamide (prepared from benzoyl chloride) on heating with phosphorus pentoxide.

$$\underset{\text{Benzamide}}{C_6H_5\ CONH_2} \xrightarrow[-H_2O]{P_2O_5} \underset{\text{Benzonitrile}}{C_6H_5C{\equiv}N}$$

Step 1 Benzamide

Benzoyl chloride : 15g
Conc. ammonia solution: 50 ml
(Sp. gr 0.88)

$$C_6H_5COCl + NH_3 \longrightarrow C_6H_5CONH_2 + HCl$$

Benzoyl chloride (15 g, 12.4 ml, 0.106 mole) is added dropwise (separatory funnel) to liquor ammonia (sp. gr. 0.88, 50 ml) cooled in ice. The reaction mixture is shaken frequently. The precipitated benzamide is filtered, washed with cold water and recrystallised from hot water. The yield is 12 g (93.8%); m.p. is 129°C.

Step 2 Benzonitrile

Benzamide : 40 g
Phosphorous pentoxide : 80 g

A mixture of benzamide (40 g, 0.33 mole) and phosphorus pentoxide (80 g) is taken in a distillation flask. The apparatus for distillation under reduced pressure is fitted up. The flask is heated with a free flame until no more liquid distills over (the receiver is connected to vacuo about 100 mm Hg pressure). The distillate is washed with a little sodium carbonate solution and water, and then dried (anhydrous calcium chloride) and distilled. The yield is 13 g (38%); b.p. is 188–89°C.

27.3 Dehydration of Oximes

27.3.1 ***Veratronitrile.*** It is obtained by dehydration of the corresponding oxime with acetic anhydride.

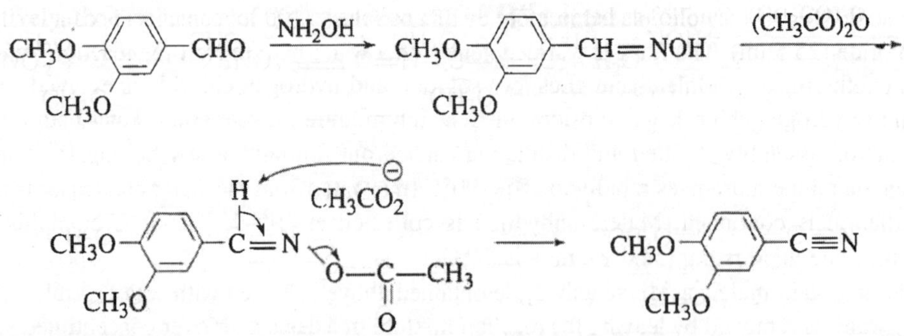

Step 1 Veratraldoxime

Veratraldehyde	: 41.5 g
Hydroxylamine hydrochloride	: 21 g
Sodium hydroxide	: 15 g

A solution of hydroxylamine hydrochloride (21 g, 0.3 mole) in water (25 ml) is added to warm solution of veratraldehyde (41.5 g, 0.25 mole) in rectified spirit (100 ml). Sodium hydroxide solution (15 g in 20 ml water) is then added and the mixture allowed to stand for 3 hours. To the mixture is added crushed ice (100 g) and it is saturated with carbon dioxide gas. The aldoxime separates as an oil, which solidifies after being left for 12 hours in an ice-chest. The solid product is filtered, washed with cold water and dried on a filter paper. The yield is 44 g (97%). It is used as such for the next step.

Step 2 Veratronitrile

Veratraldoxime	: 44 g
Acetic anhydride	: 45 ml

A mixture of dry veratraldoxime (44 g, 0.243 mole) and redistilled acetic anhydride (45 ml) is taken in a R.B. flask (100 ml capacity). The mixture is warmed (water bath) cautiously. A vigorous reaction starts. When the reaction subsides, the mixture is gently heated for 20 minutes (water bath) and poured with stirring into cold water (100 ml). Colourless crystals of veratronitrile are filtered and dried. The yield is 28 g (70%); m.p. is 67°C.

28. DEHYDROGENATION

Dehydrogenation (removal of hydrogen) can be easily and conveniently carried out by heating the compound in an inert solvent in presence of Pd-C. Thus,

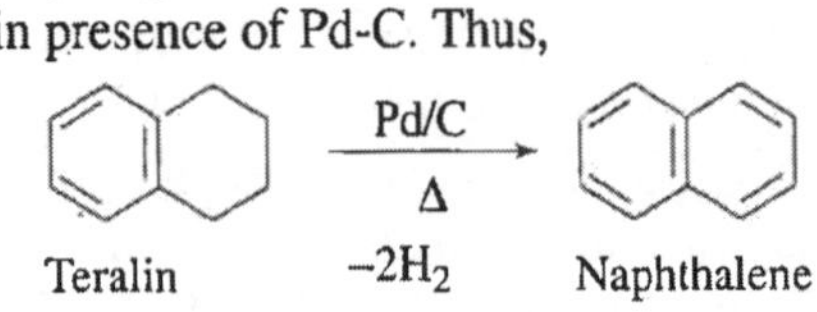

28.1 Naphthalene

Tetralin (5 ml, 0.037 mole) is refluxed for 4 hours with Pd-C (30%, 0.5 g) in a two necked R.B flask fitted with a reflux condenser. A slow current of carbon dioxide is passed through the solution. Naphthalene collects in the condenser in almost quantitative yield.

Note: In the absence of carbon dioxide the reaction takes 20–22 hours for completion.

29. DELE'PINE REACTION[1, 2]

Alkyl halides on reaction with hexamethylene tetramine (hexamine) followed by acid hydrolysis form quaternary salts.

$$\underset{\text{Alkyl halide}}{RCH_2X} + \underset{\text{Hexamine}}{C_6H_{12}N_4} \longrightarrow [RCH_2\ C_6H_{12}N_4]^+\ X^- \xrightarrow{HX} \underset{\text{(quaternary salt)}}{RCH_2N^+H_3\ X^-}$$

The same quaternary salt is an intermediate in the sommelet reaction (page 209).

30. DIAZO REACTION (GRIESS DIAZO REACTION)[3, 4]

Amines react with nitrous acid to yield diazonium salts via a number of intermediate species. This reaction is called *Diazo reaction*. Since it was discovered by Johan Peter Griess (1858) it is referred to as Griess Diazo reaction. The procedure of producing diazonium salt is referred to as diazotisation. Thus,

$$\underset{\text{Aniline}}{C_6H_5NH_2} + \underset{\text{sodium nitrite}}{NaNO_2} + HCl \xrightarrow{0-5^\circ} \underset{\text{Benzene diazonium chloride}}{C_6H_5N_2^+Cl^-} + NaCl + 2H_2O + HCl$$

Mechanism: The machanism of the diazo reaction is uncertain and somewhat complex. It is believed that the active reagent is most likely nitrous anhydride, N_2O_3 formed relatively slowly as shown here:

$$\underset{\text{Nitrous acid}}{2H{-}O{-}N{=}O} \rightleftharpoons \underset{\text{Dinitrogen trioxide}}{O{=}N{-}O{-}N{=}O} + H_2O$$

The various steps involved in the diazo reaction are as shown:

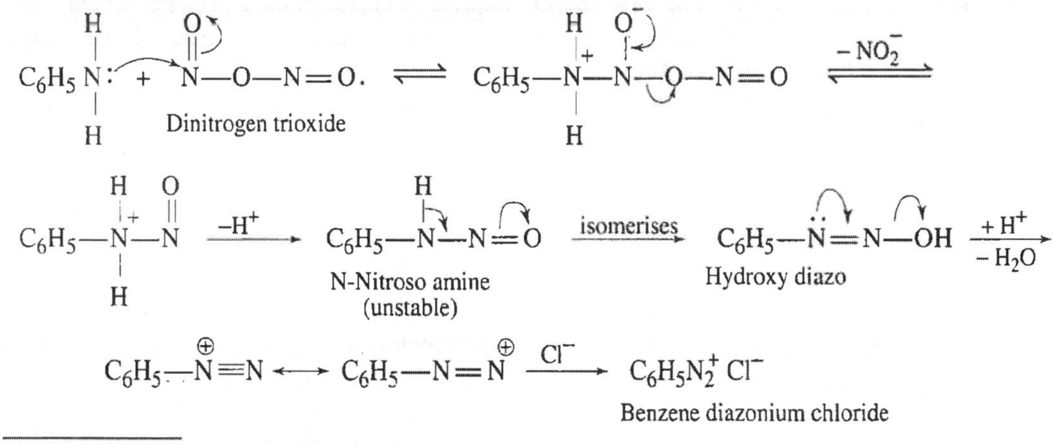

1. M. Dele'pine *Comt. Rend.* 1895, **120**, 501.
2. M. Derle'pine, *Comt. Rend.* 1897, **124**, 292.
3. P. Griess, *Ann.*, 1858, **106**, 123.
4. P. Griess, *Ann.*, 1862, **121**, 257.

Alternatively, the mechanism of diazo reaction can be represented as follows:

$$2\,NO_2^- + 2H^+ \longrightarrow 2HNO_2 \rightleftharpoons O{=}N{-}O{-}N{=}O + H_2O$$

$$O{=}N{-}O{-}N{=}O \rightleftharpoons \underset{\text{Nitrosonium ion}}{O{=}N^+} + \bar{O}{-}N{=}O$$

$$Ar\ddot{N}H_2 + \overset{\oplus}{N}{=}O \xrightarrow{-H^+} \underset{\text{(N-nitroso derivative)}}{Ar{-}\underset{H}{\underset{|}{N}}{-}N{=}O} \rightleftharpoons Ar{-}N{=}N{-}OH$$

$$Ar{-}N{=}N{-}\ddot{O}H + \overset{+}{H} \longrightarrow \underset{\text{Diazonium ion}}{Ar{-}\overset{+}{N}{\equiv}N} + H_2O$$

The diazonium salts are important intermediates in organic synthesis. Most of the reactions undergone by diazonium salts may be divided into three classes:

(i) *Replacement* In this case nitrogen is lost as N_2 and some other atom or group becomes attached to the ring in its place.

(ii) *Reactions not involving loss of nitrogen (Reduction).*

(iii) *Coupling* In this case nitrogen is retained in the product.

Typical preparations are given as follows:

30.1 Replacement of Diazo Group by Hydrogen

The amino group in an aromatic amine can be replaced by hydrogen (deamination) by boiling the diazonium salt prepared from the amine with ethyl alcohol. Thus,

$$\underset{\text{Diazonium salt}}{Ar\overset{+}{N}_2\,X^-} + CH_3CH_2OH \longrightarrow Ar{-}H + N_2 + CH_3CHO + HX$$

Mechanism: The deamination proceeds by the intermediate formation of the aryl cation, which reacts with ethanol in two ways, viz. replacement by hydrogen to form the aromatic hydrocarbon or replacement by the alkoxy group to form an ether. Thus,

$$Ar\overset{+}{N}{\equiv}N \longrightarrow \underset{\text{Aryl cation}}{Ar^+} + N_2$$

$$Ar^+ + HOCH_2CH_3 \longrightarrow Ar{-}H + CH_3CHO + H^+$$

$$Ar^+ + HOCH_2CH_3 \longrightarrow Ar{-}OCH_2CH_3 + H^+$$

The amount of ether formation depends on the nature of the substituents in the aromatic nucleus and also upon the alcohol. In general, the presence of electron withdrawing groups or atoms (e.g., halogen or nitro) favours replacement by hydrogen. Also, the replacement of the diazonium group by hydrogen can be effected by reacting with hypophosphorus acid, preferably in the presence of copper ions which appear to act catalytically.

$$ArN_2^+X^- + H_3PO_2 + H_2O \longrightarrow Ar{-}H + N_2 + H_3PO_3 + HX$$

30.1.1 *1,3,5-Tribromobenzene*

2,4,6-Tribromoaniline : 7.5 g
Absolute alcohol : 30 ml
Benzene : 7.5 ml
Conc. sulphuric acid : 2 ml
Sodium nitrite : 3 g

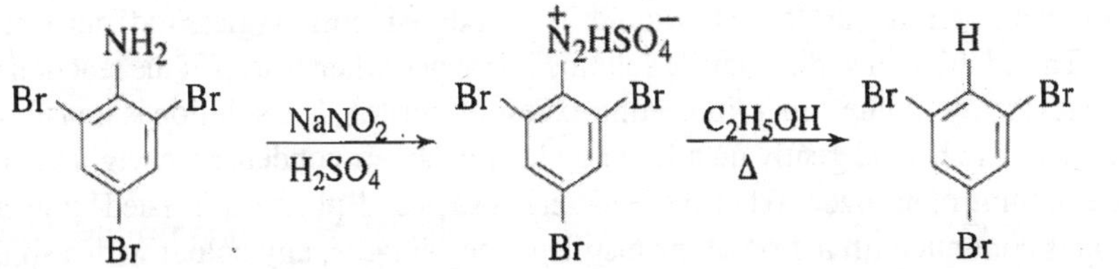

Symm.Tribromoaniline   Symm.Tribromobenzene

2,4,6-Tribromoaniline (7.5 g, 0.023 mole) (prepared as given on page 108) is dissolved in absolute alcohol (30 ml) and benzene (7.5 ml) in a round bottom flask (100 ml capacity) by heating on a water bath. The R.B. flask is fitted with a reflux condenser. Concentrated sulphuric acid (2 ml) is added to the hot solution gently, the solution being shaken. The reaction mixture is heated so that a clear solution is obtained. The flask is removed from the water bath, the reflux condenser detached and powdered sodium nitrite (3g) is added in two instalments. After each addition, the flask is fitted with the reflux condenser and shaken vigorously. The reaction mixture boils by the heat of the reaction. When the reaction has subsided, the second portion of sodium nitrite is added. The reaction mixture is heated on a boiling water bath for 1 hour, shaken at frequent intervals and allowed to cool initially at room temperature and then in an ice bath. The separated product is filtered, washed with a small quantity of alcohol and then with water (5 × 10 ml) till the sodium sulphate is removed. The crude tribromobenzene is crystallised from alcohol (charcolised). The yield is 4.5 g (63%); m.p. is 122°C.

30.2 Replacement of Diazo Group by Hydroxyl

An aqueous solution of diazonium salt on boiling or stream distillation gives phenol with the liberation of nitrogen

$$\underset{\text{Diazonium salt}}{ArN_2^+X^-} + H_2O \xrightarrow{H^+} Ar\text{–}OH + HX + N_2\uparrow$$

Mechanism:

$$Ar-\overset{+}{N}\equiv N \xrightarrow{\text{Slow}} \underset{\text{aryl cation}}{\overset{+}{Ar}} + N_2\uparrow$$

$$Ar^+ + H_2O \xrightarrow{\text{Fast}} Ar\text{–}\overset{+}{O}H_2 \longrightarrow Ar\text{–}OH + H^+$$

30.2.1 *Phenol*

Aniline : 15 g
Conc. sulphuric acid: 20 ml
Sodium nitrite : 12 g

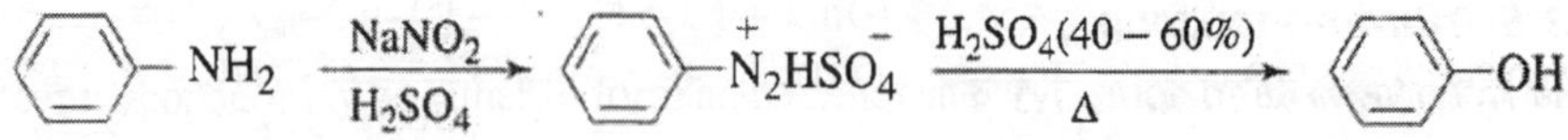

Aniline Phenol

Aniline (15 g, 0.16 mole) is added to a solution of concentrated sulphuric acid (20 ml) in water (75 ml) in a beaker (250 ml capacity). The clear solution is cooled in ice (0–5°C) and a solution of sodium nitrite (12 g in 30 ml water) is added dropwise with the help of a dropping funnel, the end of the dropping funnel being dipped under the surface of the liquid. The mixture is stirred (glass rod) during the addition of sodium nitrite. The addition of sodium nitrite solution is stopped when a drop of the reaction mixture gives a blue colour with starch-iodide paper (indicating excess of nitrite). The solution is transferred to a 1 litre round bottom flask and heated gently on a water bath using an air condenser. A vigorous reaction takes place with the evolution of nitrogen. When the reaction has subsided, phenol is isolated by steam distillation. The distillation is continued till a drop of the distillate does not give any colour with a solution of ferric chloride. The distillate (300–350 ml) is extracted with ether (3×50 ml). The combined ether extract is dried (anhydrous magnesium sulphate) and the ether distilled off (water bath). The residual phenol is purified by distillation. The yield is 10 g (66.6%); b.p. is 182°C; m.p. is 41°C.

30.2.2 m-Nitrophenol

m-Nitroaniline	: 10.5 g
Conc. sulphuric acid	: 48 ml
Sodium nitrite	: 5.4 g

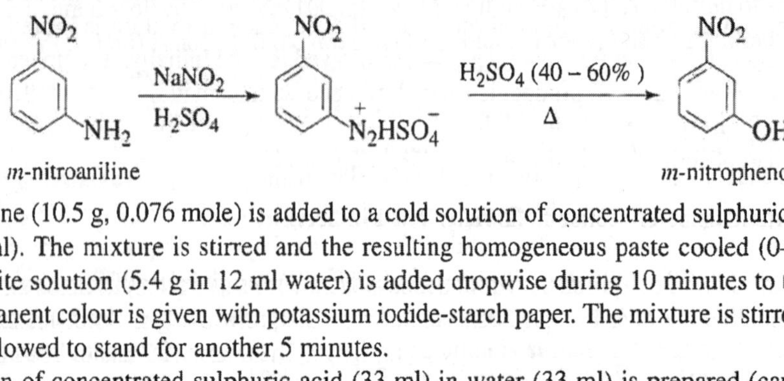

m-nitroaniline *m*-nitrophenol

m-Nitroaniline (10.5 g, 0.076 mole) is added to a cold solution of concentrated sulphuric acid (15 ml) in water (40 ml). The mixture is stirred and the resulting homogeneous paste cooled (0–5°C, ice bath). Sodium nitrite solution (5.4 g in 12 ml water) is added dropwise during 10 minutes to the stirred paste until a permanent colour is given with potassium iodide-starch paper. The mixture is stirred for 5 minutes more and allowed to stand for another 5 minutes.

A solution of concentrated sulphuric acid (33 ml) in water (33 ml) is prepared (caution) in a R.B. flask (250 ml capacity). The solution is heated to boiling and the diazonium solution (prepared above) is added in small lots (2–3 ml each time) during 20 minutes. In case a solid is separated during diazotization, the stirred suspension is added. Excessive frothing is avoided. When all the diazonium solution has been added, the mixture is boiled for 5 minutes and poured into a beaker (250 ml capacity). The solution is cooled and stirred vigorously. The separated crystals of *m*-nitrophenol are filtered, washed with a small quantity of cold water and recrystallised from dilute hydrochloric acid (1:1) as yellow crystals. The yield is 6.8 g (64.8%); m.p. is 96°C.

30.3 Replacement of Diazo Group by Iodine

Aqueous solutions of diazonium salts react readily with aqueous solution of potassium or sodium iodide to give aryl iodides

$$ArN_2^+X^- + MI \longrightarrow Ar\text{–}I + MX + N_2\uparrow$$
$$M = Na \text{ or } K.$$

Mechanism:

$$Ar\text{—}\overset{+}{N}{\equiv}N \longrightarrow Ar^+ + N_2\uparrow$$
$$Ar^+ + I^- \longrightarrow Ar\text{–}I$$

30.3.1 Iodobenzene

Aniline	: 10 g
Dil. hydrochloric acid (1:1)	: 55 ml
Sodium nitrite	: 8 g
Potassium iodide	: 18 g

$$C_6H_5N_2^+Cl^- + KI \longrightarrow C_6H_5I + KCl + N_2\uparrow$$

Sodium nitrite solution (8 g dissolved in 40 ml water) is added dropwise (using a separatory or dropping funnel with the tip of the steam below the surface of the liquid) to a cooled solution (0–5°C, ice bath) of aniline (10 g, 0.107 mole) in dilute hydrochloric acid (1:1, 55 ml). The solution is stirred or shaken during addition. The temperature is not allowed to rise above 8°C (if necessary a small amount of ice is added to the reaction mixture). The addition of sodium nitrite solution is continued till a drop of the reaction mixture gives a blue colour with potassiumiodide-starch paper. The reaction mixture should be acidic at all times (i.e., it should give a blue colour with congo red paper).

To the diazolised solution (obtained above) is added a solution of potassium iodide (18 g dissolved in 20 ml water) with shaking, when nitrogen is evolved. The mixture is allowed to stand for one hour and heated in a boiling water bath (20–30 minutes) till the evolution of nitrogen ceases. The iodobenzene separates as a dark heavy oil and is obtained by steam distillation (about 200–250 ml of distillate is collected). The distillate is extracted with ether (3 × 40 ml). The combined ether extract is washed with water and then with dilute sodium hydroxide solution till it is free from iodine. The ether extract is dried (calcium chloride) and distilled (water bath) to give iodobenzene. The yield is 18 g (82%); b.p. 188–189°C.

30.4 Replacement of Diazo Group by Chlorine or Bromine (Sandmeyers' Reaction)[1, 2]

The method described above for the preparation of iodobenzene cannot be used for the preparation of the corresponding chloro and bromo compounds. However, the diazonium group can be replaced by chlorine or bromine when an aqueous solution of the diazonium salt is warmed with cuprous chloride or bromide, dissolved in the corresponding halogen acid (viz. HCl or HBr).

$$ArN_2^+X^- \xrightarrow{CuCl/HCl} Ar\text{—}Cl + N_2\uparrow$$
$$ArN_2^+X^- \xrightarrow{CuBr/HBr} Ar\text{—}Br + N_2\uparrow$$

This reaction is known as Sandmeyer's reaction.

Mechanism:

$$H\text{–}Cl + CuCl \longrightarrow CuCl_2^- + H^+$$
$$ArN^+{\equiv}N + CuCl_2^- \longrightarrow Ar^{\cdot} + N_2\uparrow + CuCl_2$$

1. T. SANDMEYER, *Ber.*, 1884, **17**, 1663, 2650.
2. LYATTERMANN, *Ber.*, 1890, **23**, 1218.

$$Ar^{\cdot} + Cl–Cu–Cl \longrightarrow Ar–Cl + CuCl$$

The role of cupric copper may be either to form an intermediate aryl cation or an organocopper compound.

$$Ar^{\cdot} + Cu^{++} \longrightarrow \underset{\text{Aryl cation}}{Ar^{+}} + Cu^{+} \xrightarrow{Cl^{-}} ArCl + Cu^{+}$$

$$Ar^{\cdot} + Cu^{++} + Cl^{-} \longrightarrow \underset{\text{Organo copper compound}}{ArCu^{+}Cl} \longrightarrow ArCl + Cu^{+}$$

The formation of an intermediate molecular complex is supported by the fact that a molecular equivalent of the cuprous salt compound is used.

Note: For more details see Sandmeyer's reaction (page 205).

30.4.1 p-Chlorotoluene

p-Toluidine	: 15 g
Conc. hydrochloric acid	: 40 ml
Sodium nitrite	: 11 g

Cuprous chloride obtained from 45 g $CuSO_45H_2O$ (prepared as given below)

[Cuprous Chloride solution is prepared as follows: Hydrated copper sulphate (45 g) and pure sodium chloride (12.3 g) are dissolved in water (160 ml) (warmed if necessary). A solution of sodium bisulphite (11.2 g) in water (120 ml) is added to the hot solution (obtained above) during 5 minutes with constant shaking. The reaction mixture is cooled (ice-bath) and the supernatant liquid decanted. The formed cuprous chloride is washed by decantation with water containing little sulphurous acid (to prevent oxidation). The cuprous chloride is dissolved in concentrated hydrochloric acid (90 ml). The solution is kept cool (ice-salt mixture) and is used immediately after diazotization.]

p-Toluidine (15 g, 0.14 mole) is dissolved in concentrated hydrochloric acid (40 ml) and water (40 ml) in a conical flask (250 ml capacity). The solution is cooled to 0°C (ice-salt mixture) and stirred (a finely divided crystalline precipitate of the hydrochloride separates out). A solution of sodium nitrite (11 g) in water (25 ml) is added dropwise (with a separating funnel, the tip of the stem of which is dipping well below the surface of the liquid). The solution is stirred well during diazotization and the mixture is kept at 0–5°C. When the addition is complete, the separated hydrochloride salt dissolves and a soluble diazonium salt is formed (check the completion of diazotization by potassium iodide-starch paper which gives blue colour with the test solution).

The cold diazonium chloride solution (obtained above) is poured slowly with stirring in the cold cuprous chloride solution (prepared as given above). The mixture becomes viscous, due to the separation of an addition product ($CH_3\ C_6H_4\ N_2^+\ Cl^- \cdot CuCl$). The mixture is allowed to attain room temperature while shaking it occasionally. During this time (the temperature is about 15°C), the addition product breaks down with the liberation of nitrogen and *p*-chlorotoluene separates as an oily layer. The reaction mixture is warmed (60°C) to complete the decomposition, while it is shaken occasionally. The reaction mixture is steam distilled until no more oily drops come in the distillate. The *p*-chlorotoluene which is obtained as an oily product is separated (separatory funnel) and washed with sodium hydroxide solution

(2×10 ml to remove *p*-cresol which might have formed), water (2×10 ml) and dried (calcium chloride). It is finally distilled and the fraction between 160–164°C, The yield is 12 g (67.8%); b.p. is 162°C.

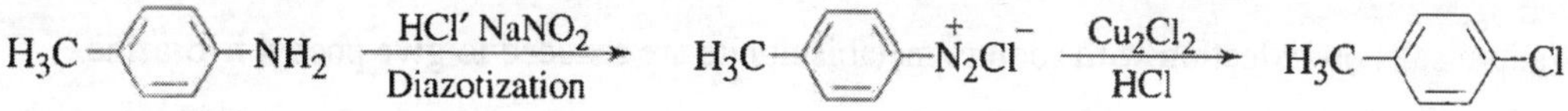

p-toluidine *p*-chlorotoluene

Note: The sodium bisulphite used for the preparation of cuprous chloride is sodium metabisulphite, $Na_2S_2O_3$, which has a purity of 90–95%. The reduction can also be carried out with an alkaline solution of sodium bisulphite prepared from 11.2 g sodium bisulphite and 10 g sodium hydroxide pellets. The reaction most likely is as follows:

$$2CuSO_4 + 4NaCl \quad Na_2S_2O_3 + H_2O \longrightarrow 2CuCl + 3Na_2SO_4 + 2HCl$$

30.4.2 p-Bromotoluene. It is obtained by the method analogous to that used for *p*-chlorotoluene, except that the amine is diazotized in sulphuric acid solution and the resulting aryl diazonium sulphate is treated with a solution of cuprous bromide in excess of hydrobromic acid.

p-Toluidine : 15.4 g
Conc. sulphuric acid : 15.4 ml
Sodium nitrite : 10.5 g
Cuprous bromide : obtained from 33.8 g $CuSO_4 \cdot 5H_2O$.
(prepared as given below)

Cuprous bromide solution is prepared as follows: Hydrated copper sulphate (33.8 g) and sodium bromide dihydrate (19.8 g) are dissolved in water (110 ml). A solution of sodium bisulphite (9 g, see page 60) in water (90 ml) is added to the hot solution (obtained above) during 5 minutes. If the blue colour is not discharged completely, add a little more sodium bisulphite. The mixture is cooled (ice-bath) and the supernatant liquid decantated with water (2×15 ml) containing a little sulphurous acid (to prevent oxidation). The remaining product is dissolved in hydrobromic acid (48% HBr, 22 ml) and the solution is cooled (ice-salt mixture) and is used immediately after diazotization.

p-Toluidine (15.4 g, 0.144 mole) is suspended in water (120 ml) and then concentrated sulphuric acid (15.4 ml) is added cautiously (R.B. flask of 250 ml capacity is used). The mixture is warmed until p-toluidine dissolves. The solution is cooled to 0–5°C (ice-salt bath) and a solution of sodium nitrite (10.5 g) in water (20 ml) is added dropwise (using a separatory funnel, the tip of the stem of the funnel dipping well below the surface of the liquid). The solution is stirred during diazotization and the mixture kept at 0–5°C. When the addition is complete, check for the completion of the reaction with potassium iodide-starch paper, which gives a blue colour with the test solution.

The cuprous bromide solution (prepared earlier) is taken in a R.B. flask (500 ml capacity) and the apparatus set up for steam distillation. The cuprous bromide solution is heated to boiling, the diazanium salt solution is added in small lots through a separatory funnel (the stem of which dips below the surface of the liquid). During the addition, steam distillation is allowed to proceed. Total time for addition is about 15–20 minutes. The steam distillation is continued till no more oily drops distil over. The steam distillate is rendered alkaline (20% NaOH solution) and the crude *p*-bromotoluene is separated (separatory funnel). It is washed with warm water (2×10 ml), sodium hydroxide solution (10 ml) and finally with water (2×10 ml). The remaining liquid (which may solidify, in which case it just should be warmed) is dried (magnesium sulphate, 1 g) and distilled. The fraction between 182–184°C is collected the yield is 12 g (48.4%), b.p. is 183°C; m.p. is 26°C.

30.5 Reactions Not Involving Loss of Nitrogen (Reduction of Diazonium Salts)

Diazonium salts on reduction with sodium metabisulphite are reduced to give phenyl hydrazine.

$$C_6H_5N_2^+Cl^- \xrightarrow{NaHSO_3} C_6H_5NHNH_2 \xrightarrow{HCl} C_6H_5NHN^+H_3Cl^-$$

Mechanism:

$$NaHSO_3 \rightleftharpoons Na^+ + S\bar{O}_3H \rightleftharpoons SO_3^{2-} + H^+$$

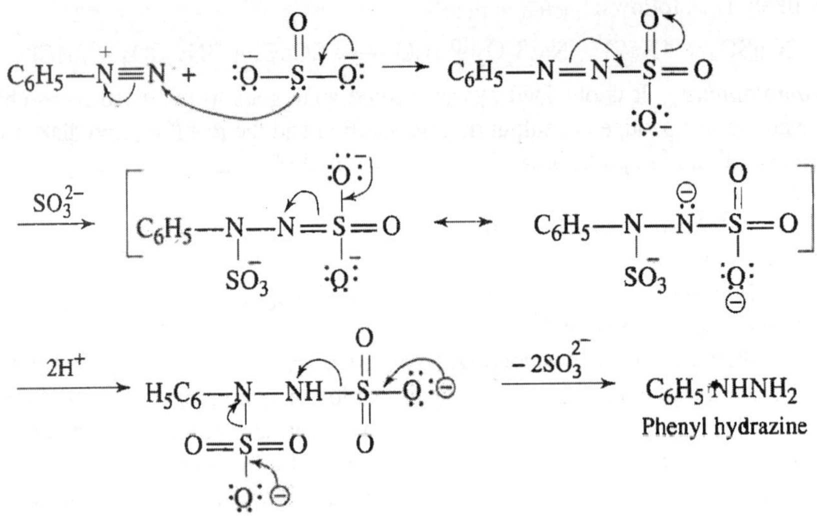

30.5.1 Phenylhydrazine

Aniline	: 23.75 g
Conc. hydrochloric acid	: 60 ml
Sodium nitrite	: 17.5 g
Sodium metabisulphite	: 67–70 g

An ice cold solution (0–3°C) of sodium nitrite (17.5g) in water (40 ml) is added dropwise through a separatory funnel (the stem of the funnel should reach the bottom of the flask) during 20–30 minutes to a vigorously stirred solution of pure aniline (23.75 g) in concentrated hydrochloric acid (60 ml) and water (30 ml) at 0°C. The solution is kept at 0°C (ice-salt bath) throughout the addition. It is ascertained that there is a slight excess of nitrous acid (potassium iodide-starch paper) at the end of 20 minutes after the last portion of sodium nitrite solution is added.

Sodium metabisulphite (56.25 g) is added to a solution of sodium hydroxide (50 g) in water (500 ml). The mixture is stirred (if possible by a mechanical stirrier or by shaking) until all the sodium metabisulphite has dissolved. The solution is cooled (25°C) and phenolphthalein indicator (few drops) is added. More sodium metabisulphite is added with stirring until the pink colour just disappears. Add little more sodium metabisulphite (6g) (the total amount of sodium metabisulphite used is approximately 67–70 g). The resulting solution is cooled to 5°C (ice bath), crushed ice (30 g) is added and the ice cold diazonium chloride solution (obtained above) is added as rapidly as possible; the reaction mixture is vigorously stirred during the addition. The mixture acquires a bright orange-red colour. It is slowly heated (water bath) to 60–70°C and kept at this temperature for 60 minutes until the colour becomes quite dark. The solution is acidified to litmus with concentrated hydrochloric acid (approx. 20–25 ml). The heating (boiling water-bath) is continued until the colour becomes lighter (4–5 hours). The solution is filtered to remove any solid product and to the hot clear filtrate is added, with stirring, concentrated hydrochloric acid (250 ml) and the solution cooled to 0°C (freezing mixture). The separated phenyl hydrazine hydrochloride is filtered and washed with dilute hydrochloric acid (1:3). This can be further purified by crystallisation from water (charcoal) (the clear solution is cooled to 0°C) as pure white crystals. The yield is 85–90%.

Phenyl hydrazine is obtained by adding sodium hydroxide solution (25%, 40 ml) to the phenylhydrazine hydrochloride. The free base is extracted with toluene (2 × 35 ml). The toluene extract is dried (sodium hydroxide pellets) and toluene distilled in vacuo. Phenylhydradine is collected at 137–138°C/8 mm. The yield is 60–65%; b.p. is 243.5°C.

Note: The sodium metabisulphite solution can also be prepared by passing sulphur dioxide gas into sodium hydroxide solution (50 g sodium hydroxide in 65 ml water and diluting to 375 ml) containing phenolphthalein indicator (few drops) till the pink colour just disappears and then for 2 minutes more.

30.6 Coupling Reactions

Diazonium salts react with aromatic amines and phenols to give azo compounds having the general formula ArN=NAr. The reaction is known as coupling reaction. Thus,

$$ArN_2^+X^- + Ar'H \longrightarrow Ar\text{–}N{=}N\text{–}Ar' + HX$$

The aromatic compound Ar′ H must contain electron releasing groups like –OH,–NH_2,–OR, NHR or –NR_2 etc. The coupling usually occurs at *para*-position with respect to the activating group. In case the *p*-position is occupied, then the coupling may occur at *ortho* position.

Mechanism: The coupling reaction is an electrophilic substitution involving the diazonium ion, which reacts at the position of greatest electron availability, i.e., the position *ortho* or *para* to the electron releasing phenoxy or amino group. Thus, the reaction of diazo compound with phenol is represented below:

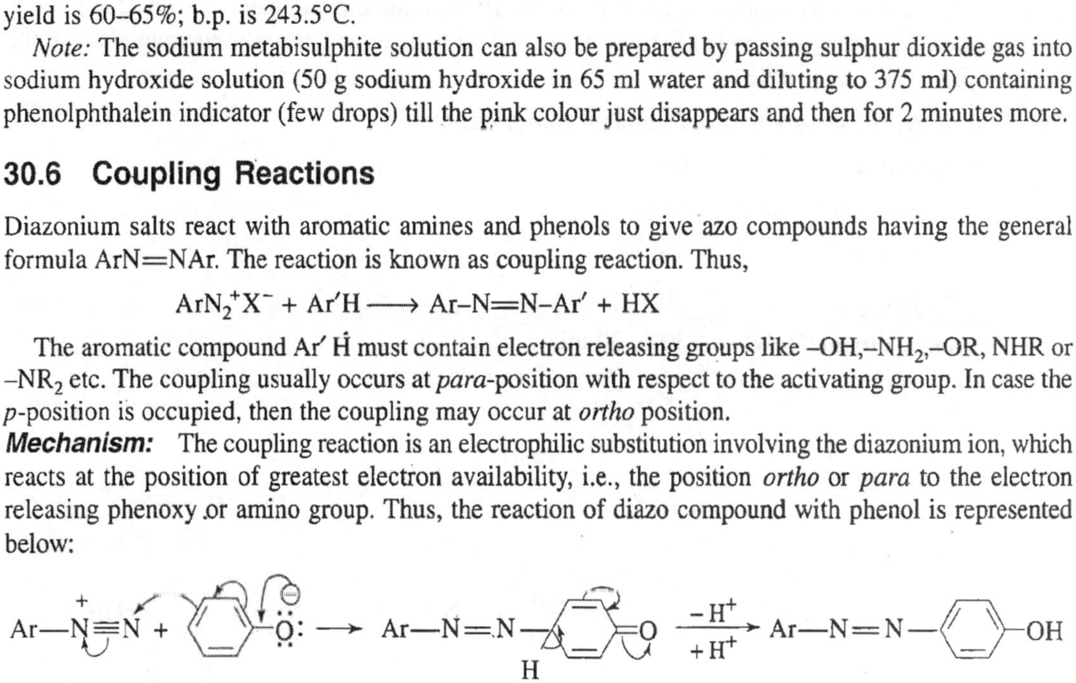

In case of 2-naphthol, coupling occurs at the more reactive 1-position. However, 1-naphthol couples exclusively in position 4.

N, N-Dialkylamines also couple in the *p*-position. Thus,

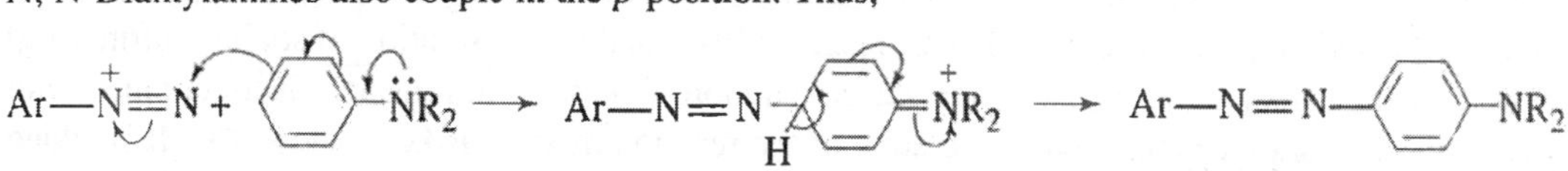

30.6.1 Phenylazo-β-naphthol

Aniline	: 4 ml
Conc. hydrochloric acid	: 13 ml
Sodium nitrite	: 3 g
β-Naphthol	: 6.4 g
Sodium hydroxide solution	: 10%, 40 ml

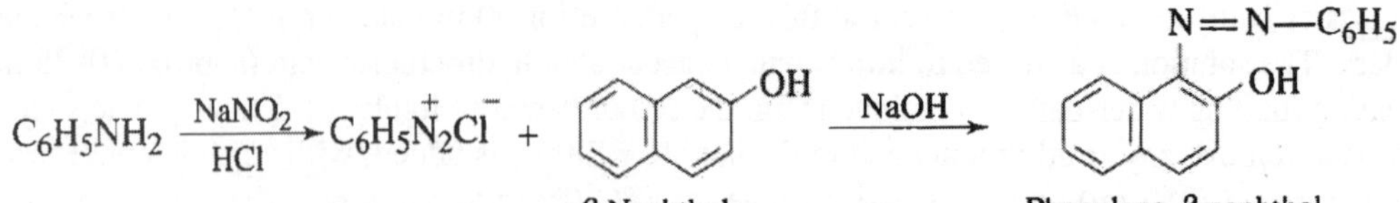

A solution of sodium nitrite (3 g in 15 ml water) is added dropwise to a stirred and cooled solution of aniline (4 ml, 0.044 mole) in a mixture of concentrated hydrochloric acid (13 ml) and water (13 ml). The reaction mixture is maintained at 0–5°C. When the addition is complete, the solution is kept for 5 minutes with occasional stirring (to complete the diazotisation) and then poured with stirring into a cold solution of β-naphthol (6.4 g, 0.044 mole) in sodium hydroxide solution (10%, 40 ml). The reaction mixture is allowed to stand in an ice-bath for 30 minutes, the mixture being stirred occasionally. The separated phenylazo-β-naphthol is filtered, washed with water and crystallised from alcohol, the yield is 8.5 g (78%); m.p. is 131°C.

30.6.2 4-(4′-Nitrobenzeneazo)-1-naphthol (Magneson II)

p-Nitroaniline	: 6 g
Conc. hydrochloric acid	: 15 ml
Sodium nitrite	: 4 g
1-Naphthol	: 5.75 g
Sodium hydroxide solution	: 30%, 30 ml

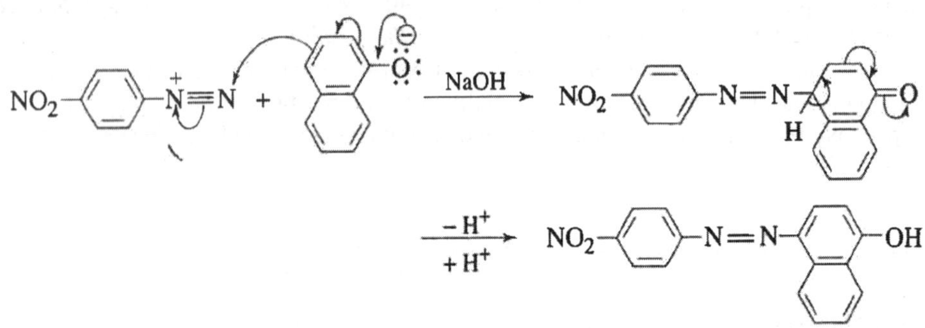

Magneson II

p-Nitroaniline (6 g, 0.043 mole) is dissolved in dilute hydrochloric acid (15 ml concentrated acid and 15 ml water). The solution is cooled (0–5°C, in an ice bath) and a cold solution of sodium nitrite (4 g) in water (10 ml) is added dropwise to the stirred solution. Excess of nitrous acid is ascertained with potassium iodide-starch paper. The temperature of the reaction mixture is kept below 8°C. It is added with stirring to a cooled (ice) solution of 1-naphthol (5.75 g, 0.04 mole) in sodium hydroxide solution (30%, 30 ml). Concentrated hydrochloric acid is then added slowly till the mixture is acidic to congo red. The colour changes from violet to dark brown. The separated product is filtered, washed with water (until free from acid) and dried on filter paper. The yield is 9.5 g (81%).

In the above preparation use of resorcinol in place of 1-naphthol gives 2,4-dihydroxy-4′-nitroazobenzene (magneson I); m.p. is 199–200°C.

30.6.3 *Diazoaminobenene.* It is obtained by the coupling of benzene diazonium chloride with aniline.

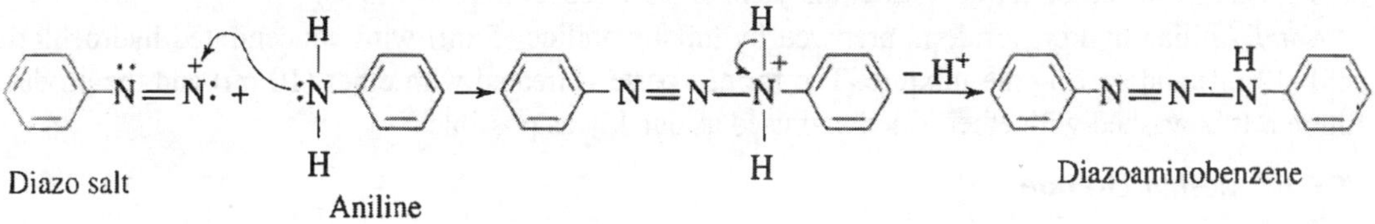

Aniline	: 10 ml
Conc. hydrochloric acid	: 11 ml
Sodium nitrite	: 3.9 g
Sodium acetate	: 27 g

A solution of sodium nitrite (3.9 g in 9 ml water) is added dropwise to a cooled (0–5°C) solution of aniline (10 ml, 0.11 mole) in dilute hydrochloric acid (11 ml concentrated hydrochloric acid and 50 ml water) with stirring during 10 minutes. The mixture is allowed to stand (0–5°C) for 15 minutes with occasional shaking. A solution of sodium acetate (27 g in 100 ml water) is added to the diazotised solution during 5 minutes. The reaction mixture is stirred for 20 minutes (the temperatures is kept below 20°C) and the separated diazoaminobenzene is filtered and washed with a little cold water. It is crystallised from alcohol. The yield is 10.5 g (97.2%); m.p. is 97°C.

Note: 1. The product is crystallised as quickly as possible to prevent it from decomposition. For this purpose the compound is dissolved in three times its volume of hot alcohol, the liquid heated for a minute and the clear solution cooled. (2) Sodium acetate diminishes the acidity of the solution to make the coupling possible.

30.6.4 p-Aminoazobenzene

Diazoaminobenzene on warming with aniline and aniline hydrochloride, rearranges to the aminoazo isomer, i.e., *p*-aminoazobenzene.

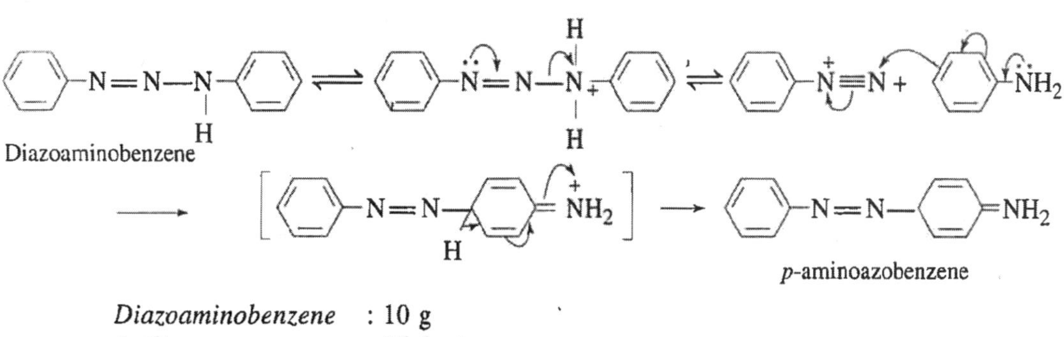

Diazoaminobenzene	: 10 g
Aniline	: 26.6 ml
Aniline hydrochloride	: 5.0 g
Glacial acetic acid	: 30 ml

A mixture of diazoaminobenzene (10 g, 0.05 mole), aniline hydrochloride (5 g) and aniline (26.6 ml) is stirred and heated in a R.B. flask at 40–45°C (water-bath) for 1 hour. The mixture is stirred frequently and allowed to stand for 30 minutes. Glacial acetic acid (30 ml) is added and the mixture stirred and poured into water (130 ml). The separated product is filtered, washed with cold water and dried. It is crystallised from carbon tetrachloride, the yield is 6.5 g (65%) m.p. is 125°C.

Note: Aniline hydrochloride is prepared by mixing aniline (5 ml) with concentrated hydrochloric acid (12 ml) and stirring the mixture. The formed paste is treated with ether (10 ml) and the product filtered. It is washed with ether, dried and used as quickly as possible.

30.6.5 Methyl Orange

Sulphanilic acid	: 10 g
Sodium carbonate (anhyd.)	: 3 g
Sodium nitrite	: 4 g
Conc. hydrochloric acid	: 10 ml
Dimethylaniline	: 6.2 ml
Acetic acid (glacial)	: 3 ml
Sodium hydroxide solution	: 20%, 40 ml
Sodium chloride	: 20 g

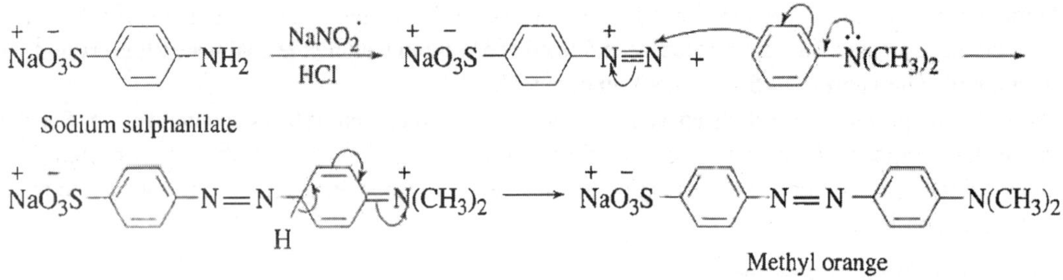

Methyl orange is more useful as an indicator than as a dye. It changes colour at pH 3.1–4.4. Treatment of a solution of methyl orange with a strong acid gives rise to a red form; this is an internal salt, which is resonance stabilized

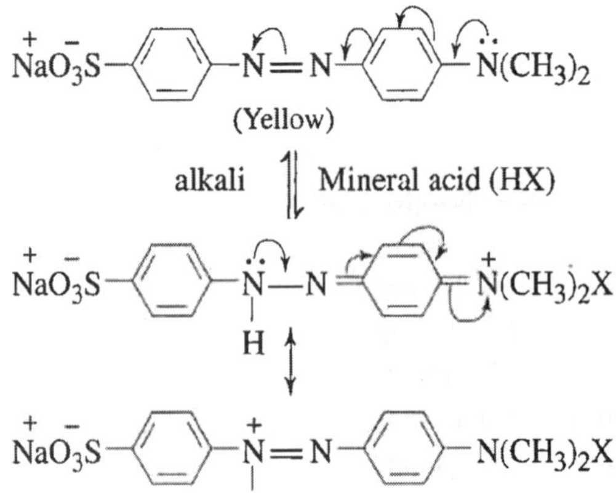

Sulphanilic acid (10 g, 0.057 mole) is dissolved in a solution of sodium carbonate (3 g) in water (100 ml). The clear solution (it may be necessary to warm the solution to obtain a clear solution) is cooled (10–15°C) and a solution of sodium nitrite (4 g in 10 ml water) is added slowly. The resulting solution is poured slowly with stirring into a beaker containing concentrated hydrochloric acid (10 ml) and crushed ice (60 g) (temperature 0–5°C). After 15 minutes the presence of free nitrous acid is tested with potassium iodide-starch paper. Fine crystals of diazobenzene sulphonate separate out. A solution of dimethylaniline (6.2 ml, 0.049 mole) in glacial acetic acid (3 ml) is added with vigorous stirring to the above suspension. The mixture is allowed to stand for 15 minutes. The red or acid form of methyl orange separates out gradually. Sodium hydroxide solution (20%, 40 ml) is added slowly with stirring to the reaction mixture. The mixture becomes orange due to the separation of sodium salt of methyl orange as fine particles. The mixture is heated almost to the boiling point when most of the dye dissolves. Add sodium chloride (about 20 g) to assist the subsequent separation of methyl orange. The mixture is warmed (80–90°C) to dissolve the salt and allowed to cool to room temperature (15 minutes) and then is cooled in ice-water. The separated methyl orange is filtered, washed with a little saturated sodium chloride solution and crystallised from hot water (100–150 ml). The yield is 13 g (81.2%). The product does not have a well defined m.p. since it is a salt.

30.6.6 Orange II (β-Naphthol orange)

Sulphanilic acid	: 10 g
Sodium carbonate (anhyd.)	: 3 g
Sodium nitrite	: 4 g
Conc. hydrochloric acid	: 10 ml
β-Naphthol	: 7.2 g
Sodium hydroxide solution	: 10%, 40 ml
Sodium chloride	: 20 g

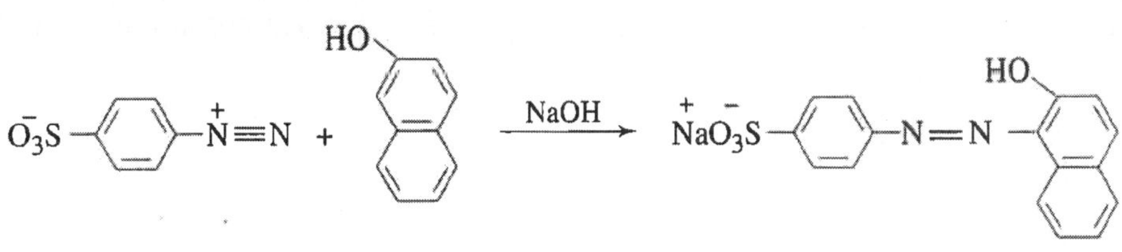

Orange II

Sulphanilic acid (10 g, 0.057 mole) is diazotised as described under methyl orange. To a cold solution (5°C) of *β*-naphthol (7.2 g, 0.05 mole) in sodium hydroxide solution (10%, 40 ml) is added with stirring the cooled and well mixed suspension of diazolised sulphanilic acid. Coupling takes place and a solid product separates. After 10–15 minutes the mixture is heated till the solid dissolves and then sodium chloride (20 g) is added. The mixture is heated to get a clear solution and then cooled in air for 1 hour and finally is cooled in ice. The separated dye is filtered, washed with a little saturated sodium chloride solution. It is recrystallised by dissolving in minimum amount of boiling water, cooled (80°C) and twice the volume of rectified spirit is added. The crystallisation occurs spontaneously. The solution is cooled and the separated orange II is filtered. The yield is 14 g (80%).

30.6.7 Methyl Red

Anthranilic acid : 10 g
Dimethylaniline : 13.8 ml
Sodium nitrite : 5.7 g
Conc. hydrochloric acid: 20 ml
Sodium acetate : 11 g
Sodium hydroxide : 20%, 8 ml
Acetic acid : 10%, 15 ml

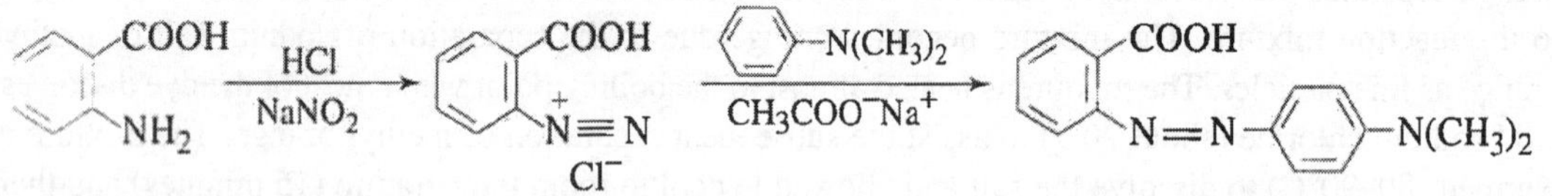

Anthranilic acid   Methyl red

Anthranilic acid (10 g, 0.073 mole) is dissolved by heating in dilute hydrochloric acid (30 ml, 1:3 v/v). The solution is cooled to 2–3°C (ice-bath) and hydrochloric acid (12 ml) is added with stirring. To the cold solution (2–3°C) is added dropwise a solution of sodium nitrite (5.7 g in 15 ml water). To the resulting solution of the diazonium salt is rapidly added dimethylaniline (13.8 ml, 0.11 mole). The mixture is stirred for 10–15 minutes (temperature < 5°C).

In another beaker, sodium acetate (11 g) is dissolved in water (15 ml). The sodium acetate solution (7 ml) is added to the reaction mixture and allowed to stand in an ice-bath for 1 hour with occasional stirring. The remaining sodium acetate solution is then added with stirring to the cooled mixture (ice-bath). It is left for 30 minutes with occasional stirring. The reaction mixture is allowed to attain room temperature and sodium hydroxide solution (about 8 ml of 20 per cent solution) is added with stirring till a distinct odour of dimethylaniline is observed. The mixture is left for 1 hour. The separated product is filtered, washed with a little water and then with dilute acetic acid (10 per cent, 15 ml) to remove dimethyl aniline; and finally is washed with water. The crude product is boiled with methanol (80 ml) for 10–15 minutes (water bath) filtered and washed with cold methanol (15 ml), the yield is 9.5 g (48%); m.p. is 181–182°C.

30.6.8 p-Nitrosodimethylaniline

Dimethylaniline : 7.5 ml
Conc. hydrochloric acid : 25 ml
Sodium nitrite : 4.5 g

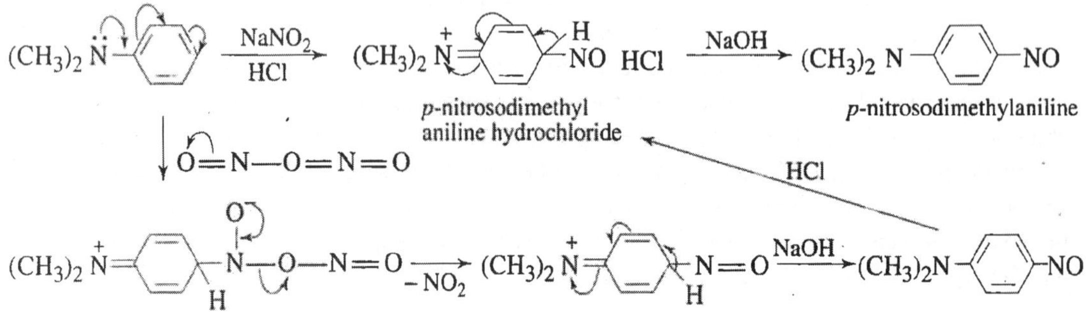

Dimethylaniline (7.5 ml, 0.059 mole) is dissolved in dilute hydrochloric acid (1:1, 50 ml). The solution is cooled to 0–5°C (ice-salt bath) and a solution of sodium nitrite (4.5 g in 10 ml water) is added in lots of 1 ml at a time with constant stirring. The completion of reaction is ascertained by testing with potassium iodide-starch paper (gives blue colour). The reaction mixture is kept for 30 minutes in an ice bath and the separated *p*-nitrosodimethylaniline hydrochloride is filtered and washed with ice-cooled hydrochloric acid. The yield is 9 g; m.p. is 176–177°C.

The hydrochloride (obtained above) is taken in a separatory funnel and treated with sodium carbonate or sodium hydroxide solution till the entire mass becomes green. It is extracted with ether (2 × 25 ml). The combined extract is dried (anhydrous potassium carbonate) and the ether solution concentrated to half the volume. It is then left in an evaporating porcelein dish for crystallisation. Brilliant green leaflets of *p*-nitrosodimethyl aniline are obtained. The yield is 7 g (79%); m.p. is 85°C.

31. DRUGS

31.1 Acetylsalicyclic Acid (Aspirin): (Analgesic Drug)

It is obtained by acetylation of salicyclic acid with acetic anhydride in presence of an acid catalyst.

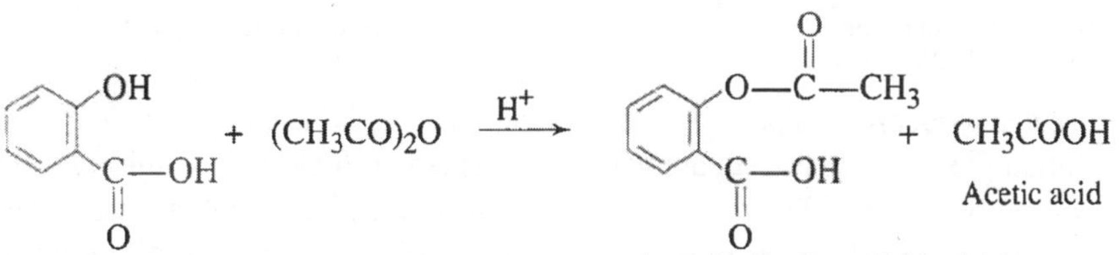

Salicylic acid — Acetic anhydride — Acetyl salicylic acid (Aspirin)

Salicylic acid : 5 g
Acetic anhydride : 10 ml
Conc. sulphuric acid : 2–3 drops

A mixture of dry salicylic acid (5 g, 0.036 mole), distilled acetic anhydride (10 ml) and conc. sulphuric acid (2-3 drops) is mixed with the help of a glass rod and the mixture warmed (water-bath) to 50 60°C for 15 minutes. The mixture is constantly stirred. It is cooled and poured into water (50 ml) and stirred. The separated aspirin is filtered. It is dissolved in hot ethanol (15 ml) and poured into warm water (40 ml). The resulting mixture is warmed to just dissolve any solid that may separate and then allowed to cool slowly. Needle like crystals of aspirin separate out. It is filtered. The yield is 5.9 g (90%); m.p. is 135–136°C.

31.2 *p*-Aminobenzenesulphonamide (Sulphanilamide)

Step 1 p-Acetamidobenzenesulphonyl chloride

Acetanilide : 10 g
Chlorosulphonic acid : 25 ml
(freshly distilled)

Acetamilide — Chlorosulphonic acid — *p*-Acetamido benzene sulphonyl chloride (sulphanilamide)

Chlorosulphonic acid (25 ml) is taken in a three necked round bottom flask, fitted with a mechanical stirrer with a mercury seal (a magnetic stirrer can also be used) and a dropping funnel. The flask is cooled by the use of an ice-bath and dry acetanilide (10 g, 0.074 mole) is added through the third neck (removing the cork when necessary) in small lots while the mixture is kept stirred. The top of the condenser is connected to a gas trap to absorb the liberated hydrogen chloride. After the addition is over, the flask is warmed on a water bath at 60°C to complete the reaction (10–15 minutes). The reaction mixture is poured into ice water (300 ml). The separated product is filtered, washed with cold water and pressed to drain out excess water. The crude product (12 g, 69%) is immediately used as such for the next stage. However, if required the crude product can be crystallised from chloroform or benzene-acetone mixture.

Note: Acetanilide is obtained as described on page 8.

Step 2 p-Acetamidobenzenesulphonamide

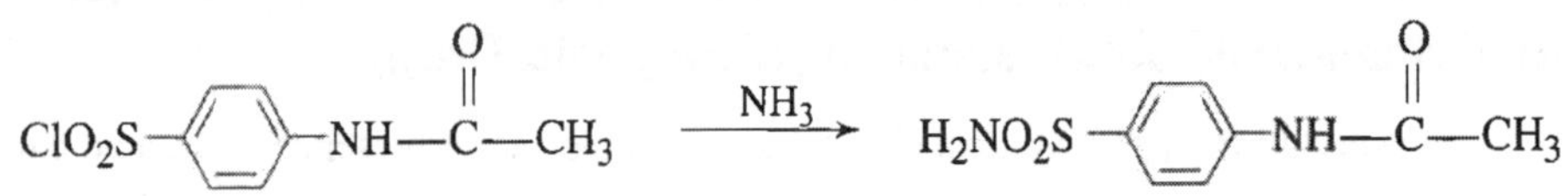

p-Acetamidobenzene sulphonyl chloride → *p*-Acetamidobenzene sulphonamide

The crude *p*-acetamidobenzenesulphonyl chloride (from step 1) is taken in a conical flask (250 ml). Ammonia solution [35 ml concentrated ammonia solution (sp. gr 0.88) and 35 ml water] is added and the contents mixed thoroughly. The mixture is heated (with occasional shaking) to 80–90°C (boiling water bath) for 10 minutes. The formed pasty suspension of sulphonamide is cooled (ice) and acidified with dilute sulphuric acid until the mixture is just acidic to congo red paper. The separated product is filtered, washed with water and the excess water drained well at the pump. It is used as such for the next step. If required, a small quantity can be crystallised from water. The m.p. is 218°C.

Step 3 p-Aminobenzene sulphonamide (sulphanilamide)

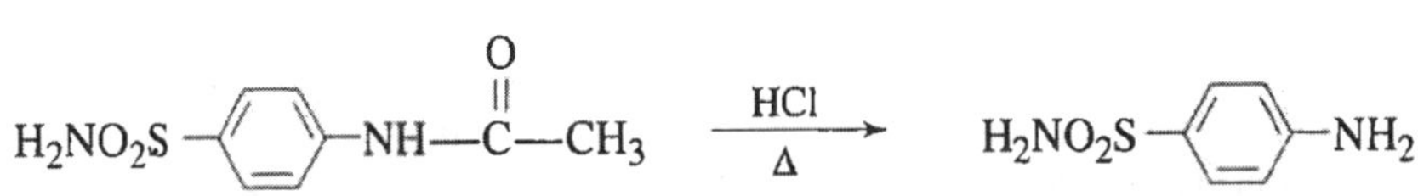

p-Acetamidobenzene sulphonamide → *p*-Aminobenzene sulphonamide (sulphanilamide)

The crude *p*-acetamidobenzene sulphonamide [obtained above in step (2)] is refluxed gently for 20–25 minutes with dilute hydrochloric acid (5 ml concentrated hydrochloric acid and 15 ml water). The solution is cooled (ice water). It is to be ascertained that no solid product separates on cooling. In case any solid separates (indicating that the hydrolysis is incomplete) the mixture is further heated for a short period. The solution is treated with decolorizing carbon, heated to boiling and filtered using a fluted filter paper. To the clear cooled solution thus obtained, is carefully added sodium bicarbonate (8 g) in small lots with stirring. The addition of sodium bicarbonate is continued till the solution does not show acidic test by litmus paper (the solution should be neutral). The mixture is cooled in ice and the separated product filtered (suction). It is recrystallised from alcohol. The yield is 6 g (68%); m.p. is 163–164°C.

31.3 Sulphaguanidine (Antibacterial Drug)

Sulphanilamide	: 10.5 g
Guanidine nitrate	: 7.5 g
Sodium hydroxide (pellets)	: 2.4 g
Methanol	: 25 ml
Cyclohexanol	: 2.1 ml
Hydrochloric acid (1.5 N)	: 68 ml

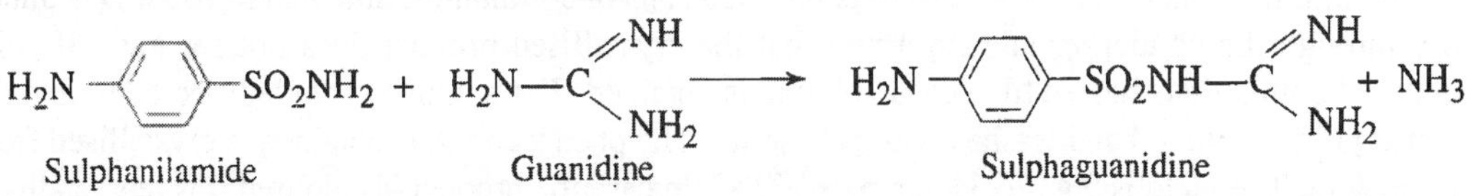

Guanidine nitrate (7.5 g, 0.06 mole) is added to methanolic sodium hydroxide solution (obtained by dissolving 2.4 g sodium hydroxide pellets into 12.5 ml methanol). The mixture is stirred at 50–60°C for 15 minutes and then cooled to about 20°C. The separated sodium nitrate is filtered and the residual salt washed with methanol (2 × 3 ml). The combined filtrate is taken in a three necked flask and sulphanilamide (10.5 g, 0.061 mole) added. The mixture is stirred for 15 minutes at 50–55°C and then methyl alcohol (15 ml) distilled off (oil bath). Cydohexanol (2.1 ml) is added to the reaction mixture and the temperature of the oil-bath raised to 180–190°C and the distillation is continued. As soon as the temperature reaches to 140–145°C, reaction commences with the evolution of ammonia. The reaction mixture is maintained at 150–160°C for 2 hours and then poured into dilute hydrochloric acid (1.5 N, 68 ml). The mixture is stirred till a clear solution is obtained. The solution is rendered alkaline by the slow addition of sodium hydroxide solution with stirring. The mixture is cooled and the separated sulphaguanidine filtered, washed with cold water and crystallised from hot water. The yield is 7 g (51%); m.p. is 189–190°C.

31.4 *p*-Ethoxyacetanilide (Phenacetin) : Analgesic Drug

p-Aminophenol	: 11 g
Acetic anhydride	: 12 ml
Sodium metal	: 1.6 g
Absolute alcohol	: 40 ml
Ethyl iodide	: 8 ml

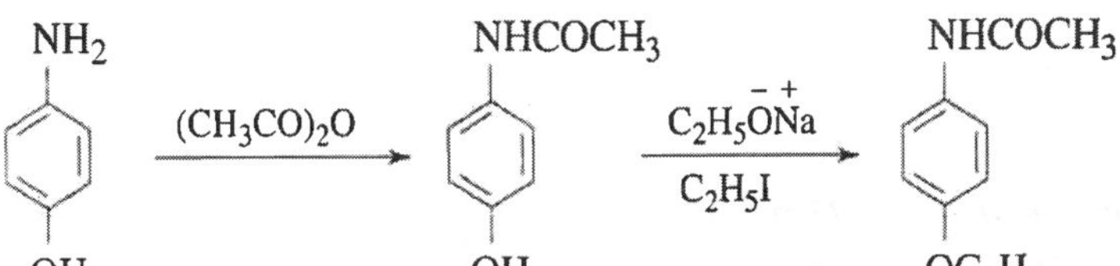

p-Aminophenol *p*-Acetyl aminophenol *p*-Ethoxyacetanilide (phenacetin)

Acetic anhyhdride (12 ml, 0.13 mole) is added with vigorous stirring (or shaking) to a suspension of *p*-aminophenol (11 g, 0.1 mole) in water (30 ml) contained in a conical flask. The mixture is then warmed in a water bath till the solid dissolves. After 10 minutes the solution is cooled, and the *p*-acetylaminophenol which separates, is filtered, washed with a little cold water and recrystallised from hot water (30–35) ml. The yield is 14 g (92.7%); m.p. is 169°C. If the compound is impure, as seen by the m.p. determination, the product is purified by dissolving in cold dilute alkali and the clear alkaline solution (obtained by filtration or ether extraction to remove any diacetate which might have formed) is acidified.

Sodium pieces (clean, 1.6 g, 0.07 mole) are added to absolute alcohol (40 ml) contained in a round bottom flask fitted with a reflux condenser (anhydrous conditions). If all the sodium has not dissolved, the flask is warmed on a water bath till a complete solution is obtained. The mixture is cooled and *p*-acetylaminophenol (10 g, 0.066 mole) is added followed by slow addition of ethyl iodide (8 ml, 0.1 mole) through the condenser. The mixture is refluxed for 50–60 minutes, and water (100 ml) is added slowly through the condenser at such a rate that the crystallised product does not separate. If solid separates, the mixture is heated till a clear solution is obtained. The solution is cooled (ice-bath) and the separated product filtered and washed with cold water. The phenacetin thus obtained is crystallised from dilute alcohol. The yield is 9 g (76.3%) m.p. is 137°C. In case the product is coloured it is recrystallised (charcoalized).

32. DUFF REACTION[1-3]

Ortho-formylation of phenols or *para*-formylation of an aromatic amine with hexamethylene tetramine (hexamine) in presence of an acid catalyst (acetic acid) is known as Duff reaction.

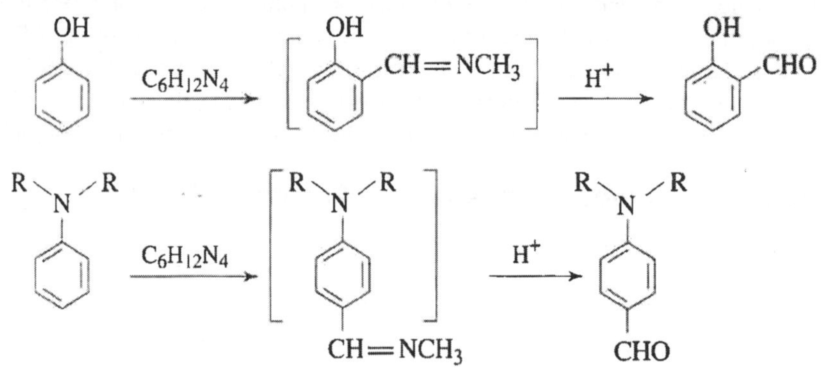

32.1 *p*-Dimethylaminobenzaldehyde

N,N-dimethylaniline : 15 g
Hexamethylenetetramine : 20 g
Ethyl alcohol (95%) : 10 ml
Glacial acetic acid : 25 ml
Formic acid (90%) : 25 ml

1. J.C. DUFF and E. BILLS, *J. Chem. Soc.*, 1932, 1987.
2. J.C. DUFF and E. BILLS, *J. Chem. Soc.*, 1934, 1305.
3. J.C. DUFF and E. BILLS, *J. Chem. Soc.*, 1945, 276.

N, *N*-Dimethylaniline (15 g, 0.124 mole), hexamethylenetetramine (20 g, 0.143 mole) and ethanol (95%, 10 ml) are placed in a 250 ml R.B. flask and is refluxed on a boiling water bath. A mixture of glacial acetic acid (25 ml) and formic acid (90%, 25 ml) is taken and 20 ml of this mixture is added to the refluxing mixture at once, and the remainder is added in 5 ml portions at 30 minutes intervals. After being heated for 5 hours, the hot mixture is poured into hydrochloric acid (0.5 N, 300 ml) with stirring and left overnight. The aldehyde is separated, which is crystallized from alcohol (30%). The yield is 7 g, (38%); m.p. is 41°C.

33. DYES

33.1 Azo Dyes

These are obtained by coupling of diazotised aromatic amines with phenols and *N*, *N*-disubstituted amines. The preparation and the mechanism of their formation has already been discussed in the coupling reactions of diazo compounds (see page 63). The methods of preparation of the following azo dyes has already been given:

Phenylazo-*β*-naphthol (Page 63)
4-(4′-Nitrobenzeneazo)-1-naphthol (magneson-II) (Page 64)
Diazoaminobenzene (Page 65)
p-Aminoazobenzene (Page 65)
Methyl orange (Page 66)
Orange II (*β*-Naphthol Orange)(Page 67)
Methyl red (Page 68)

33.2 Dyes Obtained by Other Methods

33.2.1 Fluorescein

Phthalic anhydride : 7.5 g
Resorcinol : 11 g
Zinc chloride (Anhyd.) : 3.5 g

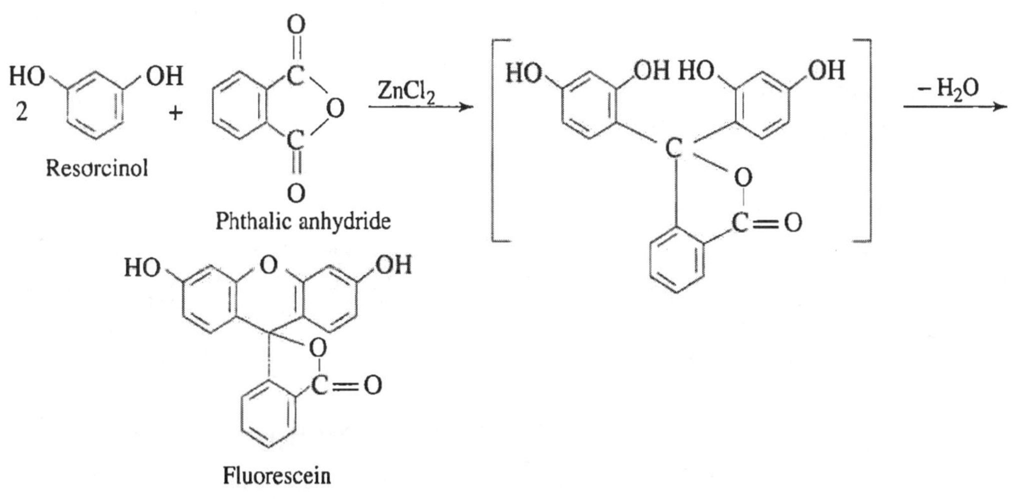

Fluorescein

An intimate mixture of anhydrous phthalic anhydride (7.5 g, 0.05 mole) and dry resorcinol (11 g, 0.1 mole) is heated to 180°C (internal temperature) (oil bath) in a round bottom flask (500 ml capacity). *Freshly fused and powdered zinc chloride (3.5 g) is added in small portions during 30–40 minutes, the mixture being stirred all the time with a thermometer. At the end the mass becomes so viscous that the stirring is not possible. The oil bath is cooled (90°C) and dilute hydrochloric acid (100 ml water and 5 ml concentrated hydrochloric acid) added to the reaction mixture. The temperature of the oil bath is raised till the water boils. The mixture is stirred so that the mass disintegrates and the zinc salts dissolve. The solid product is filtered. It is ground with water in a mortar and filtered again. The yield is 14 g (84%).

Note: (1) Dry resorcinol is obtained by drying ordinary resorcinol in an oven 120–130°C till it melts. It is cooled and powdered in a mortar. (2) Fused zinc chloride is obtained as follows: Zinc chloride is heated in a porcelein dish over a free flame using a wire guage. It melts after some time. The molten zinc chloride is stirred gently with a glass rod. When it starts fuming, it is poured in a dry mortar, powdered and kept in a tightly stoppered bottle.

33.2.2 Eosin (tetrabromo fluorescein)

Fluorescein : 8.3 g
Bromine : 6 ml

Bromine (6 ml) is added dropwise (separatory funnel) during 15–20 minutes to a stirred suspension of fluorescein (8.3 g, 0.025 mole) in alcohol (40 ml). When half the bromine has been added and the fluorescein has been converted into the dibromo compound, all the solid material dissolves, since the dibromo derivative is soluble in alcohol. Further addition of bromine gives the tetrabromo derivative, which separates out since it is sparingly soluble in alcohol. The mixture is allowed to stand for one hour. The separated orange coloured eosin is filtered, washed with cold alcohol and dried (100°C). The yield is 12 g (74%).

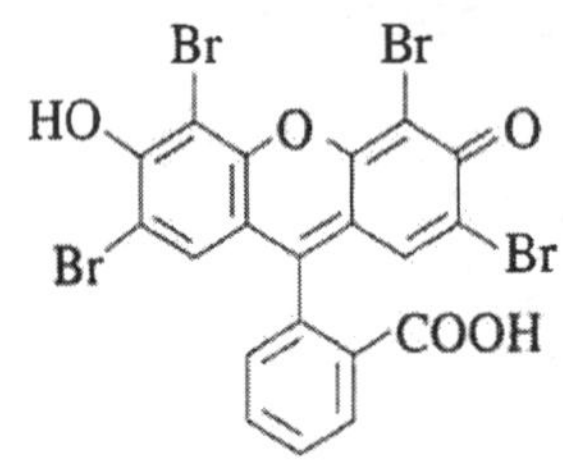

Eosin

33.2.3 Malachite green

Dimethylaniline : 9 g
Benzaldehyde : 3.5 g
Conc. hydrochloric acid: 8 ml

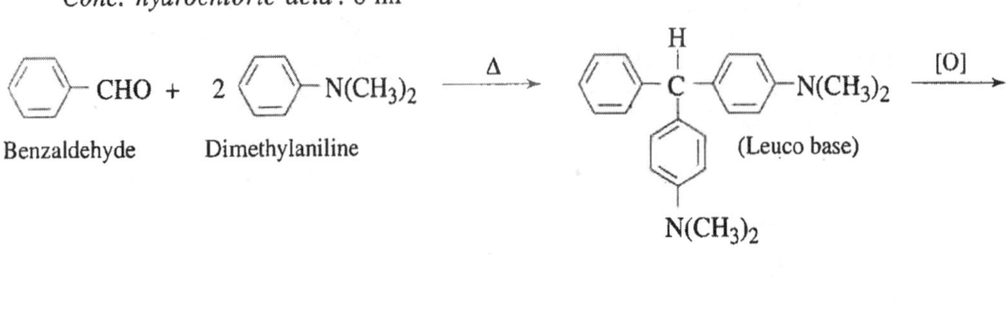

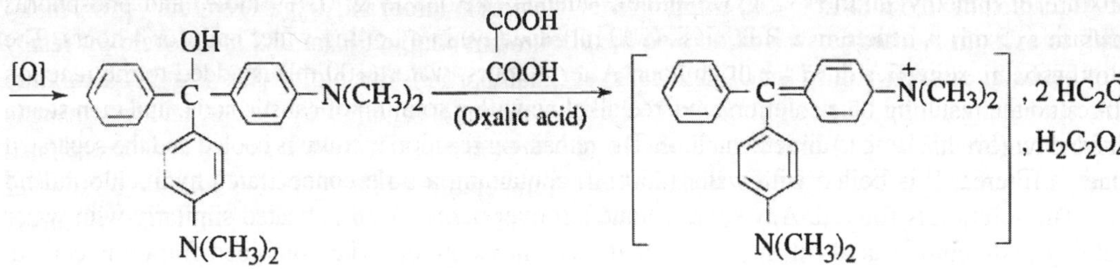

Malchite green (oxalate salt)

A mixture of dimethylaniline (9 g, 0.074 mole), benzaldehyde (3.5 g, 0.033 mole) and concentrated hydrochloric acid (8 ml) is refluxed for 12–14 hours in a R.B. flask (250 ml capacity) fitted with a reflux condenser. The reaction mixture is rendered alkaline with sodium hydroxide solution and steam distilled to remove the unreacted benzaldehyde and dimethylaniline. The residue is poured into water (500 ml). The separated leuco base is filtered, washed with water and dried. It is crystallised from hot ethyl alcohol.

The leuco base (obtained above) is converted into the dye by oxidation. The base (10 g) is dissolved (gentle warming) in dilute hydrochloric acid containing 1.35 g. hydrogen chloride (ascertained by titration with alkali) and glacial acetic acid (92 ml). The liquid is diluted to 200 ml, and cooled by adding crushed ice. Lead peroxide (3.7 g) is added in the form of a thin paste in water and the resultant mixture shaken frequently for about 10 minutes. A saturated solution of Sodium sulphate (10 g) is then added and the mixed precipitate of lead sulphate and chloride filtered off. The filtrate is heated to boiling and the free base liberated by adding sodium hydroxide. It is filtered, washed and dried. The free base is heated in 3–4 times its bulk of water and a calculated amount of oxalic acid (3 g). The solution is filtered and cooled. The oxalate of malachite green is filtered. The yield is 7 g (46%).

The hydrochloride of malachite green can be obtained by treatment of the free base with hydrogen chloride.

33.2.4 Crystal violet (hexamethyl pararosaniline)

Dimethylaniline : 12 g
Michlers ketone : 5 g
Phosphorus oxychloride : 5 ml

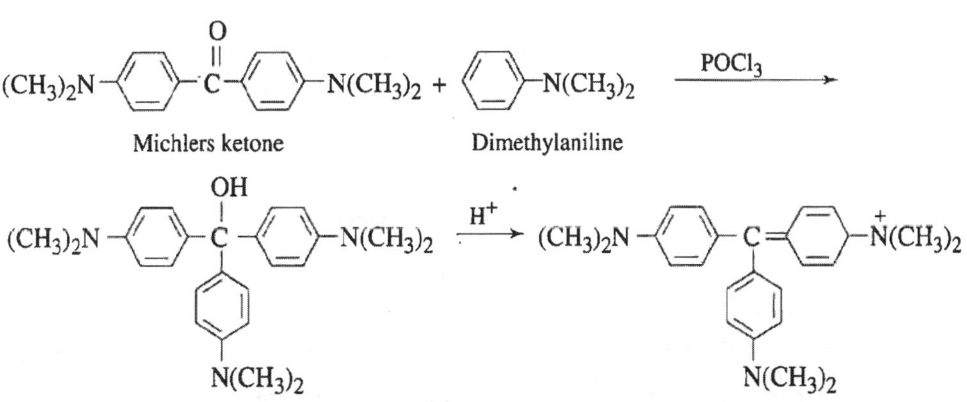

Crystal violet

A mixture of dimethylaniline (12 g, 0.1 mole), Michlers ketone (5 g, 0.019 mole) and phosphorus oxychloride (5 ml) is heated in a R.B. flask (500 ml capacity) in a boiling water bath for 4 hours. The reaction mixture solidifies during half an hour. After 4 hours, water (100 ml) is added to the reaction mixture and the resulting blue solution rendered alkaline with a solution of caustic soda, and then steam distilled to remove unchanged dimethylaniline. The remaining reaction mixture is cooled and the separated product is filtered. It is boiled with water (250 ml) containing a little concentrated hydrochloric acid (2 ml). The solution is filtered. Any solid residue left over undissolved is heated similarly with water containing hydrochloric acid till the whole of it goes into solution. The combined filtrate is cooled. Sodium chloride (50 g) is added in lots, while stirring well, until the dye is precipitated. It is filtered and crystallised from water (5–10 ml). Crystal violet separates as fine crystals having a bronze-like lusture. It is filtered and dried. The yield is 6g (79%).

34. DIENES SYNTHESIS

See Diels-Elder Reaction.

35. DOEBENER MODIFICATION OF KNOEVENAGEL CONDENSATION

See Knoevenagel condensation.

36. DIELS–ALDER REACTION[1-4]

The 4 + 2 cycloaddition reaction between a conjugated diene (4π-electron system) and a compound containing a double or triple bond, called the dienophile (2π-electron system), to form an addition product (adduct) is known as Diels-Alder reaction. Better yields are obtained if the dienophile is activated by introducing to it electron withdrawing groups such as carbonyl, nitro or nitrile. Thus,

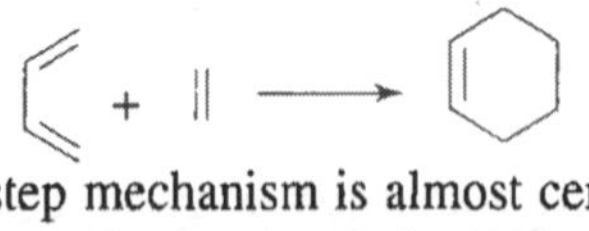

Mechanism: A concerted, single step mechanism is almost certainly involved. In this reaction two new σ-bonds are formed at the expense of two π-bonds in the starting material.

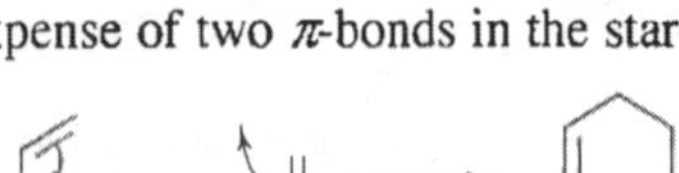

Diene Dienophile Adduct

1. O. Diels, K. Alder, *Ann.*, 1928, **460**, 98.
2. O. Diels, K. Aider, *Ann.*, 1929, **470**, 62.
3. O. Diels, K. Alder, *Ber.*, 1929, **62**, 2081, 2087.
4. L.M. Stephenson, D.E. Smith and S.P. Current, *J. Org. Chem.*, 1982, **47**, 4170.

36.1 9,10-Dihydroanthracene-endo-α, β-succinic Anhydride (Anthracene-maleic Anhydride Adduct)

Anthracene : 3 g
Maleic anhydride : 1.15 g
Xylene : 40 ml

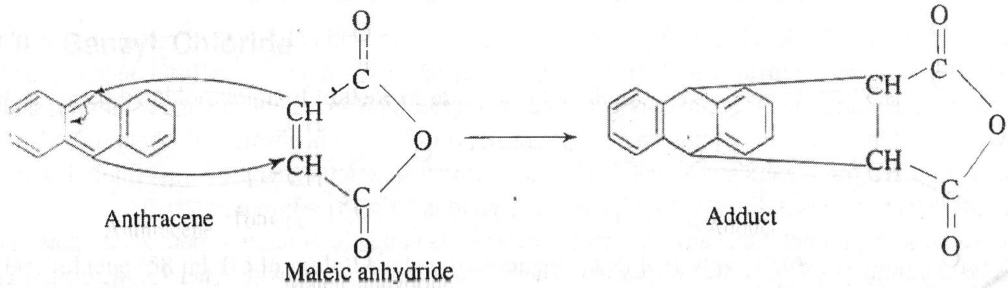

A mixture of anthracene (3 g), maleic anhydride (1.15 g) and dry xylene (40 ml) is refluxed for 15–20 minutes in a R.B. flask with occasional shaking. The clear reaction mixture is cooled and the separated adduct filtered. It is crystallised from xylene and dried in a vacuum desiccator containing paraffin wax to absorb traces of xylene. The yield is 3.2 g; m.p. is 262–263°C (decompose.).

37. DOEBNER REACTION[1-3]

The reaction of aromatic amines with aromatic aldehydes and pyruvic acid to give substituted cinchoninic acids is known as Doebner reaction. Thus,

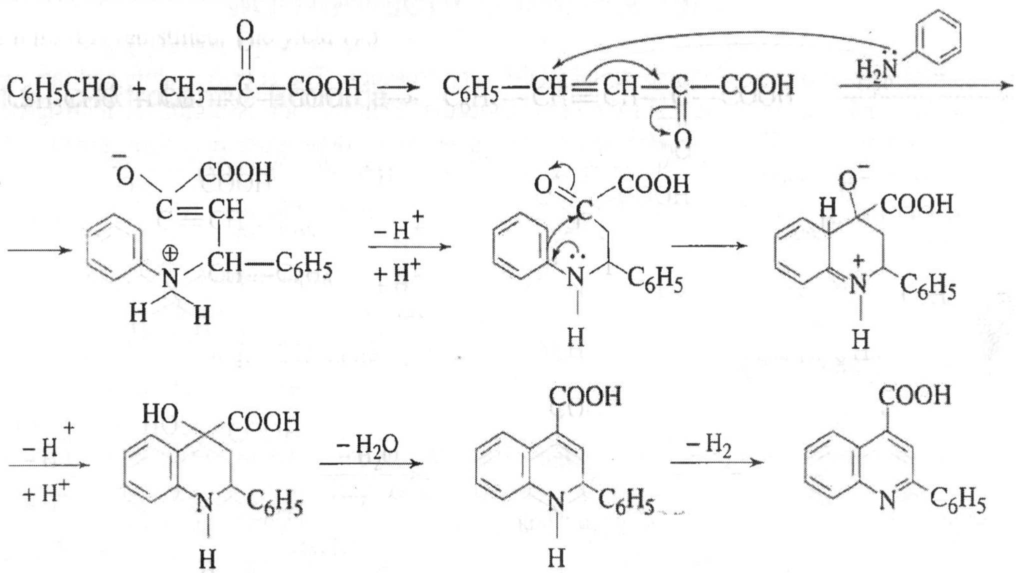

1. O. Doebner, *Ann.*, 1887, **242**, 265.
2. O. Doebner, *Ber.*, 1887, **20**, 277.
3. O. Doebner, *Ber.*, 1894, **27**, 352, 2020.

37.1 2-Phenylquinoline-4-Carboxylic Acid (Atophan)

Aniline : 11.25 ml
Benzaldehyde : 12 ml
**Pyruvic acid* : 11.5 g
Absolute alcohol : 200 ml

Pure benzaldehyde (12 ml, 0.118 mole), freshly distilled pyruvic acid (11.5 g, 0.13 mole) and absolute alcohol (100 ml) are taken in a R.B. flask (500 ml capacity) fitted with a reflux condenser and a calcium chloride guard tube. The mixture is heated on a water bath and to this boiling solution is added slowly with shaking a solution of pure aniline (11.25 ml, 0.123 mole) in absolute alcohol (100 ml). The mixture is refluxed for 3 hours and allowed to stand overnight. The separated product is filtered, washed with a little ether and crystallised from alcohol. The yield is 15 g (51%); m.p. is 210°C.

In case the solid product (atophan) does not separate on standing, the solution is poured into potassium hydroxide solution (12.5 g in 500 ml water). The mixture is extracted with ether (3 × 50 ml). The aqueous layer is treated with glacial acetic acid (35 ml) with stirring. The mixture is allowed to stand for 4–5 hours and the separated product filtered.

Note: *Pyruvic acid is obtained as follows: An intimate mixture of powdered tartaric acid (50 g, 0.33 mol) and freshly fused potassium hydrogen sulphate (75 g, 0.55 mol) is heated in a R.B. flask (500 ml capacity) fitted with a distillation arrangement. The reaction flask is heated at 210–220°C (oil-bath) until no more liquid distils over. The distillate is redistilled under reduced pressure to obtain pyruvic acid. The yield is 15 g (51%); b.p. is 75–80°C/25 mm.

38. DIECKMANN CONDENSATION

Dieckmann condensation is an intramolecular Claisen condensation. The α-carbon atom and tne ester group for the condensation come from the same molecule. This reaction is useful for making ring compounds. (Five, six and seven numbered (esters lower than adipic ester may form products by intermolecular condensation).

Diethyl adipate (diethyl hexane dioate) on heating with sodium ethoxide gives 2-oxocyclopentanecarboxylate

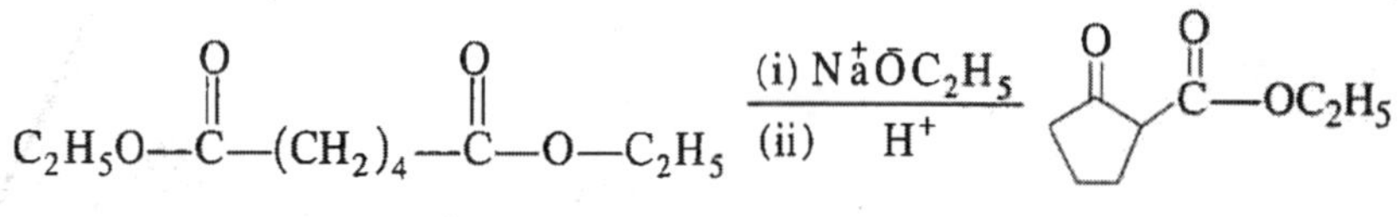

Mechanism:

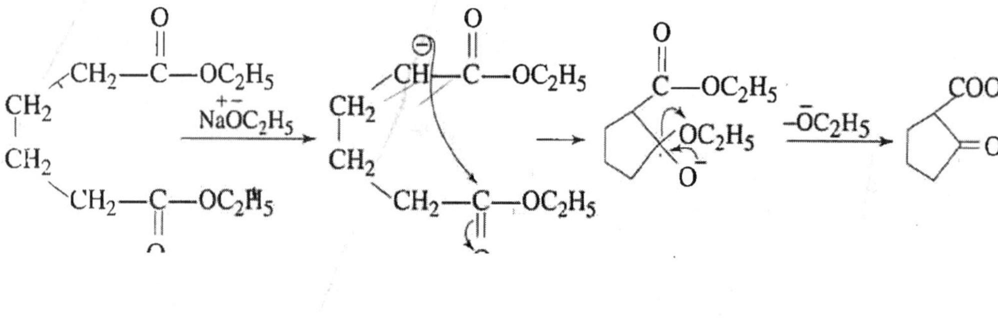

38.1 2-Ethoxycarbonylcyclopentanone (Ethyl 2-oxocyclopentane-carboxylate)

Step 1 Diethyl adipate

Adipic acid	: 73 g
Toluene	: 90 ml
Conc. sulphuric acid	: 0.75 g
Absolute ethanol	: 180 ml
Anhydrous potassium carbonate	: 75 g

Adipic acid (73 g, 0.5 mole), absolute ethanol (180 ml), toluene (90 ml) and concentrated sulphuric acid (0.75 g) are placed in a 500 ml R.B. flask fitted with a short fractionating column with a downward condenser. The mixture is heated in an oil bath at 115°C. When the acid has dissolved, an azeotropic mixture of alcohol, toluene and water commences to distil at 75°C. The distillate is collected in a flask containing anhydrous potassium carbonate (75 g) and the distillation is continued until the temperature at the top of the column rises to 78°C. The distillate is shaken thoroughly with the potassium carbonate, filtered and the filtrate is transferred to the original flask. The flask is again heated until the temperature rises to 78–80°C and is transferred to a flask of suitable size and distilled under reduced pressure. The diethyl adipate is distilled at 138°C/20 mm Hg. The yield is 95 g (94%).

Step 2 Ethoxycarbonylcyclopentanone

Pulverised sodium	: 12 g
Diethyl adipate	: 75 g
Dry benzene	: 300 ml
Absolute ethanol	: 1 ml
Hydrochloric acid (1:1)	
Sodium carbonate solution (5%)	: 50 ml

Pulverised sodium (12 g, 0.52 mole) is prepared in a 1 litre R.B. flask and is covered with dried benzene (300 ml). The flask is fitted with a reflux condenser and calcium chloride guard tube. Diethyl adipate (75 g, 0.37 mole) and absolute ethanol (1 ml) are added. The flask is heated on a water bath with continuous shaking until a vigorous reaction sets in and a cake of the sodium compound commences to separates. After the spontaneous reaction has subsided, the mixture is refluxed on a water bath overnight and is cooled in ice. The product is decomposed by acidification with ice and dilute hydrochloric acid (1:1) until the solution is just acidic to congo red paper. The benzene layer is separated and the aqueous layer is extracted with benzene (50 ml). The combined benzene layer is washed with sodium carbonate solution (5%, 50 ml) and water (100 ml) and dried over magnesium sulphate. The benzene is removed on a rotary evaporator and the residue is distilled under reduced pressure. 2-Ethoxycarbonylcyclopentanone is collected at 108–111°C/15 mm hg. The yield is 45 g (77.6%); b.p. is 102°C/11 mm Hg.

39. ELBS PERSULPHATE OXIDATION[1]

Phenols undergo Elbs persulphate oxidation. In this reaction, a monohydric phenol is oxidised to dihydric phenol by alkaline potassium persulphate. Hydroxylation normally occurs in the *p*-position. However, if the *p*-position is blocked, then hydroxylation occurs in *o*-position.

Mechanism:

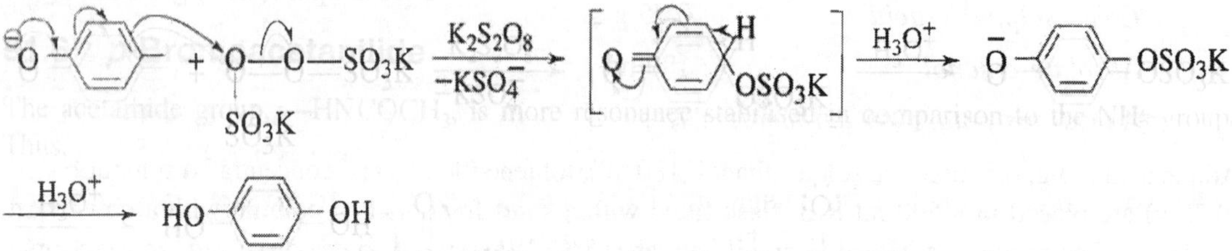

39.1 2,5–Dihydroxybenzaldehyde (gentisic aldehyde)

It is obtained by the oxidation of o-hydroxybenzaldehyde with alkaline potassium persulphate.

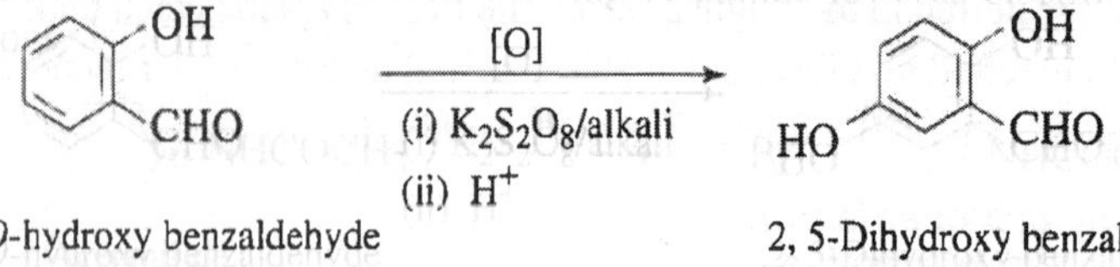

O-hydroxy benzaldehyde 2, 5-Dihydroxy benzaldehyde

o-Hydroxybenzaldehyde : 30.5 g
Sodium hydroxide solution : 20%, 100 ml
Potassium persulphate : 75 g

Aqueous potassium persulphate (75 g, 0.275 mol in 750 ml water) and sodium hydroxide solution (20%, 100 ml) are added simultaneously through two droping funnels to a solution of *o*-hydroxybenzaldehyde (30.5 g, 0.25 mol) in sodium hydroxide solution (12.5 g in 200 ml water) with vigorous stirring at 30–35°C during 90 minutes. The mixture should always remain alkaline. After the addition is over, the mixture is stirred for one hour more and kept at room temperature for 36 hours. The solution is cooled and acidified to congo red with hydrochloric acid. The unreacted *o*-hydroxybenzaldehyde (6.8 g) is extracted with ether (2 × 75 ml). The clear aqueous solution is strongly acidified with hydrochloric acid (100 ml) and heated slowly at 70°C for one hour. The separated amorphous product is filtered and the filtrate is extracted with ether (3 × 75 ml). The ether extract is dried (anhydrous sodium sulphate) and distilled. The residue is purified by column chromatography over silica gel and eluted with benzene. The yield is 6.5 g (21%); m.p. is 97-98°C.

40. ENAMINE REACTION

The enamine reaction was introduced by Stork and his co-workers and is widely used in synthesis. It provides a valuable alternative method for the selective alkylation and acylation of aldehydes and ketones.

1. J. Elbs, *J. Prakt. Chem.*, 1893, **48**, 179.

Enamines are α-, β-unsaturated amines and are obtained by the reaction of an aldehyde or a ketone with a secondary amine in presence of a dehydrating agent or in presence of a catalytic amount of *p*-toluene sulphonic acid with azeotropic removal of the water formed. The most common amines used are pyrrolidine, morpholine and piperidine.

Thus, enamine is prepared by refluxing cyclohexanone and morpholine in dry benzene using a Dean-Stark water separator.

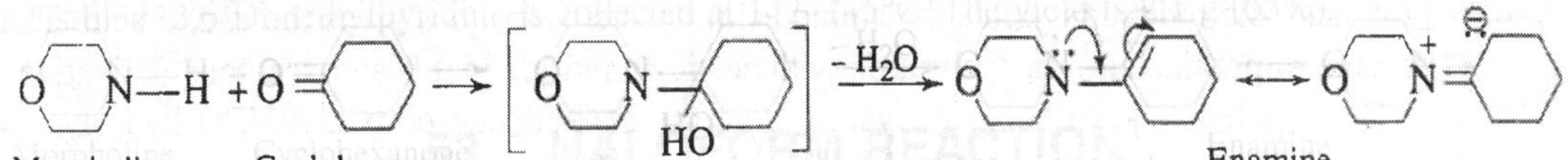

40.1 2-Acetylcyclohexanone

It is obtained by the treatment of enamine with acetyl chloride in presence of triethyl amine.

Cyclohexanone : 15.5 ml
Morpholine : 20 ml
Triethylamine : 10 ml
Acetyl chloride : 18 ml
Chloroform (dry) : 125 ml

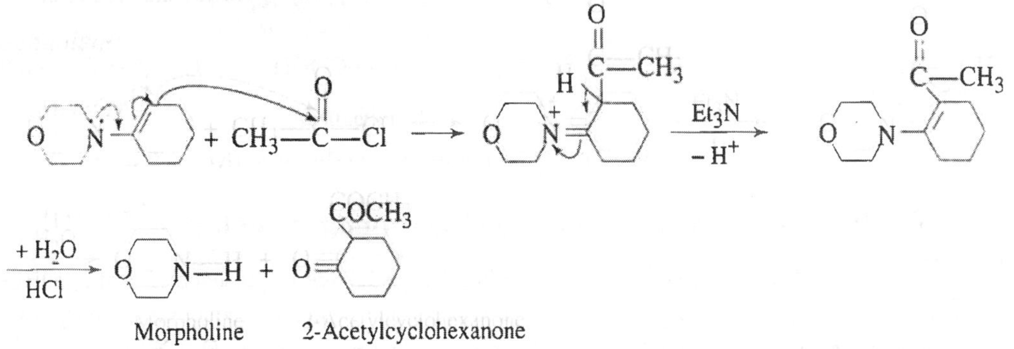

The enamine is first obtained by refluxing a solution of pure cyclohexanone (15.5 ml, 0.15 mole) and morpholine (20 ml, 0.23 mole) in dry benzene (100 ml) using a Dean-Stark water separator. When the reaction is complete (checked by TLC), the solvent is removed by distillation and the enamine is obtained by distilling it under reduced pressure at 120–130°C (15 mm). The yield is 14 g (56%).

A mixture of enamine (12 g, 0.07 mole), triethyl amine (10 ml) and dry chloroform (80 ml) is shaken to obtain a homogeneous solution (anhydrous conditions). A solution of acetyl chloride (18 ml, 0.25 mole) in dry chloroform (45 ml) is added with constant shaking to the enamine solution. The mixture is refluxed for 2 hours (boiling water bath). Water (20 ml) and concentrated hydrochloric acid (20 ml) is added to the reaction mixture and the refluxing continued for 2 hours more. The chloroform layer is separated (separatory funnel), washed with water (2 × 50 ml). The aqueous washings are mixed with an aqueous layer of the reaction mixture, the total solution is neutralised with concentrated sodium hydroxide solution keeping the mixture slightly acidic. It is extracted with chloroform (2×25 ml). The combined chloroform solution is dried (anhydrous calcium chloride), solvent distilled and the residual 2-acetylcyclohexanone purified by distillation under reduced pressure at 106–180°C (14 mm). The yield is 5.5 g (55%).

41. ERLENMYER-PLÖCHL AZLACTONE AND AMINO ACID SYNTHESIS[1, 2]

Formation of azlactone is by intramolecular condensation of acylglycine in the presence of acetic anhydride. The reaction of azlactones with carbonyl compounds, followed by hydrolysis to the unsaturated N-acylamino acid and reduction yields the amino acid; drastic hydrolysis gives the α-oxo acid. Thus, benzoyl glycine (hippuric acid) on cyclisation with acetic anhydride yields an oxazolone derivative or azlactone. This condenses with benzaldehyde to give the benzylidene derivative. The hydrolysis of benzylidene derivative to benzoyl amino cinnamic acid is achieved by refluxing in aqueous acetone. However, its hydrolysis in aqueous acid leads to phenylpyruvic acid.

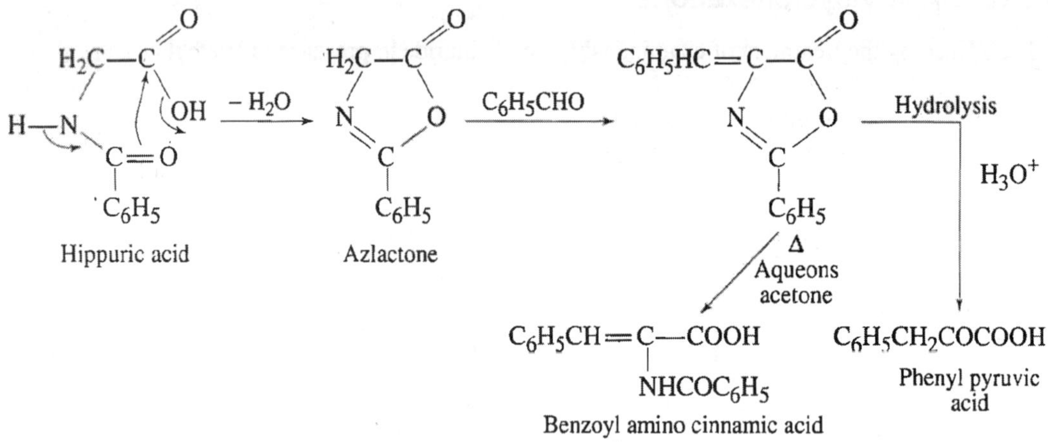

41.1. DL-Phenylalanine (2-Amino-3-phenylpropionic Acid)

It is obtained by the condensation of hippuric acid and benzaldehyde in presence of acetic anhydride and sodium acetate to give 4-benzylidene-2-phenyloxazol-5-one, which on reduction (P/HI) gives *DL*-Phenylanine.

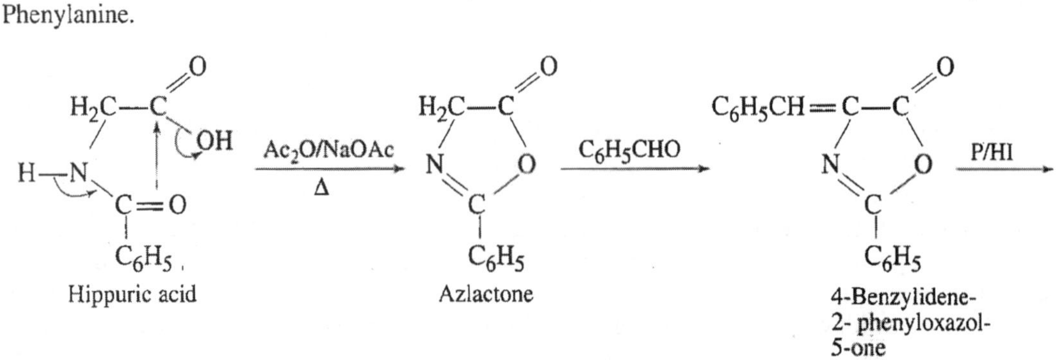

1. E. Erlenmeyer, *Ann.* 1893, **275**, 1.
2. J. Plöchl, *Ber.*, 1884, **17**, 1616.

$$\longrightarrow C_6H_5CH_2-\underset{\displaystyle NH_2}{\underset{|}{CH}}\ COOH + C_6H_5\ COOH$$

DL – Phenylalanine

Step 1 4-Benzylidene-2-phenyloxazol-5-one

Benzaldehyde	: 13 ml
Benzoylglycine (Hippuric acid)	: 22.5 g
Acetic anhydride	: 35.75 ml
Fused sodium acetate	: 10.25 g

A mixture of benzaldehyde (13 ml, 0.125 mol), benzoylglycine (p 00) (22.5 g, 0.125 mol), acetic anhydride (35.75 ml, 0.75 mol) and freshly fused and powdered sodium acetate (10.25 g, 0.125 mol) is heated in a boiling water bath with constant shaking. After the mixture has become liquid, it is heated for two more hours. Alcohol (50 ml) is slowly added and the mixture shaken and allowed to stand overnight. The separated oxazolone is filtered, washed with ice-cold alcohol (2 × 10 ml) and then with boiling water (2 × 20 ml). It is crystallised from benzene. The yield is 20 g (64%); m.p. is 167–168°C.

Step 2 DL-Phenylalanine

4-Benzylidene-2-phenyloxazol-5-one	: 12.5 g
Purified red phosphorus	: 10 g
Acetic anhydride	: 62. 5 ml
Hydriodic acid (d 1.56, 50%)	: 63 ml

A mixture of 4-benzylidene-2-phenyloxazol-5-one (12.5 g, 0.05 mol), purified red phosphorus (10 g, 0.325 mol) and acetic anhydride (62.5 ml, 0.66 mol) is placed in a three necked R.B. flask (500 ml capacity) fitted with a sealed stirrer, a reflux condenser and a dropping funnel. The mixture is stirred and hydriodic acid (63 ml, d 1.56, 50%) is added during one hour through a dropping funnel. The mixture is refluxed for 3 hours cooled and filtered, the unreacted phosphorus is washed with glacial acetic acid (2 × 5 ml). The combined filtrate is evaporated to dryness (rotary evaporator). Water (50 ml) is added to the dried residue and the evaporation to dryness is done again. The residue is dissolved in water (75 ml) and the aqueous solution extracted with ether (3 × 50 ml). The aqueous solution is heated with decolourising charcoal (water bath) and a trace of sodium sulphite and the solution is filtered. The filtrate is heated to boiling and neutralised to congo red with ammonia solution (d.0.88, about 12-13 ml is required). The solution is cooled and the separated *DL*-phenylalanine is filtered and washed with cold water and a little of ethanol. The yield is 5.5 g (67%); m.p. is 284–288°C (decomposes on rapid heating).

41.2 3-Hydroxycoumarin

It is obtained by the condensation of hippuric acid and salicylic aldehyde in acetic anhydride and fused sodium acetate. The formed azalactone is hydrolysed with aqueous alkali followed by acidification.

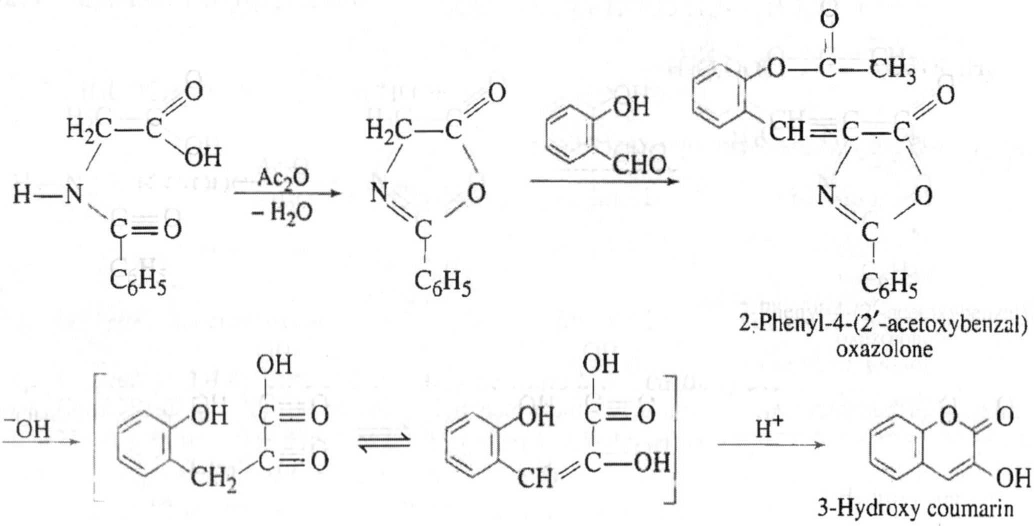

Step 1 2-Phenyl-4-(2′-acetoxybenzal) oxazolone

Salicylic aldehyde	: 25 g
Hippuric acid	: 42 g
Fused sodium acetate	: 15 g
Acetic anhydride	: 60 ml

A mixture of salicyclic aldehyde (25 g, 0.2 mol), hippuric acid (42 g, 0.278 mol, see page 207), fused sodium acetate (15 g) and acetic anhydride (60 ml) is heated in a boiling water bath for 30 minutes under anhydrous conditions. A bright yellow crystalline solid thus formed redissolves on further heating. Alcohol (20 ml) is added to the reaction mixture and allowed to stand for 6 hours. The separated azalactone is filtered, washed with cold alcohol and hot water till the aqueous washings are colourless. It is crystallised from alcohol. The yield is 40 g (63.6%). It is used as such for the subsequent step.

Step 2 3-Hydroxycoumarin

The above oxazolone (40 g, 0.13 mol) is refluxed with aqueous sodium hydroxide (20g in 100 ml water) till no more ammonia is evolved (2 hours). The resulting solution is diluted with water (100 ml) and saturated with sulphur dioxide gas with cooling. The reaction mixture is left for 12 hours in an ice bath and the separated benzoic acid filtered. The clear filtrate is heated on a boiling water bath with concentrated hydrochloric acid (60 ml) and the solution is cooled. The separated 3-hydroxycoumarin is filtered and crystallised from alcohol. The yield is 13 g (61.6%); m.p. is 151–152°C.

42. ESTERIFICATION

Esterification involves interaction between an acid and an alcohol. Thus,

$$\underset{\text{Acetic Acid}}{CH_3COOH} + \underset{n\text{-Butyl alcohol}}{CH_3CH_2CH_2CH_2OH} \rightleftharpoons \underset{n\text{-Butyl acetate}}{CH_3COOCH_2CH_2CH_2CH_3} + H_2O$$

This reaction is reversible and proceeds very slowly. However, in presence of concentrated sulphruic acid (about 3 per cent of the weight of alcohol) or dry hydrogen chloride the reaction is completed in few hours. One of the reactant is used in excess. In case the acid is costly a large excess of alcohol is used. This method of esterification gives good yields with primary alcohols.

Mechanism:

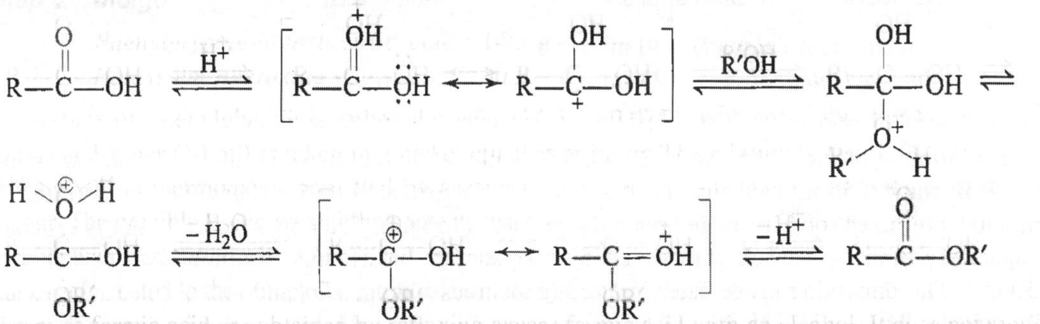

Esters of formic acid are obtained by refluxing excess formic acid with an alcohol. It does not require any sulphuric acid, since it causes decomposition of formic acid to carbon monoxide. In this case, formic acid being a comparatively strong acid is itself an acid catalyst.

Sometimes it is convenient to use benzene as solvent in the esterification reaction. This method is used for the preparation of ethyl esters of dicarboxylic acids.

A convenient and recent process for the preparation of methyl esters involves refluxing the carboxylic acid with methanol and 2, 2-dimethoxy propane in presence of *p*-toluene sulphonic acid as the catalyst. The water produced in the esterification process is removed effectively with the ketal to give methanol and acetone. Thus,

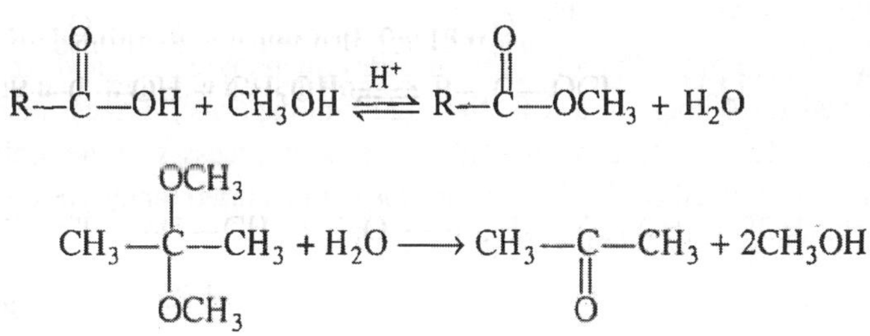

Esters are also obtained by alcoholysis of nitriles. In this process, the nitrile is heated with an alcohol in presence of concentrated sulphuric acid. Thus,

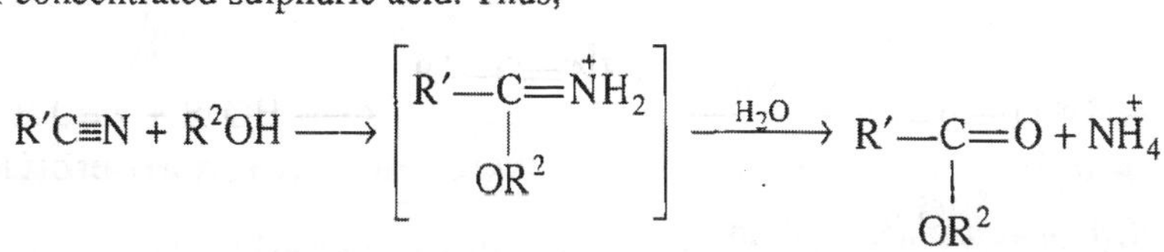

Esterification of acids by refluxing with excess alcohol in the presence of hydrogen chloride gas is known as Fisher-Speier esterification method.[1]

1. E. Fischer and A. Speier, *Ber*, 1895, **28**, 3252.

42.1 Concentrated Sulphuric Acid Method

42.1.1 n-Butyl acetate

$$CH_3COOH + CH_3CH_2CH_2CH_2OH \underset{}{\overset{H_2SO_4}{\rightleftharpoons}} CH_3COOCH_2CH_2CH_2CH_3 + H_2O$$

Acetic acid   *n*-Butyl alcohol   *n*-Butyl acetate

Acetic acid (glacial) : 30 ml
n-Butyl alcohol : 23 ml
Conc.sulphuric acid : 0.5 ml

To a solution of glacial acetic acid (30 ml, 0.52 mole) and *n*-butanol (23 ml, 0.25 mole) contained in a R.B. flask (250 ml capacity) is added conc. sulphuric acid (0.5 ml) using extra caution. The mixture is refluxed for 6 hours and poured into water (150 ml). The upper layer of ester is separated (separatory funnel), washed with water (50 ml), sodium bicarbonate solution (20 ml) and water (30 ml). The remaining crude ester is dried (anhydrous sodium sulphate) and distilled. The yield is 20 g (69%); b.p. is 124–125 °C.

42.1.2 Ethyl benzoate

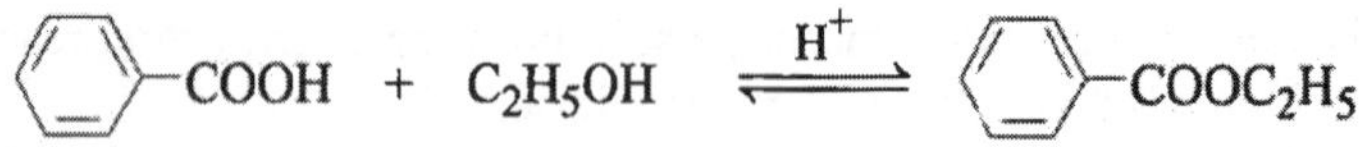

Benzoic acid   Ethyl alcohol   Ethylbenzoate

Benzoic acid : 15 g

Ethyl alcohol (absolute): 32 ml

Conc. sulphuric acid : 1.5 ml

A mixture of benzoic acid (15 g, 0.123 mole), absolute ethyl alcohol (32 ml) and concentrated sulphuric acid (1.5 ml) is refluxed for 4 hours in a R.B. flask (250 ml capacity). Excess alcohol is distilled off (steam-bath) and the residual mixture poured into water and excess acid neutralised (solid sodium bicarbonate). It is extracted with ether (2 × 50 ml), the ether extract dried (anhydrous potassium carbonate) and ether distilled. The yield is 15 g (81%); b.p. is 213°C.

42.1.3 Methyl salicylate

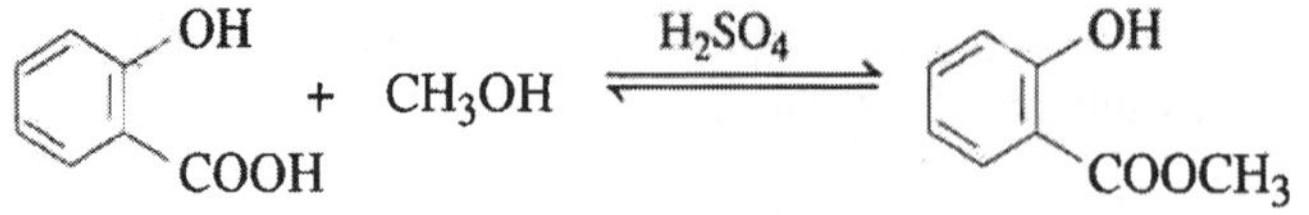

Salicylic acid   Methyl salicylate

Salicylic acid : 17 g

Methyl alcohol (absolute) : 30 ml

Conc. sulphuric acid : 5 ml

A solution of salicylic acid (17 g, 0.123 mole), absolute methyl alcohol (30 ml) and concentrated sulphuric acid (5 ml, added dropwise with caution) is refluxed for 4 hours (water bath) under anhydrous conditions. Excess methanol is removed by distillation (steam bath) and the residual product poured into water

(50 ml). It is extracted with ether (2 × 50 ml), the ether extract washed with sodium bicarbonate solution (till free of acid) and finally with water. It is dried (anhydrous sodium sulphate) and distilled. Methyl salicyllate is obtained as a colourless liquid with a fragrant smell. The b.p. is 223–225°C; yield is 16 g (85.6%).

42.2 Hydrogen Chloride Method

42.2.1 *sec. Butyl acetate*

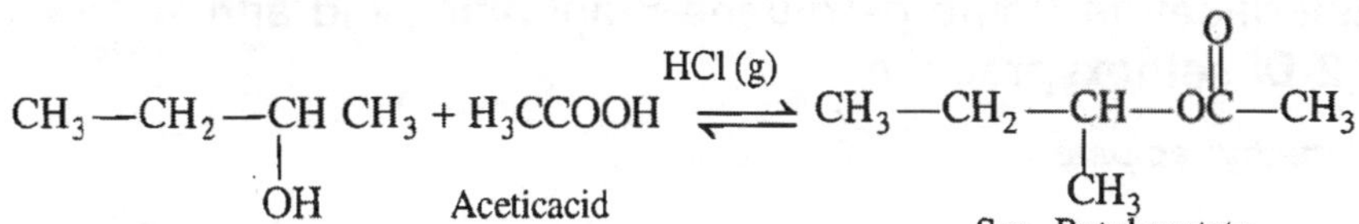

Butan - 2 - Ol Aceticacid Sec. Butyl acetate

Butan-2-ol : 23 ml
Glacial acetic acid : 30 ml
Hydrogen chloride gas

Dry hydrogen chloride gas is passed into dry butan-2-ol (23 ml, 0.25 mole) until 0.7 g is absorbed. To this solution is added glacial acetic acid (30 ml) and the solution refluxed for 10 hours. The reaction mixture is worked up as for *n*-butyl acetate. The yield is 17 g (58.6%); b.p. is 111–112°C.

42.2.2 *Ethyl p-aminobenzoate*

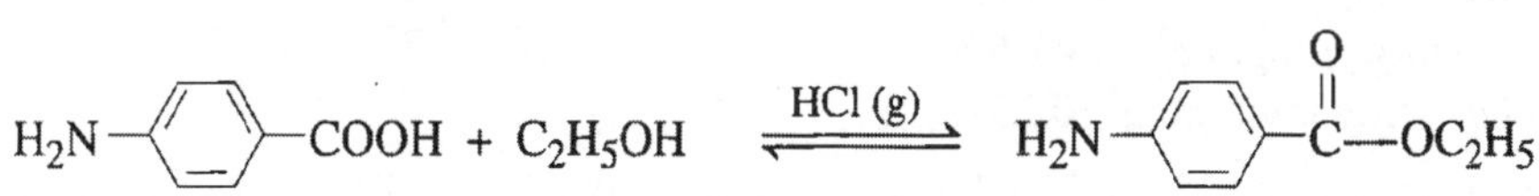

p-aminobenzoic acid Ethyl *p*-aminobenzoate

p-Aminobenzoic acid : 12 g
Ethyl alcohol (absolute): 80 ml
Hydrogen chloride gas

Dry hydrogen chloride gas is passed into absolute ethyl alcohol (80 ml) till the alcohol is completely saturated. *p*-Aminobenzoic acid (12 g, 0.088 mole) is added and the mixture refluxed (water bath) for 2 hours under anhydrous conditions. The hot solution is poured into excess water (150 ml) and sodium carbonate is added until the solution is neutral to litmus. The precipitated ethyl *p*-aminobenzoate is filtered and crystallised from rectified spirit. The yield is 10 g (68.9%); m.p. is 91°C.

42.3 Azeotropic Mixture Method

42.3.1 *Diethyl adipate*

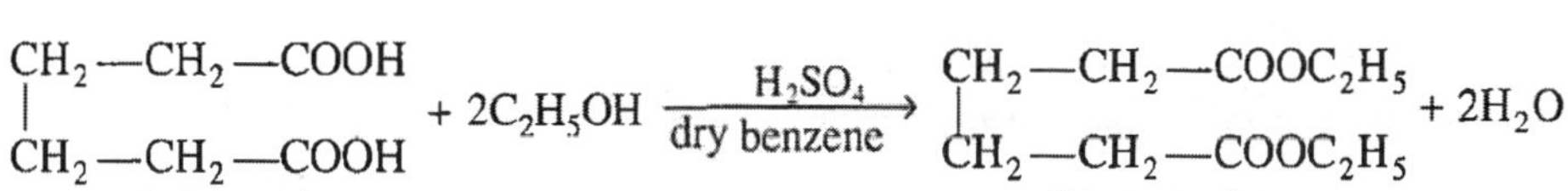

Adipic acid Diethyl adipate

Adipic acid : 20 g
Ethyl alcohol (absolute): 25 ml
Dry benzene : 50 ml
Conc. sulphuric acid : 4.5 ml

To a mixture of adipic acid (20 g, 0.137 mole) absolute alcohol (25 ml) and dry benzene (50 ml) is added concentrated sulphuric acid (4.5 ml), with caution, in a R.B. flask (200 ml capacity). The mixture is refluxed (water bath) for 5–6 hours and poured into water (150 ml). The benzene layer is separated and the aqueous layer extracted with benzene (50 ml). The combined benzene extract is dried (anhydrous calcium chloride) and distilled. The ester is finally distilled under reduced pressure. The yield is 26 g (94%); b.p. is 134–135°C/17 mm.

42.4 Esterification Using p-Toluene Sulphonic Acid and 2,2-Dimethoxypropane

42.4.1 *Dimethyl adipate*

$$\underset{\text{Adipic acid}}{HOOC(CH_2)_4COOH} + 2CH_3OH \xrightleftharpoons{H^+} \underset{\text{Dimethyl adipate}}{H_3COOC(CH_2)_4\ COOCH_3} + 2H_2O$$

$$CH_3C(OCH_3)_2CH_3 + H_2O \longrightarrow CH_3—\underset{\underset{O}{\|}}{C}—CH_3 + 2CH_3OH$$

2,2-Dimethoxypropane : 41.6 g

Adipic acid : 29.25 g
Dry methanol : 10 ml
2,2-Dimethoxypropane : 41.6 g
p-Toluene sulphonic acid : 0.25 g

A mixture of adipic acid (29.25 g, 0.2 mole), dry methanol (10 ml), 2, 2-dimethoxypropane (41.6 g, 0.4 mole) and *p*-toluene sulphonic acid (0.25 g) is stirred (magnetic stirrer) in a R.B. flask (250 ml capacity) fitted with a reflux condenser. The mixture is kept at 45°C (water bath) for 4 hours. The formed acetone and methanol are distilled (water bath). The residual ester is distilled under reduced pressure. The yield is 29 g (83.3%); b.p. is 130°C/25 mm.

42.5 Alcoholysis of Cyanides

42.5.1 *Ethyl phenylacetate*

$$\underset{\text{Benzyl cyanide}}{C_6H_5CH_2CN} + C_2H_5OH + H_2SO_4 \longrightarrow \underset{\text{Ethyl phenylacetate}}{C_6H_5CH_2—\overset{\overset{O}{\|}}{C}—OC_2H_5} + NH_4HSO_4$$

Benzyl cyanide : 14.8 g
Rectified spirit : 30.0 ml
Conc. sulphuric acid : 13.6 ml

A mixture of benzyl cyanide (14.8 ml, 0.13 mole) rectified spirit (30 ml) and concentrated sulphuric acid (13.6 ml) is refluxed in a R.B. flask (100 ml capacity) fitted with a reflux condenser (water bath). The refluxing is continued for 4 hours (two layers are separated) and poured into water (80 ml). The upper layer of the ester is separated (separatory funnel). It is dissolved in ether (20 ml) and is first washed with sodium bicarbonate solution (5 × 10 ml) until effervescence ceases, and then with water. The remaining solution is dried (anhydrous magnesium sulphate) and distilled. The yield is 15.6 g (74%); b.p. is 227°C.

43. ETARD REACTION[1-5]

Chromyl chloride is used in the Etard reaction for the oxidation of aromatic methyl derivatives to aldehydes.

$$C_6H_5CH_3 \xrightarrow{CrO_2Cl_2} C_6H_5CHO$$

43.1 o-Chlorobenzaldehyde

o-Chlorotoluene : 10 g
Carbon disulphide : 50 ml
Chromium oxychloride : 25 g
(Chromyl chloride)

o-Chlorotoluene (10 g, 9.2 ml, 0.079 mole) is dissolved in carbon disulphide (50 ml) in a 250 ml flask. Chromium oxychloride (25 g) is added to the mixture. The mixture is kept cool by a stream of water (cold water bath) and is allowed to stand for three days. The solid separated is filtered and treated with water (100 ml). A heavy oil is separated, which on distillation at 205–220°C gives *o*-chlorobenzaldehyde. The yield is 5.5 g (50%); b.p. is 209–215°C.

44. FAVORSKII REARRANGEMENT[6, 7]

The reaction of *α*-haloketones (chloro or bromo) with alkoxide ions to give rearranged esters is called Favorskii rearrangement. Thus,

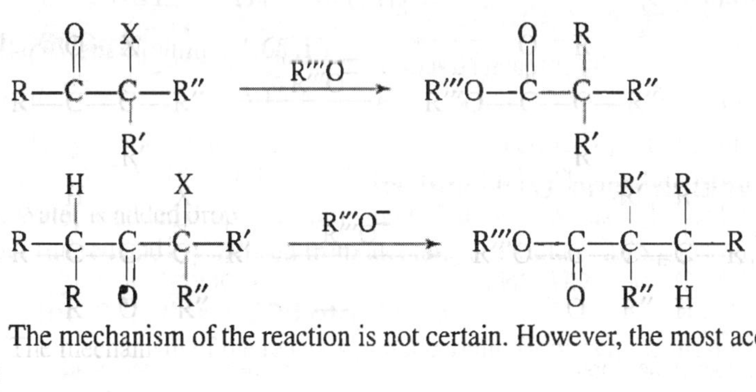

Mechanism: The mechanism of the reaction is not certain. However, the most acceptable mechanism is given below:

1. A.L. ETARD, *Compd. Rend.*, 1880, **90**, 534.
2. A.L. ETART, *Ann. Chem. Phys.*, 1881, **22**, 218.
3. H.D. LAW and F.M. PERKIN, *J. Chem. Soc.*, 1907, **91**, 191.
4. H.H. SISLER, *Inorg. Syn*, 1946, **2**, 205.
5. L.N. FERGUSON, *Chem. Rev.*, 1946, **38**, 237.
6. A.E. FAVORSKII, *J.Prakt. Chem.*, 1913, **88(2)**, 658.
7. O. WALLACH, *Ann.*, 1918, **414**, 296.

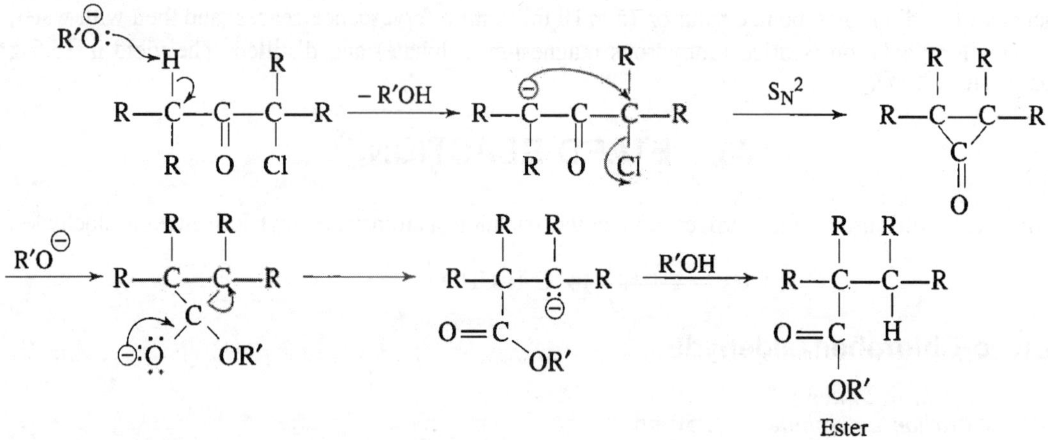

The α-halogeno cyclohexanones also undergo Favorskii rearrangement on treatment with alkali, i.e., the six-membered ring changes to a five membered ring carboxylic acid (ring contraction takes place).

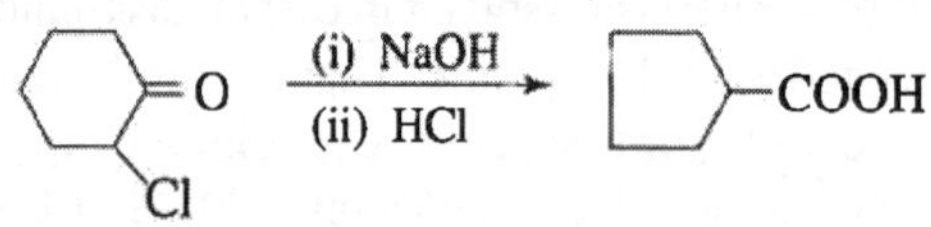

Mechanism:

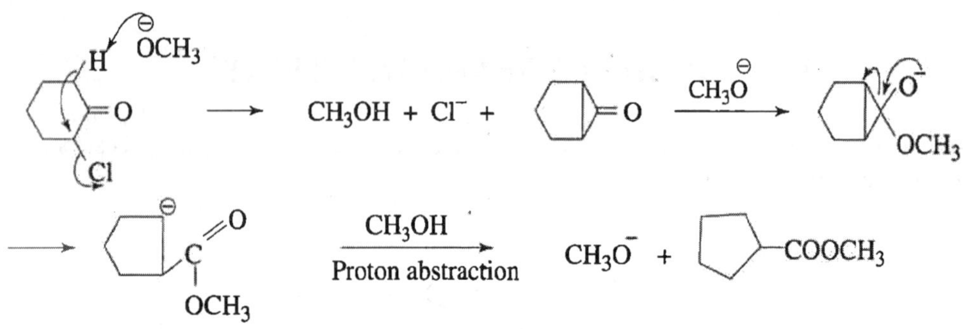

44.1 Methyl Cyclopentane Carboxylate

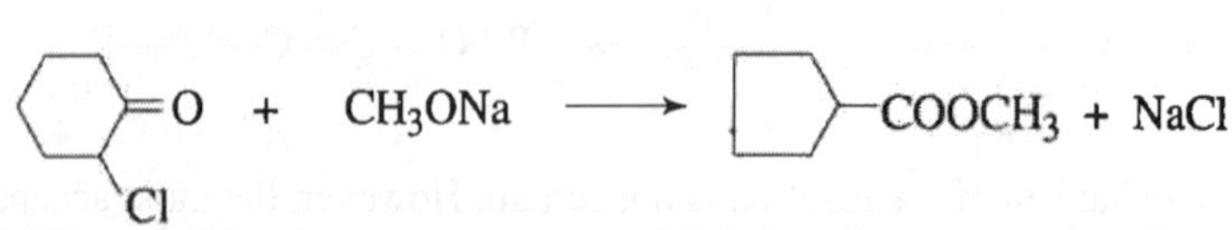

2-Chlorocyclohexanone Methyl cyclopentane carboxylate

2-Chlorocyclohexanone : 16.5 g
Sodium methoxide : 7.25 g
Dry ether : 55 ml

Sodium methoxide (7.25 g, 0.134 mole) and dry ether (40 ml) are taken in a dry three necked R.B. flask fitted with a mercury seal stirrer, a reflux condenser and a dropping funnel (all fitted with calcium chloride drying tubes). The suspension is stirred and 2-chlorocyclohexanone (16.5 g, 0.124 mole) in dry ether (15 ml) is added dropwise during 40–45 minutes. The stirring is continued and the mixture refluxed for 2 hours (water bath). It is cooled and water (40 ml) is added. The ether layer is separated (separatory funnel) and the aqueous layer saturated with sodium chloride extracted with ether (2 × 20 ml). The combined ether solution is washed successively with hydrochloric acid (1:1, 15 ml), sodium bicarbonate solution (5%, 15 ml) and saturated sodium chloride solution (15 ml). It is dried (anhydrous magnesium sulphate) and the ether removed by distillation. The residual crude ester is distilled under reduced pressure. The yield is 9 g (57%); b.p. is 70–73°C/48 mm.

45. FISCHER INDOLE SYNTHESIS[1]

The formation of an indole derivative by heating aryl hydrazones of suitable aldehydes (RCH_2CHO), ketones or ketonic acids in presence of an acid catalyst (zinc chloride, polyphosphoric acid, sulphuric acid) is known as Fischer Indole synthesis. A number of mechanisms have been proposed for this reaction. A widely accepted one is that suggested by Robinson (1918) and modified by subsequent workers.

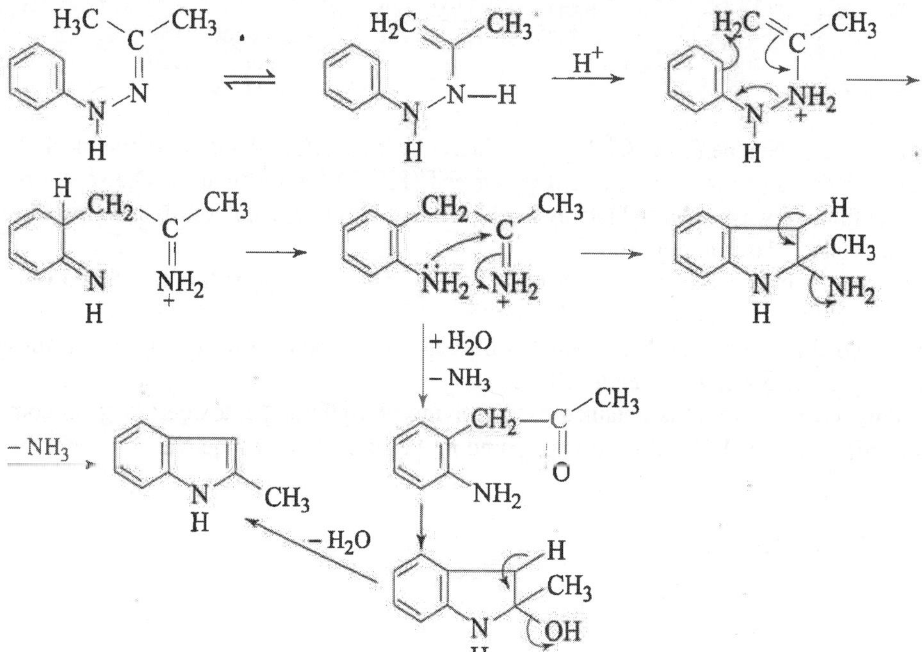

45.1 2-Phenylindole

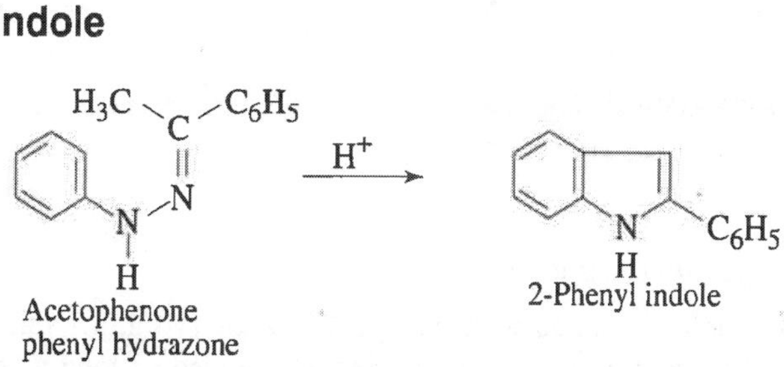

1. E. Fischer and F. Jourdan, *Ber.*, 1883, **16**, 2241.

Acetophenone	: 10 g
Phenyl hydrazine	: 9 g
Ethanol	: 30 ml
Polyphosphoric acid	: 40 g

Mechanism:

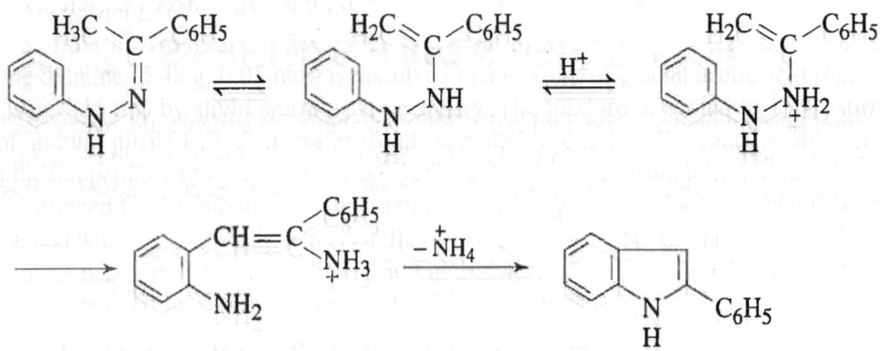

A mixture of acetophenone (10 g, 0.835 mol), phenyl hydrazine (9g, 0.835 mol) and alcohol (30 ml) containing 2–3 drops of glacial acetic acid is warmed (100°C) for 15 minutes. The separated phenyl hydrazone is filtered, washed with dilute hydrochloric acid (5 ml), ethanol (5 ml) and crystallised from ethanol. The yield is 11 g; m.p. is 106°C.

The above acetophenone phenyl hydrazone (7 g) is heated with polyphosphoric acid (40 g) on a boiling water bath in a conical flask. The mixture is heated at 100–120°C for 10 minutes. Water (100 ml) is added and the separated product is filtered and washed with water. It is crystallised from ethanol. The yield is 5 g (79%); m.p. is 187°C.

Note: Polyphosphoric acid is obtained by dissolving phosphorus pentoxide (25 g) in commercial orthophosphoric acid (d: 1.75, 15 g). It is prepared in the flask in which the reaction has to be carried out.

45.2 2-Methylindole

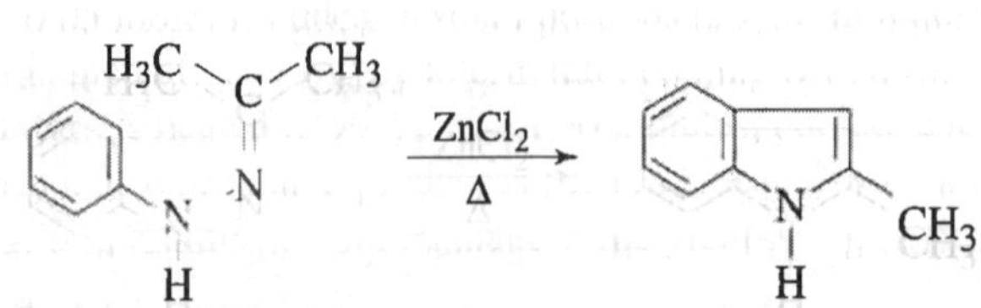

Phenyl hydrazine	: 10 g
Acetone	: 6 g
Zinc chloride (anhydrous)	: 70 g

Acetone (6 g, 7.6 ml, 0.1 mole) is mixed with phenyl hydrazine (10 g, 9.1 ml, 0.092 mole) and the mixture heated (water bath) for 15 minutes. More acetone is added as long as testing with Fehlings solution shows reduction. The crude oily acetone phenyl hydrazone is heated in a water bath to remove excess of acetone. Alternatively, the phenylhydrazone is isolated and crystallised from alcohol. The phenyl hydrazone is heated at 180°C (oil bath) with anhydrous zinc chloride (70 g) till the reaction mixture becomes completely dark. Hot water (200 ml) is added and then little hydrochloric acid. The reaction mixture is steam distilled. The oily product obtained in the steam distillate solidifies. The yield is 6 g (50%), m.p. is 95°C; b.p. is 272°C/750 mm.

45.3 1,2,3,4-Tetrahydrocarbazole

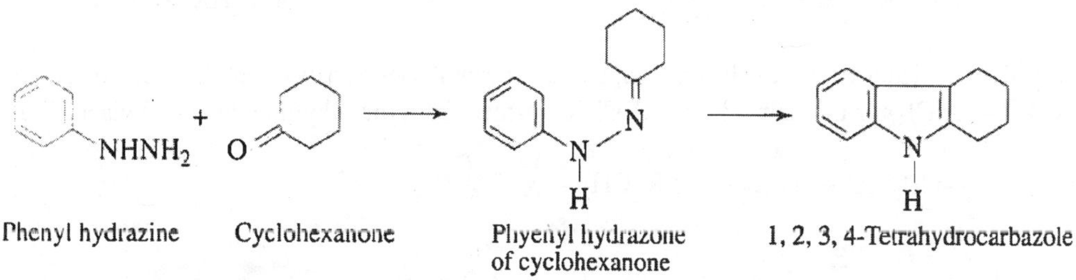

Cyclohexanone : 4.9 g
Acetic acid (glacial) : 18 ml
Phenyl hydrazine : 4.9 ml

A mixture of cyclohexanone (4.9 g, 0.05 mol) and glacial acetic acid (18 ml) is refluxed in a R.B. flask (100 ml capacity). Phenylhydrazine (4.9 ml, 0.05 mol) is added slowly through the condenser and the mixture refluxed for 30 minutes. The solution is cooled (ice-salt bath) with stirring. The separated product is filtered, washed with water and finally with dilute alcohol (1:1). It is crystallised from methanol. The yield is 6.5 g (76%); m.p. is 116–117°C.

Note: Phenyl hydrazine is highly poisonous. It should not be inhaled. Even contact with skin should avoided. In case it comes in contact with the skin, wash with 2 per cent acetic acid and then with soap and water.

46. FRIEDEL-CRAFTS REACTION[1]

This reaction involves alkylation or acylation of organic compounds in presence of aluminium chloride or other lewis acid catalysts. Thus,

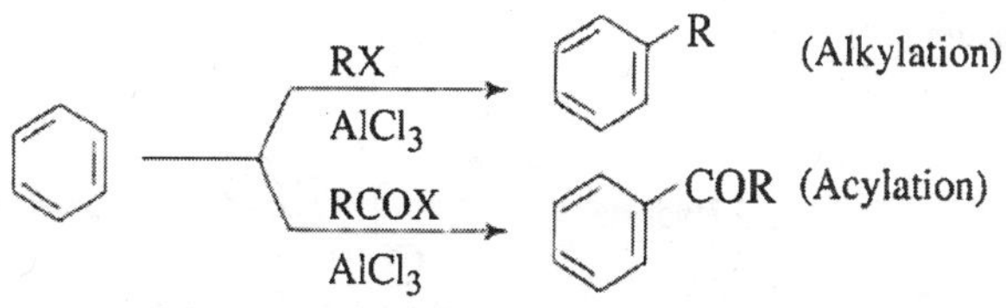

In the alkylation, the reaction does not stop at the stage of mono-substitution, since the alkyl benzene (Ar-H) produced undergoes alkylation more easily than the original hydrocarbon, owing to the electron releasing effect of the alkyl group. Thus, a mixture of substances result, which are difficult to purify.

1. C. Friedel and J.M. Crafts, *Compt. Rend.*, 1877, **84**, 1392, 1450.

However, some mono alkyl benzenes are prepared by using excess of hydrocarbon, which also acts as diluent in moderating the violence of the reaction.

The alkylation reaction is considered to procceed by way of carbonium ions, which attack the aromatic nucleus:

$$R—X: + AlCl_3 \rightleftharpoons R^+ [XAlCl_3]^- \rightleftharpoons R^+ + [XAlCl_3]^-$$

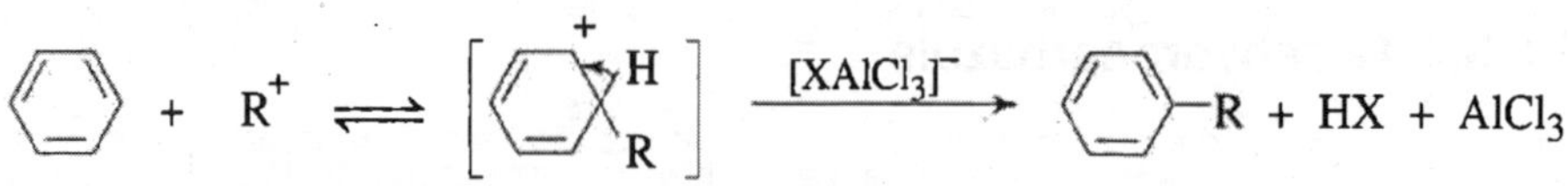

Alternatively, Friedel-Crafts alkylation is supposed to occur through an electrophilic attack of a complex (e.g. $R–X^{\delta+}–^{\delta-}AlCl_3$) or ion pair (eg, $R^+ Al^-Cl_3X$) formed from the alkyl halide and lewis acid. Thus,

$$R–CH_2–X: + AlCl_3 \rightleftharpoons R–CH_2 - \overset{\delta+}{X} - \overset{\delta-}{AlCl_3}$$

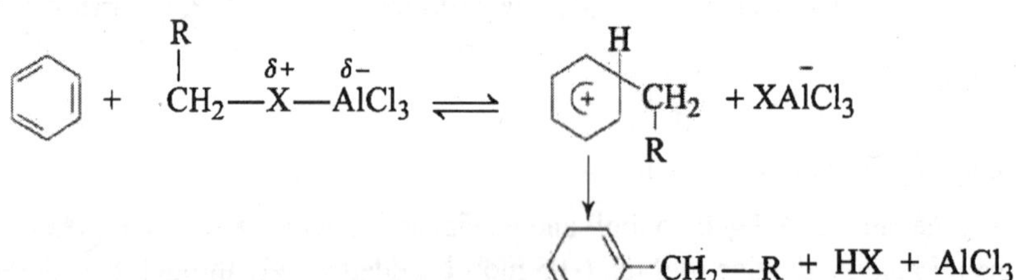

The exact pathway by which the reaction proceeds depends on the nature of the alkyl halide. With secondary and tertiary alkyl halides the reaction appears to occur through the formation of a carbonium ion.

In many cases, the alkyl group may undergo skeletal rearrangement under the reaction conditions. For example, alkylation of benzene with propyl chloride in presence of aluminium chloride gives a mixture of propyl benzene and isopropyl benzene (1 : 2). However, *iso*-butyl bromide or chloride with benzene in presence of aluminium chloride gives exclusively *t*-butylbenzene. The rearrangement process involve a 1, 2- shift of a hydride ion to form a more stable secondary or tertiary carbonium ion.

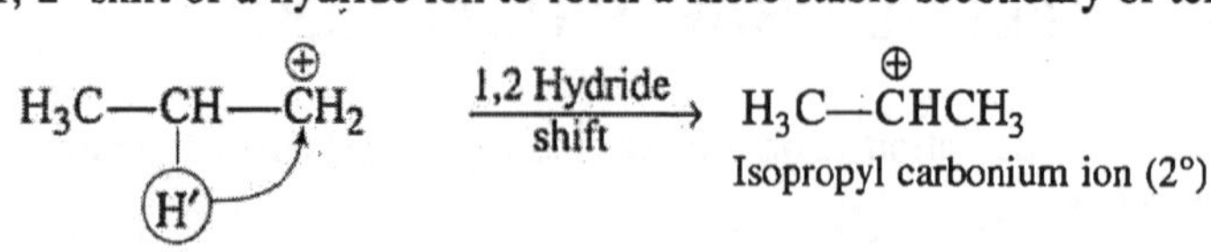

Isopropyl carbonium ion (2°)

n-propyl carbocation(1°)

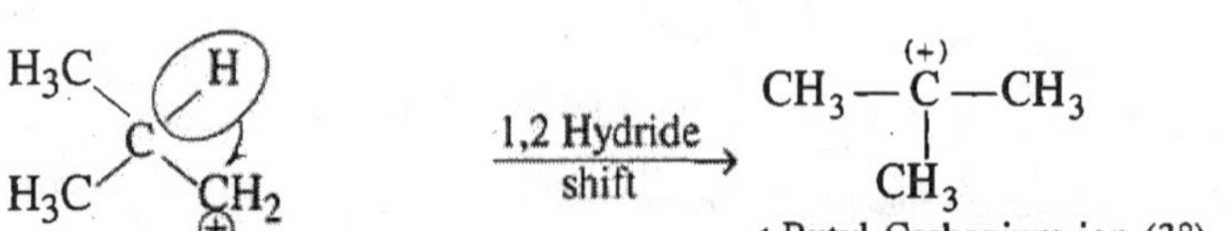

t-Butyl Carbonium ion (3°)

Typical examples of Friedel-Crafts acylation are preparation of tert. butyl benzene and diphenyl methane.

The Friedel-Crafts alkylation requires catalytic quantities of aluminium chloride, but for acylation with an acid chloride at least one molecular equivalent of aluminium chloride is necessary, since much of the catalyst is effectively removed by complex formation. When a carboxylic anhydride is employed as the acylating agent at least two moles of aluminium chloride is necessary.

The Friedel-Crafts acylation reaction is extremely useful for the preparation of *o*-hydroxy ketones. In this method, the complete O-methyl ether of a reactive phenol on reaction with an acyl chloride in presence of aluminium chloride using ether as solvent, gives rise to *o*-hydroxyketones. There is simultaneous demethylation of one of the methoxyl groups and acylation. This reaction is illustrated by the preparation of 2-hydroxy-4-methoxy benzophenone.

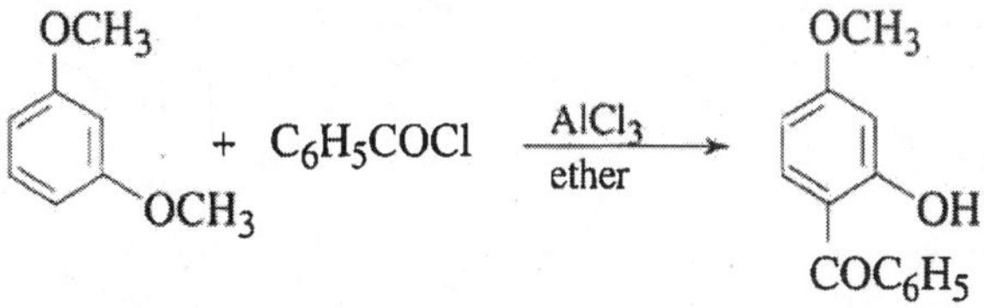

The mechanism of acylation with acid halides is usually regarded as involving the acyl cation. Thus,

$$RCOCl + AlCl_3 \longrightarrow [RCO]^+ [AlCl_4]^- \rightleftharpoons [RCO]^+ + [AlCl_4]^-$$

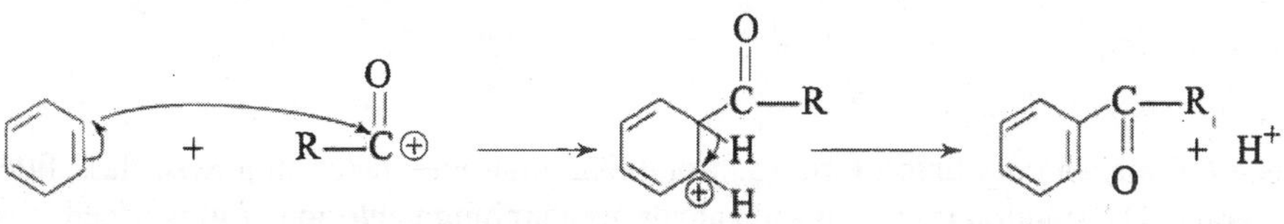

$$H^+ + [AlCl_4]^- \longrightarrow HCl + AlCl_3$$

Typical examples of Friedel-Crafts acylation are preparation of acetophenone, β-benzoylpropionic acid and *o*-benzoylbenzoic acid.

46.1 Alkylation

46.1.1 tert. Butylbenzene

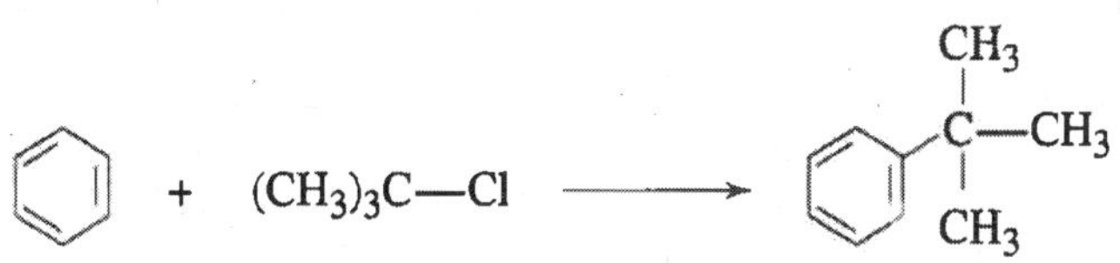

Tert. butyl chloride | Tert. butyl benzene

Benzene (dry) : 100 ml

Aluminium chloride (anhyd) : 11.25 g

tert. butyl chloride : 13.2 ml

Anhydrous aluminium chloride (11.25 g) and pure dry benzene (100 ml) are taken in a three necked R.B. flask (250 ml capacity) equipped with a mercury sealed stirrer, a reflux condenser and a dropping funnel (anhydrous conditions). The mixture is stirred, cooled (crushed-ice bath) and tert. butyl chloride (13.2 ml, 0.12 mole) is added dropwise (dropping funnel) during 2 hours. The temperature is maintained between 0–5°C during the addition. After all the tert. butyl chloride has been added, the stirring is continued for 25–30 minutes and treated with crushed ice (40 g) added in small portions with stirring. The mixture is finally treated with cold water (20 ml).

The resulting reaction mixture is steam distilled till no more oily drops distil over. The upper organic layer is separated (separatory funnel) from the distillate. It is washed with water, dried (anhydrous magnesium sulphate) and benzene removed by distillation (water bath). The residue is finally distilled. The yield is 10 g (61.3%); b.p. is 168.5°C.

46.1.2 Triphenylmethane

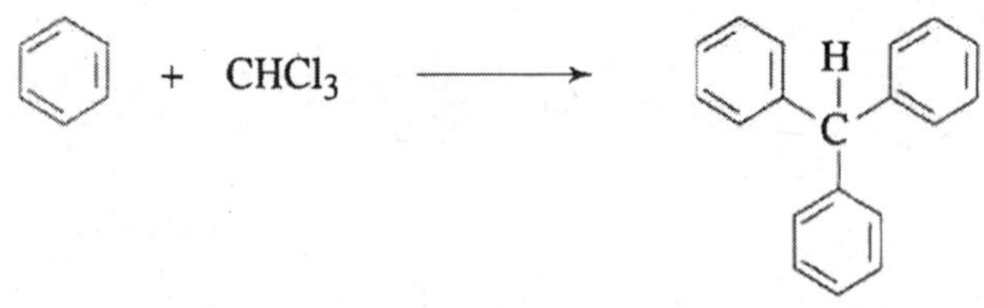

Triphenyl methane

Benzene (dry) : 50 ml

Chloroform (dry) : 5.2 ml

Aluminium chloride (anhyd.) : 7 g

Dry benzene (50 ml) and dry chloroform (5.2 ml, 0.065 mole) are taken in a R.B. flask fitted with a reflux condenser. The solution is shaken and anhydrous aluminium chloride (7 g) is added in small lots (2–3 g each) at intervals of 3–5 minutes. A brisk reaction sets in and the liquid boils with the evolution of hydrogen chloride gas. The mixture is refluxed (sand bath) for 25–30 minutes, cooled and poured into ice-cooled water (50 ml). The benzene layer is separated (separatory funnel), dried (anhydrous calcium chloride) and the benzene removed by distillation (steam bath). The residue is finally distilled under reduced pressure and the fraction between 190–215°C at 10 mm is collected. The yield is 6 g (37.7 %); m.p. is 92°C.

46.2 Acylation

46.2.1 Acetophenone

$$(CH_3CO)_2O + AlCl_3 \rightleftharpoons CH_3\overset{+}{C}O + \overline{A}lCl_3CH_3COO$$

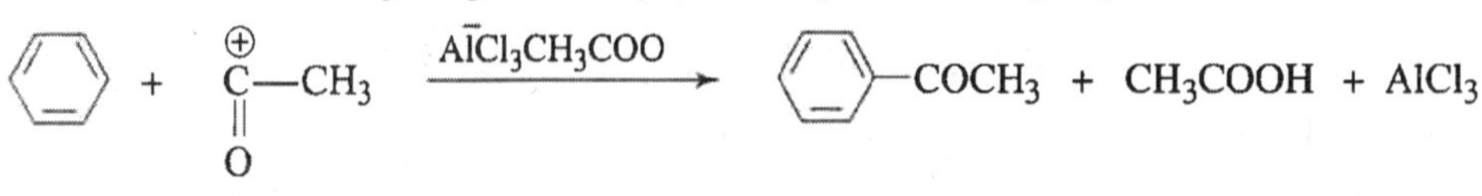

Benzene (dry) : 40 ml

Aluminium chloride (anhyd.): 15 g

Acetic anhydride : 14 ml

Dry benzene (40 ml) and finely powdered anhydrous aluminium chloride (15 g, 0.112 mole) are taken in a two necked R.B. flask (250 ml capacity) fitted with a reflux condenser and a dropping funnel (anhydrous conditions). Acetic anhydride (14 ml, 0.148 mole) is added dropwise (dropping funnel) while shaking the reaction vessel. Considerable amount of heat is evolved during the addition with vigorous evolution of hydrogen chloride gas. If necessary, the reaction mixture is cooled (cold water bath). After the addition of acetic anhydride, the reaction mixture is heated at 60°C (water bath) for 40–45 minutes, shaking the contents occasionally. The mixture is cooled and poured with stirring into a beaker containing concentrated hydrochloric acid (100 ml) and crushed ice (150 g). The aluminium chloride complex decomposes. The upper benzene layer is separated (separatory funnel) and the aqueous layer extracted with carbon tetrachloride (35 ml). The combined benzene and carbon tetrachloride layer is washed with dilute sodium hydroxide (25 ml) and then with water. It is dried (anhydrous magnesium sulphate) and distilled to remove the solvent (water bath). The residue is finally distilled to obtain acetophenone. The yield is 10 g (56%); b.p. is 202°C.

46.2.2 β-Benzoyl propionic acid

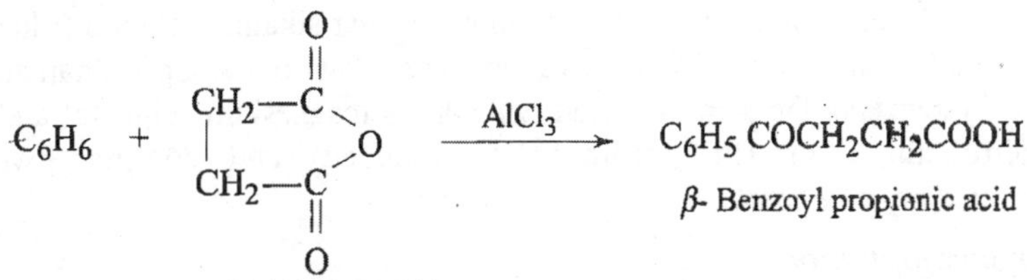

Benzene   Succinic anhydride

Benzene (dry) : 65 ml
Succinic anhydride : 12.2 g
Aluminium chloride (anhyd.): 32 g

Pure succinic anhydride (12.2 g, 0.122 mole) and sodium dried pure benzene (65 ml) are taken in a three necked R.B. flask (250 ml capacity) fitted with a reflux condenser and a dropping funnel. Powdered anhydrous aluminium chloride (32 g, 0.24 mole) is added through the third neck of the flask (anhydrous conditions). The reaction mixture is shaken, the mixture becomes warm and hydrogen chloride is evolved. The reaction mixture is heated for 30 minutes (boiling water bath) with frequent shaking. The reaction mixture is cooled (ice bath) and water (40 ml) added dropwise (dropping funnel) followed by addition of concentrated hydrochloric acid (15 ml). Excess benzene is removed by steam distillation and the remaining solution cooled. The separated product is filtered, washed with cold dilute hydrochloric acid (1:3, 30 ml) and then with cold water (30 ml). The crude acid is dissolved in sodium carbonate solution (15 per cent, 70 ml) by boiling for 10 minutes. The clear alkaline solution (filter if necessary) is treated with decolourizing carbon (1 g), stirred and filtered. It is cooled and acidified with concentrated hydrochloric acid (20 ml) and cooled. The separated β-benzoylpropionic acid is filtered. The yield is 16 g (73.7%), m.p. is 115°C.

46.2.3 o-Benzoylbenzoic acid

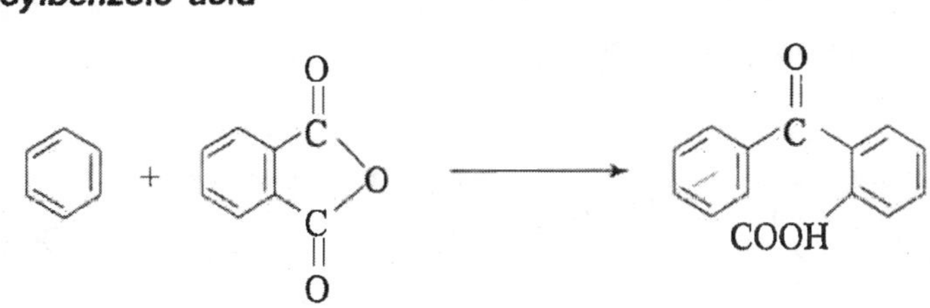

Benzene   Phthalic anhydride   *o*-Benzoyl benzoic acid

Benzene (dry) : 15 ml
Phthalic anhydride : 3 g
Aluminium chloride (anhyd.) : 6 g

Pure dry benzene (15 ml) and phthalic anhydride (3 g, 0.02 mole) are taken in a R.B. flask (100 ml capacity) fitted with a reflux condenser and a calcium chloride drying tube. Anhydrous powdered aluminium chloride (6 g) is added with shaking to the flask. If the reaction does not start, the flask is gently warmed to initiate the reaction and hydrogen chloride gas is evolved. The flask is warmed or cooled (as may be necessary) to maintain a steady evolution of hydrogen chloride. The reaction mixture is finally heated for one hour (boiling water bath). It is cooled (ice bath) and ice cold water added dropwise, while shaking the flask, to decompose the complex. When the HCl gas evolution has subsided, more water (15 ml) and concentrated hydrochloric acid (20 ml) is added until the solution becomes clear. Excess benzene is removed by steam distillation. The solid obtained on cooling the aqueous solution is filtered and washed with ice-cold water. It is purified by dissolving in sodium carbonate solution (1.5 g in 30 ml water), warming and acidifying the clear alkaline solution (filter if necessary). The yield is 3 g (65.2 %); m.p. is 93°C (monohydrate). The anhydrous sample obtained by dissolving the mono hydrate in benzene, the solution dried (anhydrous magnesium sulphate) and concentrated. Addition of petroleum ether (till turbidity is produced) and cooling gives anhydrous *o*-benzoylbenzoic acid, m.p. is 128°C.

46.2.4 p-Methylacetophenone

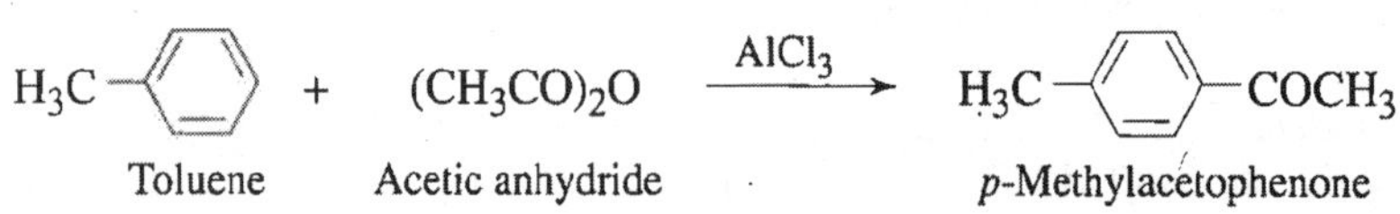

Toluene (dry)	: 70 ml
Aluminium chloride (anhyd.)	: 40 g
Acetic anhydride	: 12 ml

Sodium dried pure toluene (70 ml, 0.66 mole) and powdered anhydrous aluminium chloride (40 g) are taken in a three necked R.B. flask (500 ml capacity) fitted with a mercury sealed stirrer, a reflux condenser and a dropping funnel (anhydrous conditions). Freshly distilled acetic anhydride (12 ml, 0.127 mole) is added dropwise (dropping funnel) during 20 minutes to the stirred mixture. The temperature rises with the evolution of hydrogen chloride gas. The reaction mixture is heated (water bath) for 30 minutes till the evolution of hydrogen chloride ceases. The mixture is cooled (ice bath) and treated with crushed ice (100 g) and concentrated hydrochloric acid (100 ml). The mixture is stirred, toluene layer separated (separatory funnel) and washed with water, sodium hydroxide solution (10%, 2 × 50 ml) and then again with water (25 ml). It is dried (anhydrous sodium sulphate) and distilled at atmospheric pressure till the temperature rises to 125°C and finally under reduced pressure. The *p*-methylacetophenone distills at 93–94°C/7mm. The yield is 14 g (82.4%); b.p. is 225°C.

46.2.5 2-Hydroxy-4-methoxybenzophenone

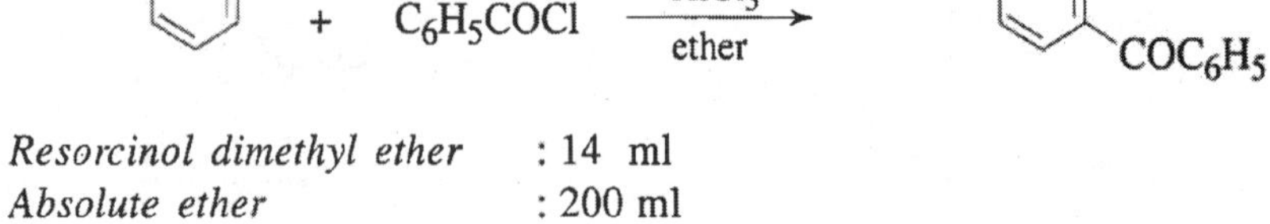

Resorcinol dimethyl ether	: 14 ml
Absolute ether	: 200 ml
Aluminium chloride (anhyd.)	: 20 g
Benzoyl chloride	: 15 ml

Anhydrous aluminium chloride (20 g) is dissolved in absolute ether (200 ml) with cooling (cold water bath) and resorcinol dimethyl ether (14 ml, 0.107 mole) added. The mixture is vigorously stirred and cooled to 0°C and freshly distilled benzoyl chloride (15 ml, 0.129 mole) is added dropwise during one hour under anhydrous conditions. The mixture is stirred for 6 hours more. An oily red aluminium chloride complex separates. The reaction mixture is kept for 48 hours at room temperature and treated with crushed ice (100 g) and concentrated hydrochloric acid (25 ml). The mixture is heated on a water bath for 30 minutes, the ether being allowed to distil away. The cooled reaction mixture is extracted with ether (2 × 100 ml). The ether extract is washed with sodium carbonate solution (5 per cent) and then extracted with sodium hydroxide solution (10 per cent, 200 ml). The cooled, clear alkaline solution is acidified with hydrochloric acid and the separated product is filtered and washed with cold water. It is crystallised from dilute alcohol. The yield is 18 g (73.8%); m.p. is 66°C.

Note: (1) Resorcinol dimethyl ether is obtained as described on p. 146(2) In a similar way, 2-hydroxy-4-methoxyacetophenone, (m.p. 50–51°C) is obtained by using acetyl chloride in place of bezoyl chloride.

47. FRIES REARRANGEMENT[1, 2]

Esters of phenols on heating with anhydrous aluminium chloride undergo rearrangement to give phenolic ketones. This is known as Fries rearrangement. Thus,

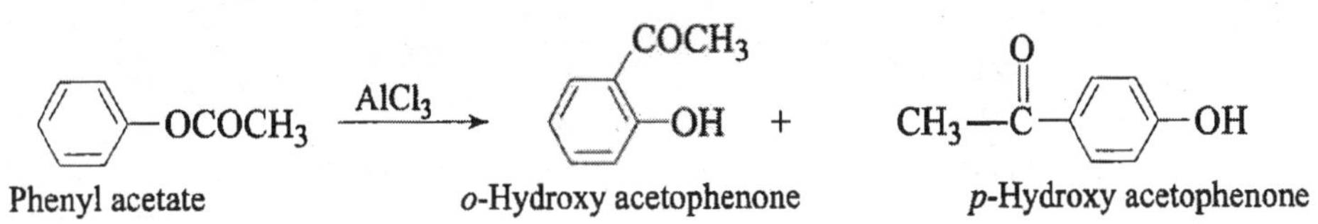

Generally, low temperatures (< 60°C) favour the formation of the *p*-isomer, whereas high temperature (≥ 160°C) favour the formation of *o*-isomer.

Mechanism: The mechanism of Fries reaction is not certain. There is evidence for an intermolecular mechanism, but recently it has been shown that the rearrangement is at least partly intramolecular. It would appear that the mechanism is best regarded as a combination of the two mechanism occuring simultaneously.

Intermolecular

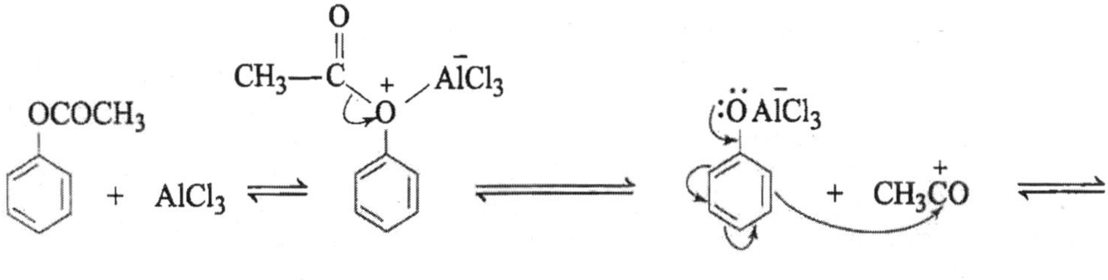

1. K. FRIES and G. F. FINK, *Ber*, 1908, **41**, 4271.
2. K. FRIES and W. PLAFFENDORF, *Ber*, 1910, **43**, 212.

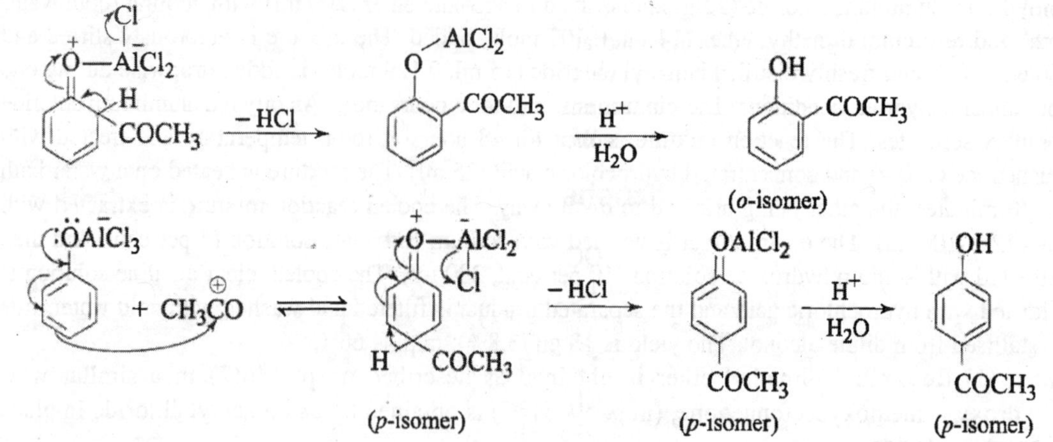

Intramolecular

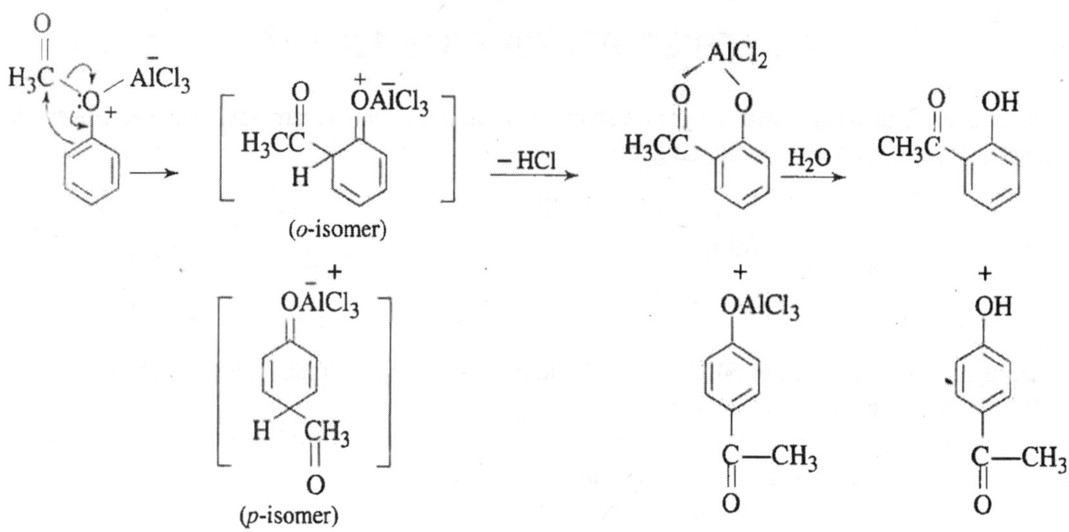

47.1 2,5-Dihydroxyacetophenone

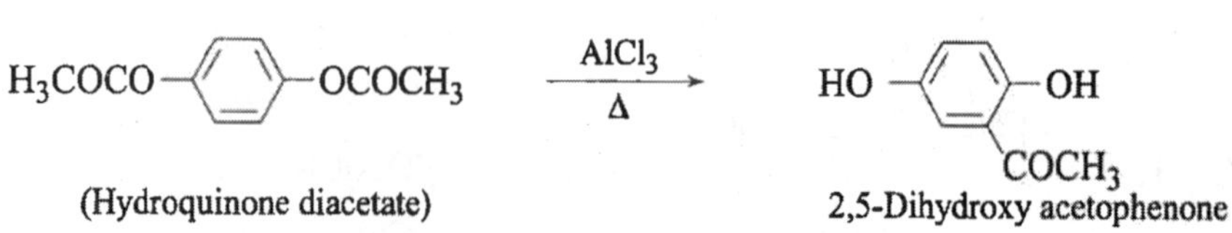

Hydroquinone diacetate : 10 g
Aluminium chloride (Anhyd.) : 21.5 g

An intimate mixture of hydroquinone diacetate (p. 4) (10 g, 0.05 mole) and anhydrous aluminium chloride (21.5 g) is heated in a R.B. flask (100 ml capacity) fitted with an air condenser and a calcium chloride drying tube. The heating is initially done slowly (oil bath) so that the temperature reaches 110–120°C during 30 minutes. The reaction starts and the evolution of hydrogen chloride begins. The temperature is slowly raised to 160–165°C and maintained at this temperature for one hour. The reaction mixture is allowed to cool slowly to room temperature. Crushed ice (60 g) and concentrated hydrochloric acid (5 ml) are added to decompose the aluminium chloride complex. The separated product is filtered, washed with water and crystallised from alcohol. The yield is 4.5 g (57.7%); m.p. is 202–203°C.

47.2 2-Hydroxy-5-methylbenzophenone

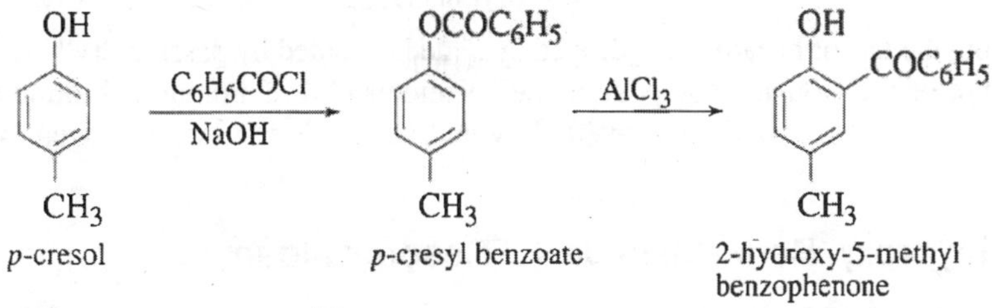

p-Cresol	: 11 g
Benzoyl chloride	: 11 ml
Sodium hydroxide (2 N)	: 22 ml
Aluminium chloride (anhyd.)	: 8 g
Conc. hydrochloric acid	: 20 ml

Step 1 p-Cresyl benzoate

To a mixture of *p*-cresol (11 g, 0.1 mole), water (100 ml) and benzoyl chloride (11 ml, 0.095 mole) in a R.B. flask is added sodium hydroxide solution (2N, 22 ml) till the mixture is strongly alkaline. The mixture is shaken vigorously while cooling under the tap. The reaction mixture should always be alkaline. If necessary, more sodium hydroxide solution is added. The mixture is poured into water. The separated solid (scratching the flask is necessary to induce crystallisation) is filtered, washed with water and crystallised from dilute alcohol (1:1). The yield is 13 g; m.p. is 71°C.

Step 2 2-Hydroxy-5-methylbenzophenone

A mixture of *p*-cresyl benzoate (10 g, 0.047 mole) and anhydrous aluminium chloride (8 g) is heated in a R.B. flask (250 ml capacity) in an oil bath at 90°C under anhydrous conditions. After the reaction mixture has melted, the temperature is raised to 120°C and then slowly to 140°C and kept for about 15 minutes at 140°C. The cooled reaction mixture is treated with crushed ice (50 g) and concentrated hydrochloric acid (20 ml). The separated product is filtered. The 2-hydroxy-5-methylbenzophenone is obtained by steam distillation of the crude product. It is finally crystallised from alcohol. The yield is 8 g (80%); m.p. is 83–84°C.

48. GATTERMANN FORMYLATION[1, 2]

The Gattermann formylation also known as Gattermann aldehyde synthesis consists in the preparation of aldehydes of phenol and phenolic ethers or of heterocyclic compounds by treatment of aromatic

1. L. GATTERMAN, *Ber.*, 1898, **31**, 1149.
2. L. GATTERMAN, *Ann.*, 1907, 313.

substrate with hydrogen cyanide and hydrogen chloride in presence of lewis acid catalyst (zinc chloride, aluminium chloride).

Mechanism: A number of mechanisms have been proposed. A widely accepted one is that imidoformyl chloride is formed as an intermediate.

$$HCN + HCl \xrightarrow{AlCl_3} HN{=}CHCl$$

$$C_6H_6 + HN{=}CHCl \xrightarrow[-HCl]{AlCl_3} \underset{\text{Aryl imine}}{C_6H_5CH{=}NH} \xrightarrow{H_2O} C_6H_5CHO + NH_3$$

In this reaction, the use of hazardous hydrogen cyanide is avoided by passing dry hydrogen chloride gas into a mixture of the aromatic compound, zinc cyanide and or aluminium chloride and a solvent (anhydrous ether). The zinc cyanide is converted by hydrogen chloride into hydrogen cyanide, which reacts in situ.

48.1 2,4-Dihydroxy Benzaldehyde (β-Resorcylaldehyde)

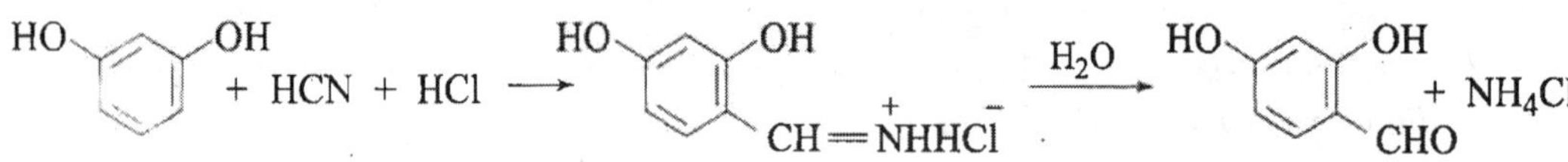

Resorcinol

2, 4 Dihydroxy benzaldehyde

Resorcinol (dry) : 10 g
Ether (anhyd.) : 150 ml
Zinc cyanide (anhydrous): 14 g

Dry resorcinol (10 g, 0.09 mole), anhydrous ether (150 ml) and zinc cyanide (dry, 14 g, 0.119 mole) are taken in a 500 ml dry filtration flask with arrangement for bubbling hydrogen chloride gas. The side tube is connected through a rubber tube to a small funnel, which is partially dipped in dilute sodium hydroxide solution to absorb excess hydrogen chloride and hydrogen cyanide gas. The flask is cooled (ice bath) and a current of dry hydrogen chloride gas is bubbled through the mixture. The zinc cyanide gradually disappears and the aldimine hydrochloride separates as a thick oil which solidifies in 40–60 minutes. A slow stream of hydrogen chloride is passed continuously at room temperature for 3–4 hours. The ether is decanted and the solid washed with dry ether (2 × 25 ml). The remaining aldimine hydrochloride is dissolved in water (20 ml) and the solution heated for 5 minutes (water bath). The solution is cooled and the separated aldehyde filtered. It is crystallised from a small amount of water and dried. The yield is 9 g (72%); m.p. is 135°C. If the product is coloured it can be removed by recrystallisation using decolourising charcoal.

48.2 *p*-Methoxybenzaldehyde (Anisaldehyde)

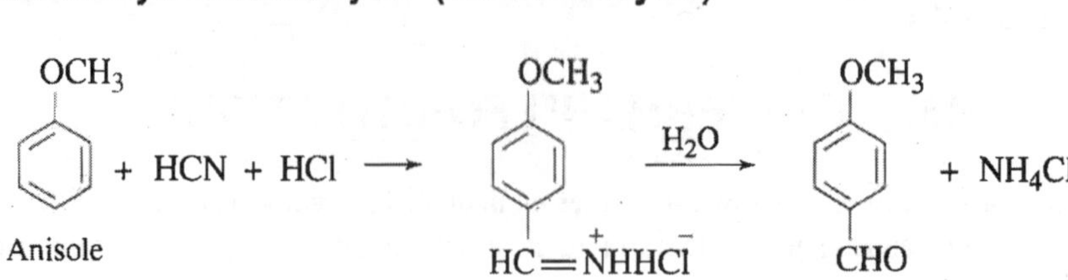

Anisole

Anisaldehyde

Anisole (dry) : 15 ml
Benzene (anhydrous) : 40 ml
Zinc cyanide : 26 g
Aluminium chloride (anhyd.): 23 g

Dry anisole (15 ml, 0.138 mole), sodium dried benzene (40 ml) and dry zinc cyanide (26 g, 0.22 mole) are taken in a 500 ml dry filtration flask with arrangement for bubbling hydrogen chloride gas and absorbing excess hydrogen chloride (as in the case of 2,4-dihydroxy benzaldehyde, page 102). The mixture is cooled (cold water bath) and a rapid stream of dry hydrogen chloride gas passed for one hour. Anhydrous powdered aluminium chloride (23 g) is added to the reaction mixture and hydrogen chloride gas passed for 3–4 hours at 40–45°C (use warm water bath). The reaction mixture is shaken occasionally. The cooled reaction mixture is poured with stirring into excess of dilute hydrochloric acid. The resulting mixture is refluxed for 30 minutes (water bath) to decompose the imine hydrochloride. The mixture is steam distilled. The organic benzene layer containing anisaldehyde is separated (separatory funnel) from the steam distillate. It is dried and distilled to remove benzene. Finally, anisaldehyde is obtained by distillation. The yield is 18 g (95.7%); b.p. is 246–248°C (134–135°C/12 mm).

49. GABRIEL SYNTHESIS[1]

This is the conversion of alkylhalides to primary amines by treatment with potassium phthalimide and hydrolysis of the formed *N*-alkyl phthalimides.

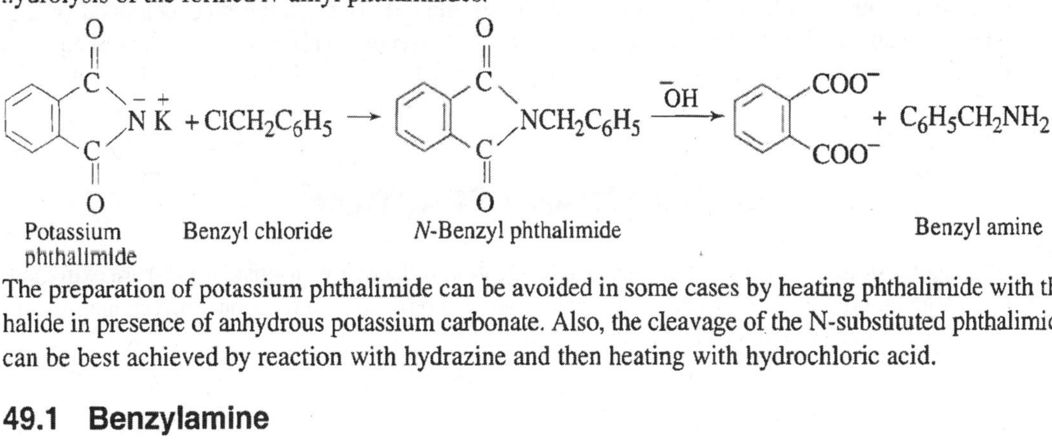

Potassium phthalimide — Benzyl chloride — *N*-Benzyl phthalimide — Benzyl amine

The preparation of potassium phthalimide can be avoided in some cases by heating phthalimide with the halide in presence of anhydrous potassium carbonate. Also, the cleavage of the N-substituted phthalimide can be best achieved by reaction with hydrazine and then heating with hydrochloric acid.

49.1 Benzylamine

Step 1 Phthalimide

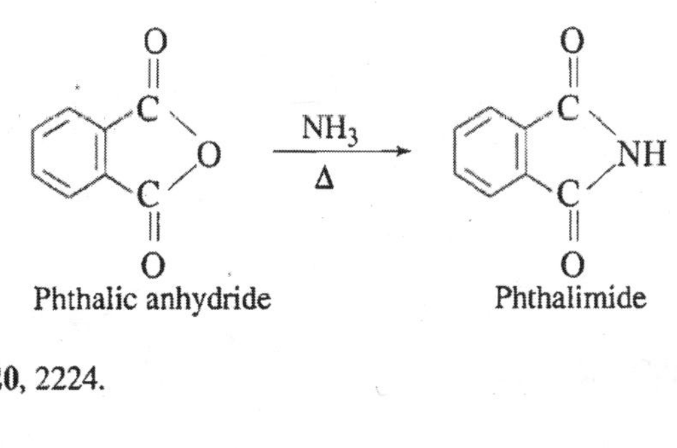

Phthalic anhydride — Phthalimide

1. S. GABRIEL, *Ber.*, 1887, **20**, 2224.

A mixture of phthalic anhydride (50 g, 0.337 mol) and concentrated ammonia (d: 0.88, 55 ml) is refluxed in a R.B. flask (500 ml capacity) fitted with an air condenser (6–8″). The mixture finally becomes a homogeneous melt. (temperature reaches 300°C in about 1.5–2 hours). The water formed is evaporated during the first hour. The contents of the flask are poured into a porcelein dish and allowed to cool. It is grinded to a fine powder in a mortar. The yield is 47.5 g (96%); m.p. is 233–234°C. It is sufficiently pure.

Step 2 N-Benzylphthalimide

Phthalimide	: 36.6 g
Potassium carbonate (Anhyd.)	: 19 g
Benzyl chloride	: 37.75 g

A finely powdered and ground mixture of phthalimide (36.6 g, 0.25 mol) and anhydrous potassium carbonate (19 g, 0.185 mol) is treated with redistilled benzyl chloride (37.75 g, 0.3 mol) in a R.B. flask (250 ml capacity) fitted with a reflux condenser. The reaction mixture is heated in an oil bath at 180–190°C for 3 hours. Excess of benzyl chloride is removed by steam distillation. The mixture is cooled and the separated product filtered and washed with water. It is recrystallised from glacial acetic acid. The yield is 45 g (76 %); m.p. is 116°C.

Step 3 Benzylamine

A suspension of the above *N*-benzyl phthalimide (29.6 g, 0.125 mol) in alcohol (20 ml) and hydrazine hydrate (100%, 6.25 g, 0.125 mol) is warmed. After 20 minutes it is heated with excess hydrochloric acid on a steam bath. The formed phthalyl hydrazide is filtered. The filtrate is concentrated (rotary evaporator) and cooled. The clear solution (filter if necessary) is rendered alkaline with excess of aqueous sodium hydroxide. The liberated benzylamine is extracted with ether. The ether solution is dried (potassium hydroxide pellets) and the solvent evaporated. The residual product is distilled. The yield is 12.25 g (94%); b.p. is 185–187°C.

50. GRIGNARD REACTION[1]

Grignard reagents or the alkylmagnesium halides, R-Mg-X introduced by Grignard (1900) are extremely valuable in organic synthesis. A Grignard reagenet is usually prepared by a reaction between magnesium (1 atom) and alkyl halide (1 mole) in dry, alcohol free ether.

$$R\text{–}X + Mg \longrightarrow RMgX$$

The etheral solution of Grignard reagent is generally used in all reactions.

Addition of organomagnesium compound (Grignard reagent) to a carbonyl containing compound is known as a Grignard reaction.

Mechanism:

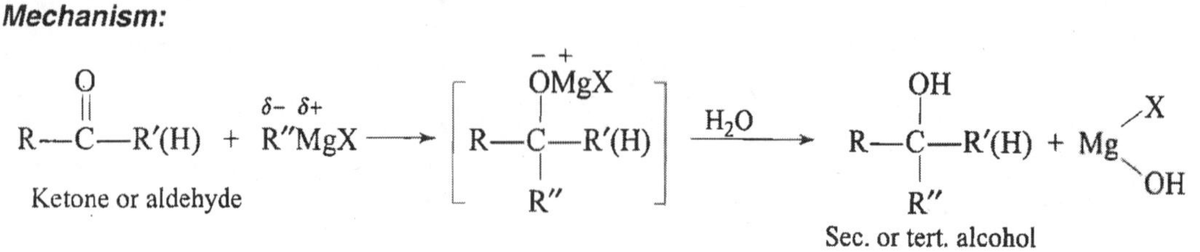

1. V. Grignard, *Compt. Rend.*, 1900, **130**, 1320.

The synthetic utility of Grignard reagents is illustrated by their carbonation (see preparation of benzoic acid from the appropriate Grignard reagent, p 10).

Another important reaction is the reaction of Grignard reagent on an ester to give *tertiary* alcohol.

50.1 Triphenylcarbinol

It is obtained by the reaction of a suitable Grignard reagent, viz. C_6H_5MgBr with methyl benzoate. Thus,

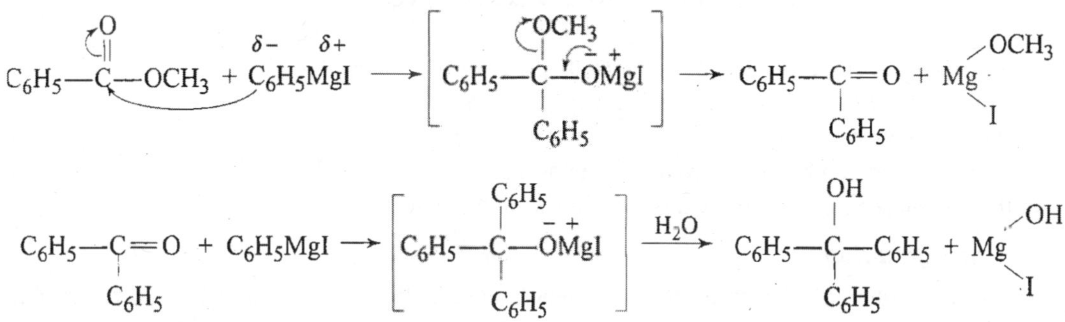

Magnesium turning	: 1.69 g
Iodine	: One crystal
Dry ether	: 50 ml
Bromobenzene	: 7.5 ml
Methyl benzoate	: 4 ml

A mixture of magnesium turnings (1.69 g), sodium dried ether (8 ml) and a crystal of iodine is taken in a three neck R.B. flask (250 ml capacity) fitted with a reflux condenser and a dropping funnel (anhydrous conditions). Bromo benzene (7.5 ml, 0.07 mole) in dry ether (30 ml) is taken in the dropping funnel. The bromobenzene solution (5 ml) is added to the reaction mixture and the flask shaken. In case the reaction does not commence in 2–3 minutes (as indicated by the disappearance of iodine colour, appearance of turbidity and boiling of ether), the reaction mixture is gently warmed (warm water bath) until the reaction commences. When the reaction has started, the remaining bromobenzene solution is added dropwise at such a rate that the ether boils gently without any external heating. The mixture is finally refluxed gently (water bath) for 15 minutes.

The formed Grignard reagent is cooled (ice-water) and a solution of methyl benzoate (4 ml, 0.032 mole) in dry ether (10 ml) is added dropwise (dropping funnel). The reaction mixture is shaken occasionally. The rate of addition is so adjusted that the mixture refluxes gently. Finally, the reaction mixture is refluxed for 10 minutes (water bath) and the cooled reaction mixture poured cautiously into crushed ice (80 g) and dilute sulphuric acid (10%, 40 ml). The mixture is stirred for 10–15 minutes so that the magnesium compound is decomposed. The ether layer is separated (separatory funnel) and the aqueous solution extracted with ether (2 × 25 ml). The combined ether solution is washed with dilute sulphuric acid (10%, 2 × 15 ml), water (20 ml) and finally with water (15 ml) containing sodium bisulphite (0.5 g) (to remove the iodine used to start the reaction). The remaining ether solution (A) is distilled (water bath) and water (40 ml) added to the residue. The mixture is steam distilled until no more oil (unchanged reactants and diphenyl ketone obtained as byproduct) passes over. The residual solution is cooled and the separated product is filtered and dried. It is crystallised from methylated spirit or from benzene. The yield is 5 g; m.p. is 162°C. Alternatively, the ether solution (A) obtained above is dried (anhydrous sodium sulphate), petroleum ether (25 ml) added and the resulting solution is concentrated slowly (water bath) until crystals of triphenyl carbinol begin to separate. The mixture is first cooled to room temperature and then at 0°. The product obtained is crystallised as given above.

51. HALOGENATION

Aromatic halogen compounds can be obtained by the following methods.

1. Replacement of diazo group by a halogen (see p 59).
2. Direct halogenation.
3. Replacement of a hydroxyl group by a halogen.

In the present section, only the last two methods are discussed.

Direct halogenation

The nuclear substitution of benzene with chlorine or bromine in the cold is not possible. However, in the presence of aluminium amalgam, pyridine or iron the reaction takes place readily to give the mono halogenated derivatives as the main product. Di-substituted products (mainly the *para* isomer) are obtained if the proportion of halogen is increased. Typical procedures are illustrated by the synthesis of chloro- and bromo-benzenes.

Mechanism: The catalyst increases the electrophilic reactivity of the halogen. The mechanism of chlorination is illustrated below:

$$AlCl_3 + Cl—Cl \longrightarrow \overset{\delta-}{AlCl_3 \ldots Cl} \ldots \overset{\delta+}{Cl}$$

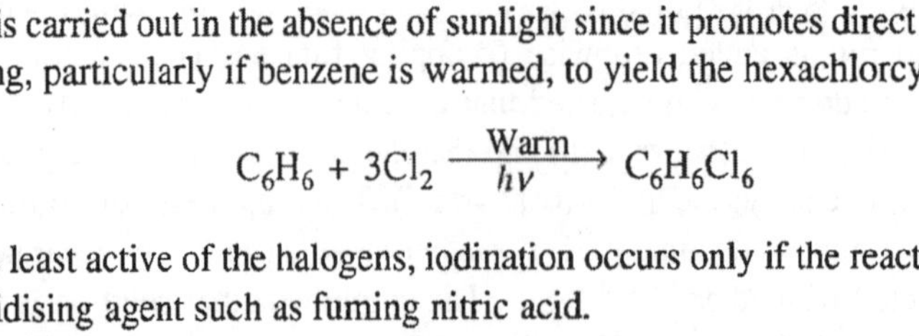

The halogenation is carried out in the absence of sunlight since it promotes direct addition of the halogen to the aromatic ring, particularly if benzene is warmed, to yield the hexachlorcyclohexane

$$C_6H_6 + 3Cl_2 \xrightarrow[h\nu]{\text{Warm}} C_6H_6Cl_6$$

Since iodine is the least active of the halogens, iodination occurs only if the reaction is carried out in the presence of an oxidising agent such as fuming nitric acid.

In case some groups like NH_2, OH or $NHCOCH_3$ are present, facile halogenation (particularly bromination) takes place. This is illustrated by the preparation of 2,4,6-tribromoaniline, 2,4,6-tribromophenol, *p*-bromo acetanilide and 1-bromo-2-naphthol.

Side chain halogenation in the absence of catalysts: For example, treatment of toluene with chlorine or bromine at boiling point preferably in presence of sunlight results in halogenation of the side chain. This is illustrated by the preparation of benzyl chloride and *p*-nitrobenzyl bromide.

The replacement of a hydroxyl group by an halogen is illustrated by the preparation of ethyl bromide and ethyl iodide. However, chloroform is obtained from acetone by reaction with bleaching powder.

51.1 Chlorobenzene

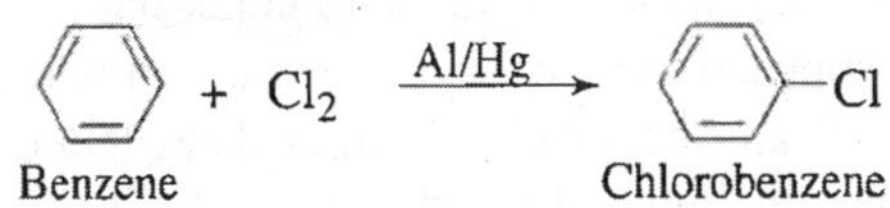

Benzene (dry) : 50 ml

Aluminium amalgam : 0.3 g

Chlorine gas

Benzene (dry, 50 ml, 0.56 mole) and aluminium amalgam (0.3 g) are taken in a two necked R.B. flask (200 ml capacity) fitted with a reflux condenser, a gas inlet tube going up to the bottom of the flask and an arrangement for absorbing the hydrogen chloride gas evolved during the reaction. Dry chlorine from a cylinder is passed through benzene. An exothermic reaction occurs and hydrogen chloride gas is evolved. The chlorination is continued by cooling the reaction mixture (cold water) until there is an increase in weight by 17 g. This is ascertained by weighing the flask before starting chlorination and after passing the gas. The liquid is poured into cold water (70 ml). The lower layer of chlorobenzene is separated (separatory funnel), washed with dilute sodium hydroxide solution and water. It is dried (anhydrous calcium chloride) and distilled using a small fractionating column. Chlorobenzene is obtained in the fraction 127–135°C. The yield is 30 g (47.6%), b.p. is 131–132°C. The high boiling fraction (residue) consists of *p*-dichlorobenzene (b.p. 175°C, m.p. 53°C) and *o*-dichlorobenzene (b.p. 179°C) The *p*-isomer separates on cooling in ice.

**Note:* The aluminium-amalgam is obtained as follows: Aluminium foil (10g) is cut into thin strips (10–15 cm long), loosely folded and place in a conical flask (250 ml capacity) and covered with sodium hydroxide solution (10%). The contents are heated (water bath) until vigorous evolution of hydrogen takes place. The foil is washed thoroughly with water and alcohol. Mercuric chloride solution (2%) is added to cover the foil and allowed to stand for 2 minutes. Excess solution is poured and the amalgam washed with water, alcohol and covered with dry benzene.

51.2 Bromobenzene

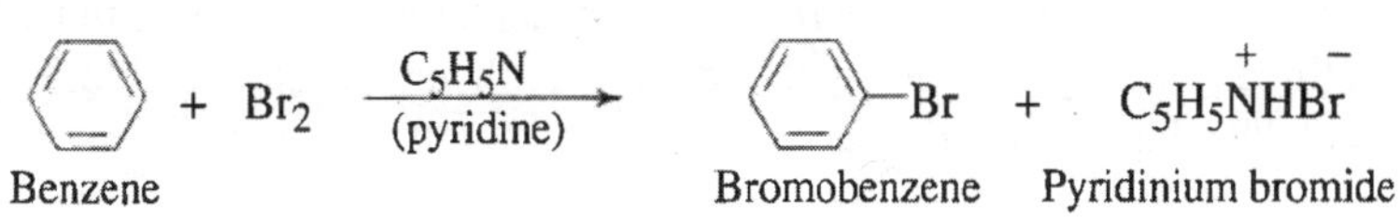

Benzene (dry) : 50 ml

Pyridine (dry) : 0.5 ml

Bromine : 35 ml

Dry benzene (50 ml, 0.56 mole) and dry pyridine (0.5 ml) are taken in a three necked R.B. flask (250 ml capacity) fitted with a reflux condenser, a dropping funnel and a device for absorbing the hydrogen bromide gas evolved during the reaction. The flask is cooled (cold water) and bromine (35 ml) added carefully through the dropping funnel. A vigorous reaction occurs and hydrogen bromide is evolved. The reaction mixture is warmed to 25–30°C (water bath) for one hour and the temperature is then raised to 65–70°C (water bath) for 40–45 minutes. During this time all bromine disappears and evolution of hydrogen bromide ceases. The reaction mixture is washed (separatory funnel) with sodium hydroxide solution (10%) till the washings are alkaline to litmus and is finally washed with water. It is dried (anhydrous magnesium sulphate) and distilled. The fraction boiling at 150–170°C is collected. The liquid is redistilled using a small fractionating column. The yield is 30 g (34%); b.p. is 156°C.

The residual product consists of *p*-dibromobenzene. It is crystallised from alcohol. The yield is 15g; m.p. is 89°C.

Note: Other halogen carriers e.g., 1–2 g iron fillings or 1 g aluminium can also be used. In such a case bromine should be added slowly.

51.3 Iodobenzene

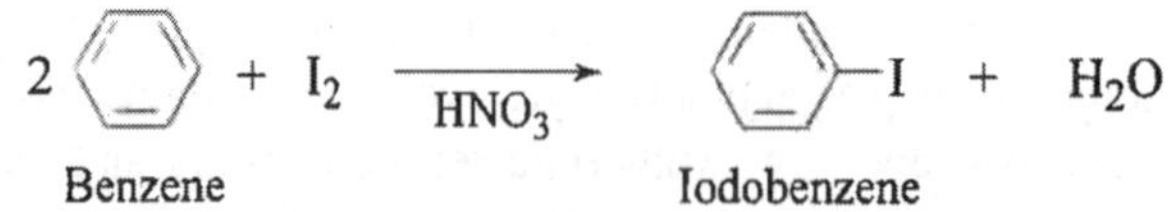

Benzene (dry) : 50 ml
Iodine : 42.3 g
Fuming nitric acid : 30.3 ml
(d. 1.50)

Dry benzene (50 ml, 0.56 mole) and iodine (42.3 g, 0.167 mole) are taken in a three necked R.B. flask (250 ml capacity) fitted with a sealed mechanical stirrer, a reflux condenser and a dropping funnel. The mixture is stirred, heated to 50°C (water bath) and fuming nitric acid (d: 1.50, 30.3 ml) is added (dropping funnel) slowly during 25–30 minutes. Oxides of nitrogen are evolved, temperature rises and the mixture boils. After the addition of fuming nitric acid is complete, the mixture is refluxed gently for 15–20 minutes. If some iodine is still present, more nitric acid is added to the warm solution until the purple colour due to iodine changes to brownish red. The lower oily layer is separated (separatory funnel), dilute sodium hydroxide solution (10 per cent, 60 ml) added and the mixture steam distilled till no more oil passes over. Some yellow solid consisting of nitro compounds may collect at the end of the distillation. This is removed by stirring (mechanical stirring) of the oil for 3 hours with concentrated hydrochloric acid (6 ml), water (90 ml) and iron filings (30 g) in a three necked flask fitted with a reflux condenser. The cooled mixture is rendered distinctly acidic to congo red with hydrochloric acid and again steam distilled. The oily layer is separated, dried (anhydrous calcium chloride) and distilled. The yield is 55 g (80.9%); b.p. is 184-186°C.

51.4 2,4,6-Tribromoaniline

The amino group in aniline activates the nucleus to such an extent that substitution takes place at the two *ortho* and one *para* position. Thus, reaction with bromine gives 2,4,6-tribromoaniline

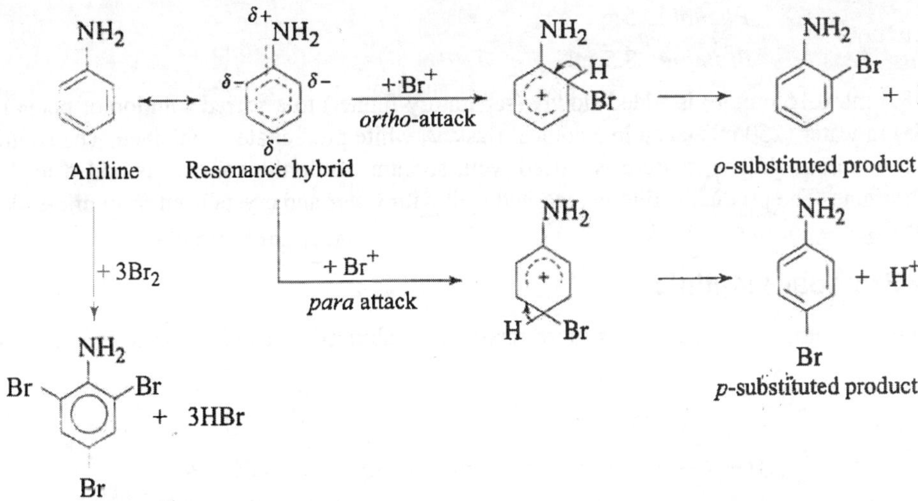

2, 4, 6-Tribromoaniline

Aniline : 5 ml
Acetic acid (glacial) : 40 ml
Bromine : 8.4 ml

A solution of bromine (8.4 ml, 0.163 mole) in glacial acetic acid (20 ml) is added dropwise (dropping funnel) to a solution of redistilled aniline (5 ml, 0.055 mole) in glacial acetic acid (20 ml) taken in a conical flask (100 ml capacity). During the addition, the flask is shaken and cooled in ice. A yellow coloured pasty mass is obtained. It is poured into excess of water. The separated product is filtered, washed with water and crystallised from rectified spirit. The yield is 10 g (55.6%); m.p. is 120°C.

51.5 2,4,6-Tribromophenol

The phenolic group, like the amino group activates the benzene ring towards electrophilic substitution reactions. Like aniline, in the case of phenol also the substitution takes place in the two *ortho* and one *para* position. Thus, reaction with bromine gives 2,4,6-tribromophenol.

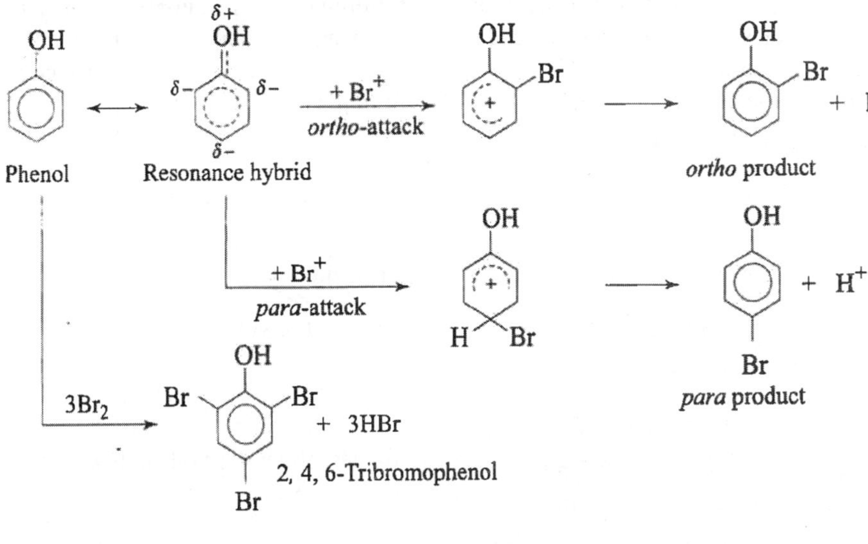

2, 4, 6-Tribromophenol

Phenol : 5 g
Bromine : 8.5 ml

Bromine (8.5 ml, 0.165 mole) is added slowly (separatory funnel) to a stirred solution of phenol (5 g, 0.053 mole) in water (250 ml) taken in a conical flask. A white precipitate is obtained. The solution is finally yellow coloured. The mixture is stirred well, sodium bisulphite solution is added to destroy excess of bromine. The product is filtered, washed well with water and crystallised from dilute alcohol, the yield is 6 g (34%); m.p. is 95°C.

51.6 *p*-Bromoacetanilide

The acetamide group, —$HNCOCH_3$, is more resonance stabilised in comparison to the NH_2 group. Thus,

This means that the unshared electron pair on the *N* atom of the acetamide group is not available for donation to the ring by a resonance mechanism, as is the pair of electrons on the amino group. Hence, the ring of acetanilide is less active than the ring of aniline towards electrophilic reagents and give mono substitutions.

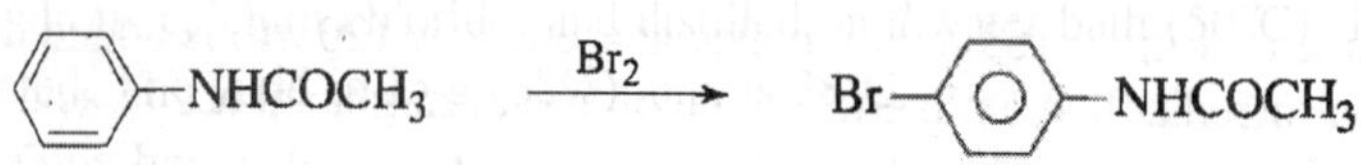

Acetanilide   *p*-Bromoacetanilide

Acetanilide : 7 g
Bromine : 2.7 ml
Acetic acid (glacial) : 65 ml

A solution of bromine (2.7 ml, 0.052 mole) in glacial acetic acid (30 ml) is added (dropping funnel) slowly to a stirred solution of acetanilide (7 g, 0.051 mole) in glacial acetic acid (35 ml). During the addition, the reaction mixture is kept in a cold water bath. After the addition is over, the reaction mixture will be orange coloured due to slight excess of bromine. The reaction mixture is allowed to stand for 15–20 minutes at room temperature and poured into water (200 ml). Sodium bisulphite solution is added just to remove orange colour. The separated product is filtered, washed with cold water and crystallised from dilute alcohol. The yield is 8 g (72.7%); m.p. is 167°C.

51.7 1-Bromo-2-naphthol

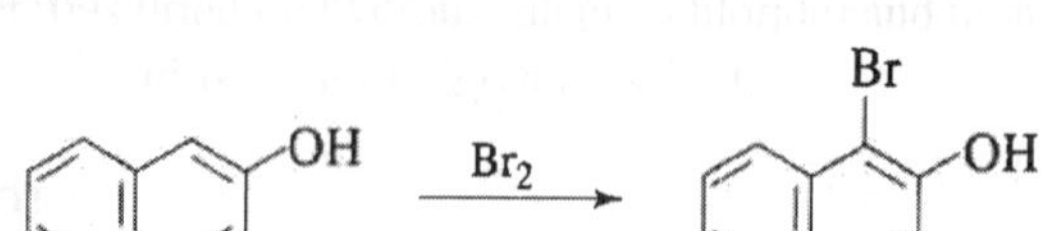

β-Naphthol   1-Bromo-2-naphthol

β-Naphthol (2-Naphthol)

: 7.2 g
Bromine : 2.5 ml
Acetic acid (glacial) : 40 ml

A solution of bromine (2.5 ml, 0.049 mole) in glacial acetic acid (10 ml) is added dropwise (dropping funnel) to a stirred solution of β-naphthol (7.2 g 0.05 mole) in glacial acetic acid (30 ml). During the addition, the reaction mixture is stirred and kept in a cold water bath. The resulting reaction mixture is poured into ice cold water and sodium bisulphite solution is added to destroy excess of bromine. The separated product is filtered, washed with water and crystallised from alcohol. The yield is 10 g (92%); m.p. is 80–81°C.

51.8 Benzyl Chloride

It is obtained by chlorination of toluene in presence of sunlight.

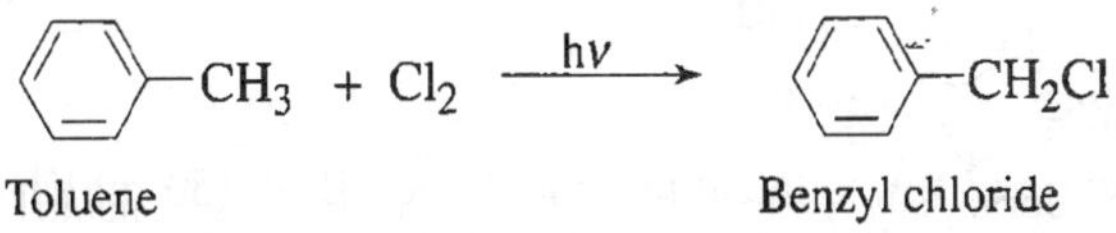

Dry toluene (58 ml, 0.546 mole) is taken in a three necked R.B. flask (250 ml capacity) fitted with a reflux condenser, an inlet tube extending to the bottom of the flask for passing chlorine gas and a thermometer (the bulb of the thermometer is adjusted within 2 cm of the bottom). The top of the reflux condenser is connected to a calcium chloride guard tube and an arrangement for absorbing excess of chlorine gas into sodium hydroxide solution. Toluene is boiled gently and a stream of chlorine gas passed into it until the temperature reaches 157–158°C.

The completion of the reaction is determined by weighing the flask and toluene before and after passing the chlorine gas so that the increase in weight is 18.5 g. The reaction time is considerably reduced by exposure of the mixture to bright sunlight or a mercury-vapour lamp. After the reaction has completed, the mixture is distilled at atmospheric pressure until the temperature reaches 135–140°C. The distillation is subsequently continued under reduced pressure. Benzyl chloride distils at 64–69°C/12 mm. It is redistilled. The yield is 50 g (72.5%); b.p. is 63–65°C/12 mm (179°C).

51.9 *p*-Nitrobenzyl Bromide

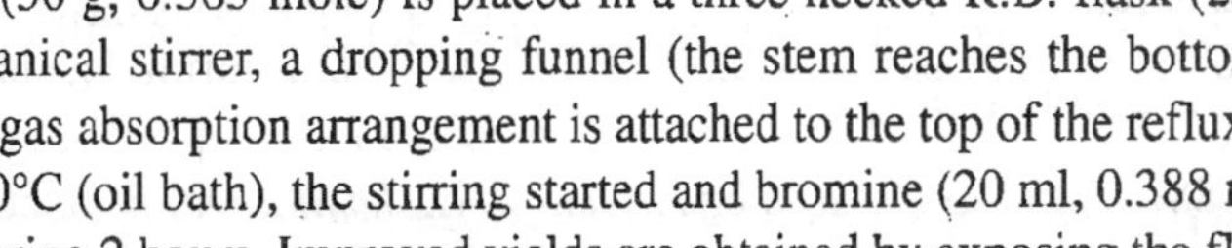

p-Nitrotoluene : 50 g
Bromine : 20 ml

Pure *p*-nitrotoluene (50 g, 0.365 mole) is placed in a three necked R.B. flask (250 ml capacity) fitted with a sealed mechanical stirrer, a dropping funnel (the stem reaches the bottom of the flask) and a reflux condenser. A gas absorption arrangement is attached to the top of the reflux condenser. The flask is heated to 145–150°C (oil bath), the stirring started and bromine (20 ml, 0.388 mole) added dropwise (dropping funnel) during 2 hours. Improved yields are obtained by exposing the flask to the light of two 300 watt tungsten lamps during the addition of bromine. After the addition of bromine has been completed, the reaction mixture is poured into hot petroleum ether (b.p. 80–100°C, 750 ml) in a R.B. flask. The mixture is heated (water bath) till the material dissolves. Decolourising charcoal (5 g) is added, heating continued for 10–15 minutes and filtered. The filtrate is cooled and the separated *p*-nitrobenzyl bromide is filtered. The yield is 45 g (57%); m.p. is 98–99°C.

Note: (1) *p*-Nitrobenzyl bromide is lachrymatory. If it comes in contact with skin wash immediately with alcohol. (2) It can also be obtained by the reaction of *p*-nitro toluene with an equimolar amount of NBS (refluxing 5–6 hours in $CHCl_3$ solution).

51.10 Ethyl Bromide

$$CH_3CH_2OH + H_2SO_4 \longrightarrow CH_3\overset{+}{C}H_2\bar{O}SO_3H + H_2O$$

$$CH_3CH_2\overset{+}{O}S\bar{O}_3H + KBr \longrightarrow C_2H_5Br + KHSO_4$$

Ethyl alcohol (absolute) : 18 ml
Conc. sulphuric acid : 13 ml
Potassium bromide : 24 g

Concentrated sulphuric acid (13 ml) is taken in a distillation flask (50 ml capacity) and absolute ethyl alcohol (18 ml, 0.3 mole) added dropwise, the flask being cooled under a tap during the addition. Powdered potassium bromide (24 g, 0.2 mole) is then added in small portions (3–4 g each time), the reaction mixture being shaken occasionally. The distillation flask is stoppered, distillation assembly fixed and cold water passed through the condenser and the receiver cooled in an ice-bath. The mixture is distilled gently on a sand bath till no more oily drops pass over. The lower layer of ethyl bromide is separated (separatory funnel), washed with dilute sodium carbonate (10 per cent), and then with cold water, dried (anhydrous calcium chloride) and distilled on a water bath (50°C). The fraction between 37–39°C is collected. The yield is 15 g (68%); b.p. is 38°C.

51.11 Ethyl Iodide

$$2P + I_2 \longrightarrow 2PI_3$$

$$6C_2H_5OH + 2PI_3 \longrightarrow 6C_2H_5I + 2H_3PO_3$$

Iodine : 25 g
Red phosphorus : 2.5 g
Alcohol : 16 ml

Powdered iodine (25 g) is added in small lots at a time to a mixture of red phosphorus (2.5 g) and alcohol (16 ml, 0.273 mole) placed in a R.B. flask (100 ml capacity), the flask being cooled under a tap during the addition. A reflux condenser is attached to the R.B. flask and left overnight. The mixture is refluxed for 2 hours and the ethyl iodide is distilled off on a boiling water bath. The distillate (usually, coloured due to the presence of free iodine) is washed (separatory funnel) with dilute sodium hydroxide (2*N*) and then with water. It is dried (anhydrous calcium chloride) and then distilled, the fraction between 70–74°C is collected. The yield is 18 g (43%); b.p. is 72°C.

51.12 Chloroform

$$CH_3COCH_3 + 3Cl_2 \longrightarrow CH_3COCCl_3 + 3HCl$$

$$CH_3COCCl_3 + Ca(OH)_2 \longrightarrow 2CHCl_3 + (CH_3COO)_2\,Ca$$

Bleaching powder : 100 g
Acetone : 25 ml

Acetone (25 ml) is added slowly to a fine suspension of finely powdered bleaching powder (100 g) and water (350 ml) contained in a R.B. flask (750 ml capacity) fitted with a reflux condenser. When the vigorous reaction has subsided, the same flask is fitted with a distillation arrangement. The flask is heated (sand bath) till no more chloroform passes over (1–1.5 hours). The chloroform obtained is washed (separatory funnel) with sodium hydroxide solution (2*N*) and water. It is dried (fused calcium chloride) and distilled. The fraction between 60–61°C is collected. The yield is 15 g; b.p. is 61°C.

52. HANTZSCH PYRIDINE SYNTHESIS[1-3]

The reaction involves the condensation of two moles of a β-dicarbonyl compound (viz. ethyl acetoacetate) with one mole of an aldehyde in the presence of ammonia. The resulting dihydropyridine is dehydrogenated with an oxidizing agent.

The reaction mechanism depends on the reaction conditions. In case ethylacetoacetate and the aldehyde are allowed to react in presence of a basic catalyst (as in the synthesis of 2,6-dimethyl pyridine), an alkylidene-bis-acetoacetate is formed by successive Knoevenagal and Michael reactions. Cyclisation of the 1,5-dione with ammonia gives the dihydropyridine derivative. This on oxidation with a concentrated nitric acid-sulphuric acid mixture gives the pyridine derivative.

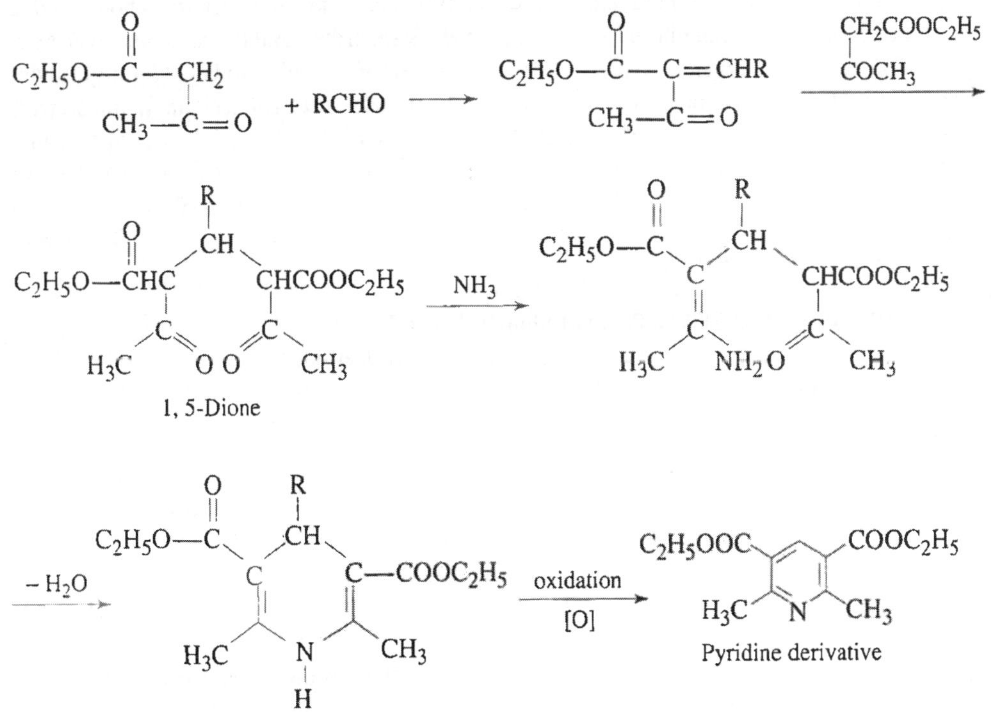

1. A. Hantzsch, *Ann.*, 1982, **215**, 1,72.
2. A. Hantzsch, *Ber.*, 1885, **18**, 1744.
3. A. Hantzsch, *Ber.*, 1886, **19**, 289.

52.1 2,6-Dimethylpyridine

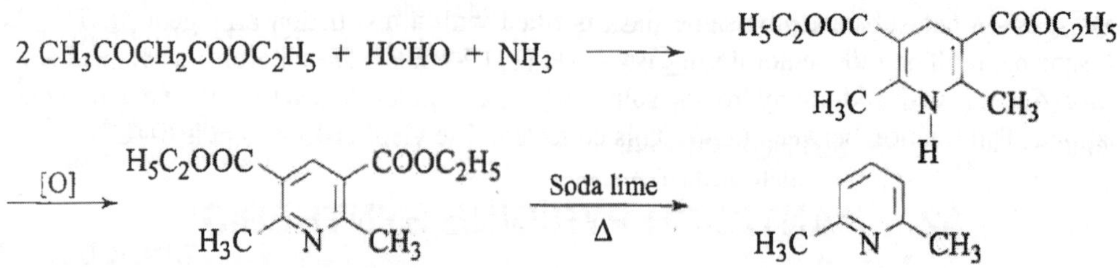

Diethyl 2,6-dimethylpyridine-3,5-dicarboxylate

2,6-Dimethyl pyridine

Step 1 Diethyl 1,4-dihydro-2, 6-dimethylpyridine-3, 5-dicarboxylate

Ethyl acetoactate : 25.5 ml
Formalin (40%) : 7.5 ml
Diethylamine : 0.5 ml

Formalin (40%, 7.5 ml) is added to cooled (0°C) ethyl acetoacetate (26g, 25.5 ml, 0.2 mole) followed by the addition of a few drops of diethylamine as a catalyst. The mixture is kept for 6 hours at 0°C and then for 40 hours at room temperature. The organic layer is separated and the aqueous layer extracted with ether. The combined organic layer and ether extract is dried (anhydrous calcium chloride) and ether removed under reduced pressure (rotary evaporator). The resultant residue is transferred in a reagent bottle with an equal volume of alcohol. A steady stream of ammonia gas (from a cylinder) is passed into the cooled solution (0°C) for one hour. The mouth of the bottle is stoppered and allowed to stay at room temperature for 40 hours. The solution is filtered and alcohol removed (boiling water bath) in a china dish. The product separated on cooling is filtered and crystallised from alcohol (200 ml). The yield is 18 g (71%); m.p. is 182–183°C.

Step 2 Diethyl 2,6-dimethylpyridine-3,5-dicarboxylate

Diethyl 1,4-dihydro-2,6-dimethylpyridine-3,5-dicarboxylate : 12 g
Conc. nitric acid (d: 1.42) : 3 ml
Conc. sulphuric acid : 2.8 ml

To a suspension of diethyl 1,4-dihydro-2,6-dimethylpyridine-3,5-dicarboxylate (12 g, 0.338 mole) in cold water (25 ml) is carefully added concentrated nitric acid (d: 1.42, 3 ml) and concentrated sulphuric acid (2.8 ml). The reaction mixture is shaken and heated cautiously on a boiling water bath until a vigorous reaction sets in. When the reaction has moderated, the mixture is heated cautiously for 15 minutes more until the oxidation is complete and a deep-red solution is obtained. To the cooled reaction mixture is added water (50 ml) and crushed ice (about 50 g) and the mixture rendered distinctly alkaline with concentrated aqueous ammonia solution (d: 0.88). The separated product is filtered, washed with cold water and crystallised from dilute alcohol. The yield is 7.5 g (64%); m.p. is 71–72°C.

Step 3 2,6-Dimethylpyridine

Diethyl 2,6-dimethylpyridine-3,5-dicarboxylate : 7.5 g
Soda lime : 45 g

An intimate mixture of diethyl 2,6-dimethylpyridine-3,5-dicarboxylate (7.5 g, 0.03 mole) and soda lime (10–14 mesh, 45 g) is heated at 250°C (oil bath) in a R.B. flask (100 ml capacity) fitted with an arrangement for distillation. The temperature of the oil bath is maintained at 250°C till no further material distils below 105°C (about 2 hours). The oil bath is removed and the heating is continued with a strong direct bunsen burner. The product distilled by direct heating is collected. It is treated with potassium hydroxide pellets and extracted with ether. The ether is removed by distillation and the residue is distilled. 2,6-Dimethylpyridine is collected at 142–145°C. The yield is 2.1 g (65%).

53. HALOFORM REACTION[1]

The haloform reaction, also known as Lieben Iodoform reaction, consists in the cleavage of methyl ketones (CH_3—CO—R), acetaldehyde, ethanol and secondary methyl carbinols (CH_3CHOHR) with halogens (mostly iodine) and a base to give haloform (iodoform if iodine is used) and carboxylic acids.

$$RCOCH_3 + 3NaOI \longrightarrow RCOCI_3 + 3NaOH$$

$$RCOCI_3 + NaOH \longrightarrow CHI_3 + RCO\overset{-}{O}\overset{+}{N}a$$

This reaction is best carried out by dissolving the compound in dioxan, adding dilute sodium hydroxide, then a slight excess of iodine in potassium iodide solution, warming and finally adding water. If the compound contains the acetyl group, iodoform is precipitated.

Mechanism:

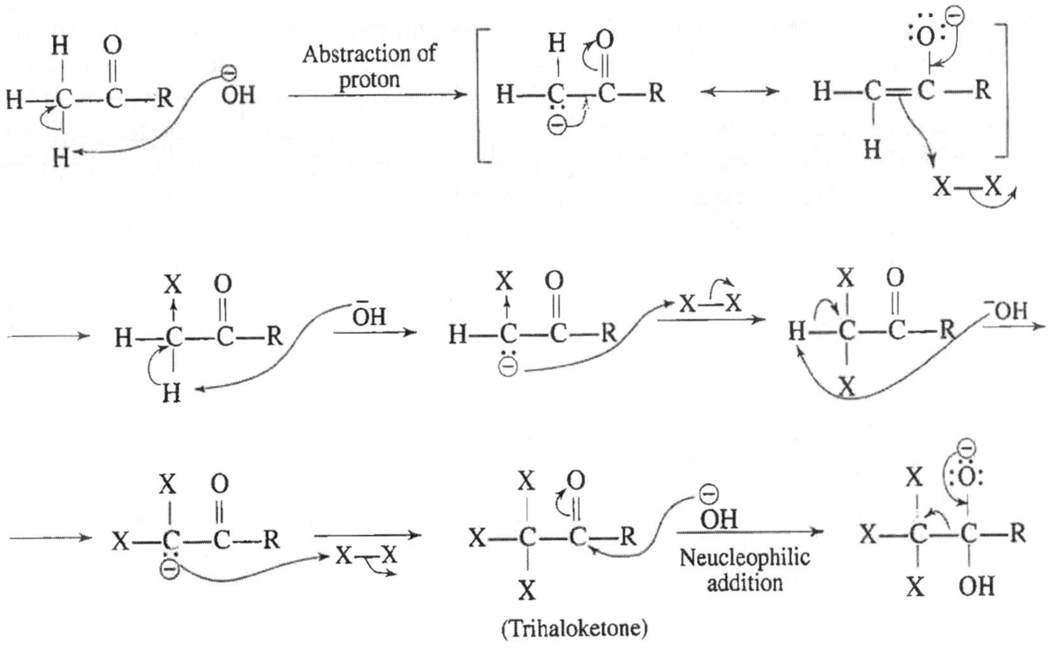

1. A. Lieben, *Ann.* (suppl.), 1870, **7**, 218.

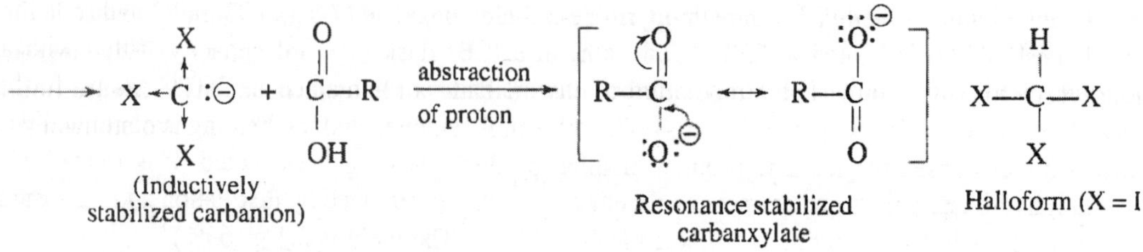

53.1 Iodoform

From acetone

$$CH_3COCH_3 \xrightarrow{\overset{(-)}{O}H} CH_3CO\overset{(-)}{\ddot{C}}H_2 \xrightarrow{I_2} CH_3COCH_2I \xrightarrow{\bar{O}H/I_2} CH_3COCHI_2 \xrightarrow{\bar{O}H/I_2}$$

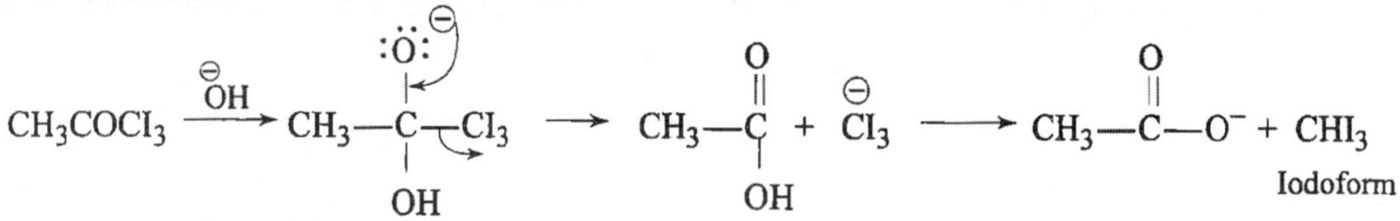

Acetone	: 3 ml
Sodium hydroxide solution	: 10%, 15 ml
Iodine solution	: 12.5 g. iodine dissolved in a solution of 25 g KI in 50 ml water.

To a solution of acetone (3 ml, 0.04 mole) in water (30 ml) and sodium hydroxide solution (10%, 15 ml) is added iodine solution (12.5 g iodine dissolved in a solution of 25 g potassium iodide in 50 ml water) dropwise with constant shaking till the colour of iodine persists. The mixture is heated at 60°C (water bath). More iodine solution is added if the colour disappears. The mixture is heated (water bath) till the precipitate settles down. The yellow precipitate of iodoform is filtered and crystallised from dilute methanol (1:1). The yield is 5 g (31.2%); m.p. is 119°C.

From alcohol

$$C_2H_5OH + 4I_2 + 3K_2CO_3 \longrightarrow CHI_3 + HCO\overset{-}{O}\overset{+}{K} + 5KI + 3\ CO_2 + 2H_2O$$

Alcohol	: 5 ml
Potassium carbonate	: 10 g
Iodine (powdered)	: 10 g
Potassium iodide	: 20 g

Potassium carbonate (10 g) is dissolved in water (40 ml) in a conical flask. Alcohol (5 ml, 0.085 mole) is added and then iodine solution (10 g iodine dissolved in a solution of 20 g potassium iodide in 100 ml water) is added in small quantities at a time with stirring. The mixture is gradually heated to 80°C (water bath), kept for 5 minutes and then cooled. The separated iodoform is filtered and crystalised from dilute alcohol (1:1). The yield is 2.4 g (7.2%); m.p. is 119°C.

Note: For other application of haloform reaction see page 169 to 171.

54. HETEROCYCLIC COMPOUNDS

The heterocyclic compounds are divided into the following categories.

A. Heterocyclic compounds containing nitrogen

A. 1. *Nitrogen present in five-membered ring.*

2,5-Dimethyl pyrrole (see Pall-Knorr synthesis)
3,5-Diethoxycarbonyl-2, 4-dimethylpyrrole (see Pall-knorr synthesis)
3,5-Dimethylpyrazole (see Pall-knorr synthesis)
3-Methyl-1-phenylpyrazol-5-one (see Pall-knorr synthesis)
2,3-Dimethyl-1-phenylpyrazol -5-one (antipyrin) (see Pall-knorr synthesis)
2-Phenyl indole (see Fischer indole synthesis)
2-Methyl indole (see Fischer indole synthesis)
Indigo
5,5-Diphenylhydantoin
Benzimidazoles
Benzotriazoles

A. 2. *Nitrogen present in a six-membered ring.*

Quinoline (see Skraup synthesis)
8-Hydroxy quinoline (see Skraup synthesis)
2,4,6-Trimethyl quinoline (see Combes quinoline synthesis)
4-Methyl carbostyril
Barbituric acid
Quinoxalines
Piparazine-2,5-dione

B. Heterocyclic compounds containing oxygen

B. 1 *Oxygen present in five-membered ring.*

2,5-Dimethylfuran
Benzofuran

B. 2 *Oxygen present in a six-membered ring.*

Coumalic acid
Benzopyrones
Coumarin
3-Hydroxycoumarin [See Erlenmyer-Plöchl Azlaclone Synthesis]
4-methyl-7-hydroxycoumarin
2-methyl-7-hydroxychormone (See Kostanecki acylation)
7-Hydroxyflavone (See Baker-Venkataraman reaction)
Anthraquinone
Acridone

C. Sulphur present in a five-membered ring

2,5-Dialkyl thiophens
Benzothiophens

D. Nitrogen and sulphur both present

Thiazoles

54.1 2,3-Dimethyl-1-phenylpyrazol -5-one (Antipyrin)

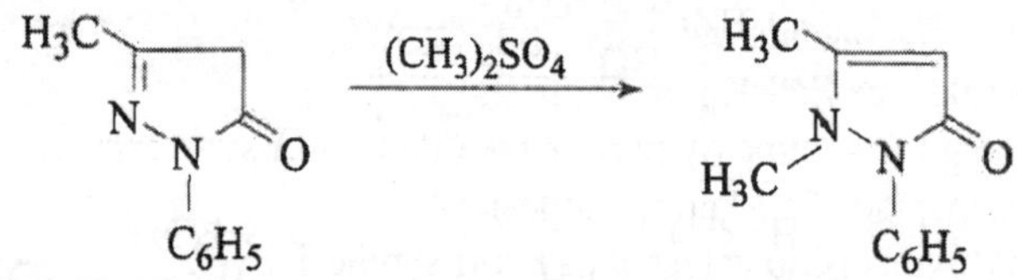

3-Methyl-1-phenylpyrazol -5-one	: 4.35 g
Sodium hydroxide	: 1 g
Dimethyl sulphate	: 3.6 g, 2.7 ml

To a solution of 3-methyl-1-phenylpyrazol-5-one (4.35 g, 0.025 mole, see page 177) in methanol (5 ml) is added sodium hydroxide solution (1 g sodium hydroxide dissolved in minimum amount of water). The mixture is stirred and dimethyl sulphate (3.6 g, 2.7 ml, 0.0285 mol) added. The mixture is refluxed (boiling water bath) and cooled. Methanol is distilled off and hot water added to the residual mixture. The clear solution (filter if necessary) is extracted with benzene (3 × 20 ml). The combined benzene solution is evaporated and the residual product crystallised from hot water (charcoal). The yield is 3.5 g (74%); m.p. is 113°C.

54.2 Indigo

It is obtained by fusing phenylglycine-*o*-carboxylic acid with potassium hydroxide to give indoxyl, which on oxidation gives indigo.

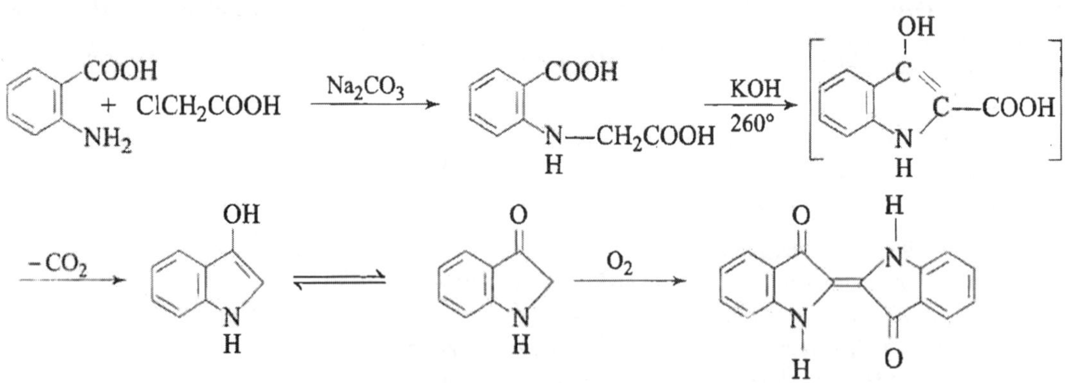

Indigo

Step 1 Phenylglycine-o-carboxylic acid

Anthranilic acid	: 30.9 g
Chloroacetic acid	: 21 g
Sodium carbonate; (anhyd)	: 48 g

A mixture of anthranilic acid (30.9 g, 0.225 mole) chloroacetic acid (21 g, 0.225 mole), anhydrous sodium carbonate (48 g) and water (450 ml) is refluxed for 3 hours in a R.B. flask fitted with a reflux condenser. The reaction mixture is cooled and acidified with concentrated hydrochloric acid to congo red. It is allowed to stand overnight and the separated product filtered and crystallised from hot water. The yield is 16.8 g, (39%); m.p. is 218–220°C.

Step 2 Indigo

Phenylglycine-o-carboxylic acid : 14.5 g
Potassium hydroxide : 50 g

A mixture of phenylglycine-*o*-carboxylic acid (14.5 g. 0.075 mol), potassium hydroxide pellets (50 g) and water (10 ml) is taken in a nickel crucible or basin. The mixture is heated to 260°C while stirring with a thermometer protected by a copper tube. The mixture melts and assumes an orange colour. The crucible is cooled and the contents dissolved in water (400 ml). The clear solution is taken in a buchner flask fitted with a one holed cork carrying a glass tube which dips into the solution and the flask is connected to the pump. Oxidation takes place with atmospheric oxygen and indigo is precipitated as a dark blue powder. It is filtered, washed with water and dried. The yield is 5.4 g (55%). Alternatively, the clear alkaline solution is acidified with concentrated hydrochloric acid and oxidised with ferric chloride solution until no more of indigo is precipitated. It is filtered, washed with water and dried.

Dyeing of cotton (vat dyeing): Indigo (0.2 g) is moistened on a watch glass with ethanol. It is suspended in water (1 ml) and aqueous sodium hydroxide solution (20%, 1 ml) added. A warm solution (50°C) of sodium dithionite (0.3 g) in water (20 ml) is added to reduce the indigo. When a clear solution (vat) is obtained, it is diluted with water (40 ml). A piece of cotton, which has been thoroughly washed with detergent and rinsed with water is suspended in the clear solution and heated (water bath) to 50°C for one hour. During this period the cotton is stirred to ensure even impregnation. The cotton is removed from the solution and allowed to stand in air for 30 minutes to develop the colour. The colour can be brightened by heating on a water bath for 15 minutes, with a dilutle solution of a mild detergent followed by rinsing and drying. The reactions taking place are:

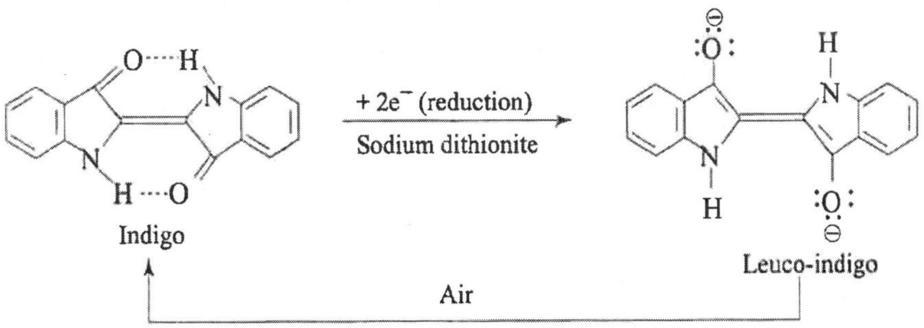

54.3 5, 5-Diphenylhydantoin

It is prepared by condensing benzil (a 1,2-diketone) with urea. The intermediate heterocyclic pinacol undergoes pinacol-pinacolone rearrangement to give 5,5-diphenylhydantoin.

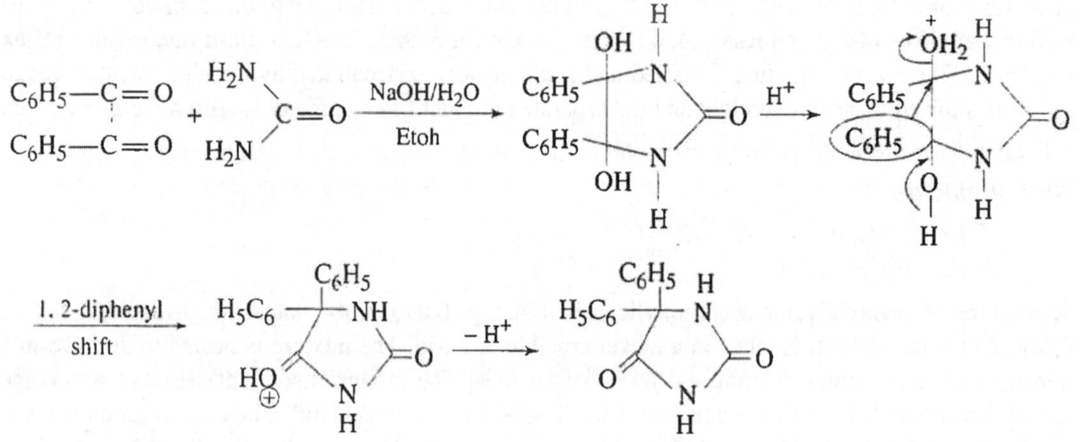

5-5 Diphenylhydantoin

Benzil	: 7.9 g
Urea	: 4.5 g
Sodium hydroxide solution (30%)	: 23 ml
Ethanol	: 100 ml

A mixture of benzil (7.9 g, 0.037 mol), urea (4.5 g, 0.075 mol), aqueous sodium hydroxide (30%, 23 ml) and ethanol (100 ml) is refluxed for 2 hours in a R.B. flask fitted with a reflux condenser. The cooled reaction mixture is poured into water (150 ml). The clear solution (filter if necessary) is acidified with concentrated hydrochloric acid and cooled (ice bath). The separated product is filtered and crystallised from alcohol, the yield 4.2 g (48.7%) m.p. 297–298°C.

54.4 Benzimidazoles

Benzimidazoles are prepared by heating *o*-diamino benzene with a carboxylic acid,

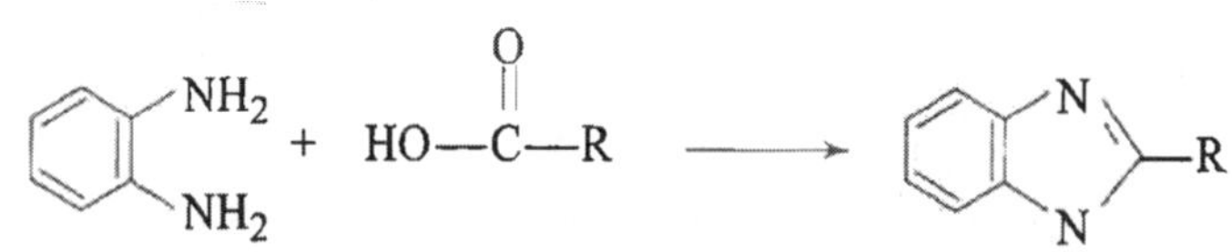

o-diaminobenzene — Benzimidazoles

54.4.1 Benzimidazole

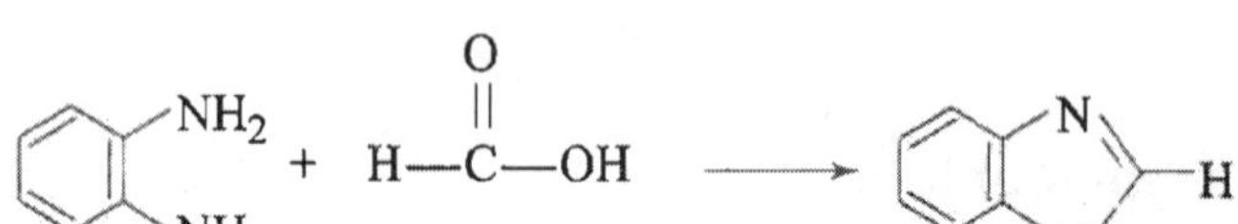

Benzimidazole

Step 1 *o*-Phenylenediamine. It is obtained by reduction of o-nitroaniline with zinc and sodium hydroxide.

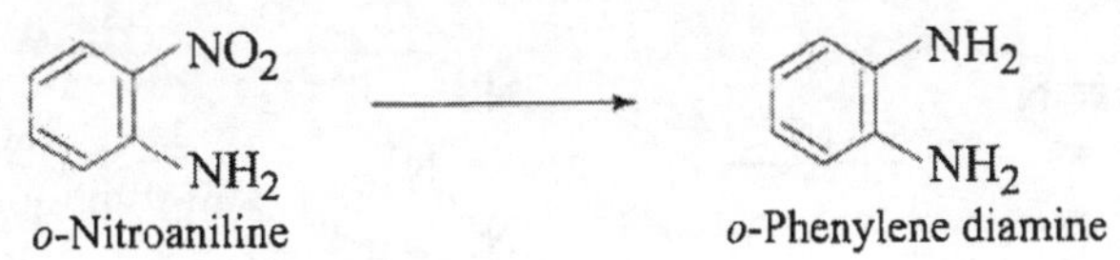

o-Nitroaniline : 23 g
Sodium hydroxide solution : 20%, 13.5 ml
Rectified spirit : 80 ml
Zinc dust : 45 g

A mixture of *o*-nitroaniline (23 g, 0.165 mole) sodium hydroxide solution (20%, 13.5 ml) and rectified spirit (80 ml) is taken in a three necked R.B. flask (500 ml capacity) fitted with a sealed stirrer and a reflux condenser. The mixture is stirred and heated to gentle boiling on a water bath. The source of heat is removed and zinc powder is added in 4-5 g portions at such a rate that the solution is kept boiling. In case the reaction stops, it is necessary to add more sodium hydroxide solution (5 ml, 20%). A total of 45 g (0.7 mole) zinc powder is added. The mixture is refluxed for one hour, the colour of the solution changes from deep red to almost colourless. The mixture is filtered while hot and the zinc residue is washed with hot rectified spirit (2 × 50 ml). To the combined filtrate is added sodium dithionite (1 g) and the solution concentrated under reduced pressure (steam bath) to 40–50 ml. The resultant solution is cooled in a freezing mixture (ice salt) and the separated product filtered, w .shed with ice water (5–10 ml) and dried in a vacuum desiccator. The crude product is crystallised from hot water containing sodium dithionite (0.5 g) (cooling in freezing mixture is necessary). The yield is 14.2 g (79%); m.p. is 100–101°C.

Step 2 Benzimidazole

o-Phenylenediamine : 13.5 g
Formic acid (90%) : 8 ml

A mixture of *o*-phenylenediamine (13.5 g, 0.125 mol) and formic acid (90%, 8 ml, 0.17 mol) is heated at 100°C (water bath) for 2 hours. The mixture is cooled and basified by addition of sodium hydroxide solution (10%). The separated benzimidazole is filtered and washed with ice-cold water. It is crystallised from boiling water (200 ml). The yield is 12.5 g, (85%); m.p. is 171–172°C.

54.4.2 2-Methylbenzimidazole

The m.p. is 176°C. This is similarly prepared from *o*-phenylenediamine and acetic acid.

54.4.3 2-Benzylbenzimidazole

This is prepared in 55 per cent yield from *o*-phenylene diamine and phenyl acetic acid. The m.p. is 190–191°C.

54.5 Benzotriazole

It is prepared by the reaction of *o*-phenylene diamine with nitrous acid.

Mechanism:

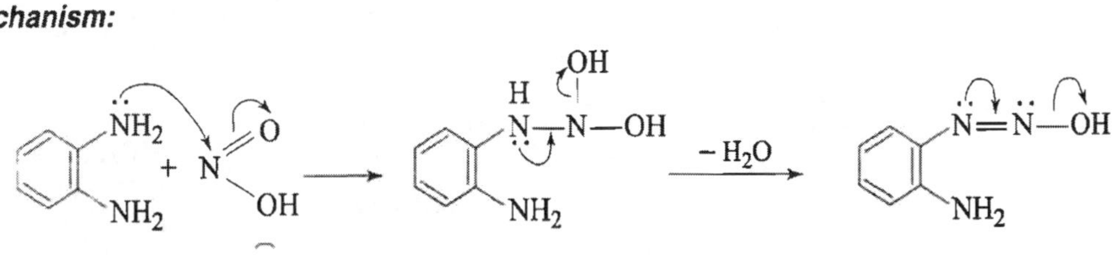

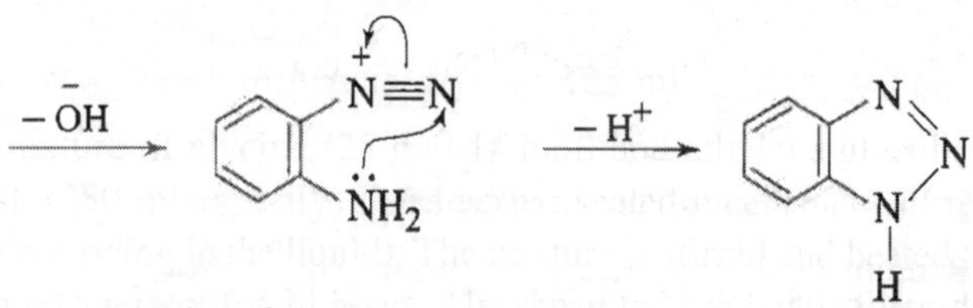

o-phenylene diamine : 5.4 g
Acetic acid (Glacial) : 5.7 ml
Sodium nitrite : 3.8 g

o-Phenylene diamine (5.48 g, 0.05 mol) is dissolved in a mixture of glacial acetic acid (6g, 5.7 ml, 0.1 mol) and water (15 ml) by slight warming if necessary. The mixture is cooled to 15°C, stirred, and a solution of sodium nitrite (3.8 g) in water (8 ml) is added in one lot. The reaction mixture becomes warm and the temperature reaches to about 80°C. The colour changes from deep red to pale brown. The stirring is continued for 15 minutes and the mixture cooled (ice-water). The separated benzotriazole is filtered, washed with ice-cold water and crystallised from hot water (60–65 ml). It may be necessary to add a crystal of benzotriazole for seeding. The yield is 4 g (67%); m.p. is 99–100°C. A pure white product can be obtained by sublimation at 90–95°C at 0.2 mm Hg.

54.6 4-Methylcarbostyril (2-Hydroxy-4-methylquinoline)

It is obtained by the cyclisation of acetoacetanilide with concentrated sulphuric acid.

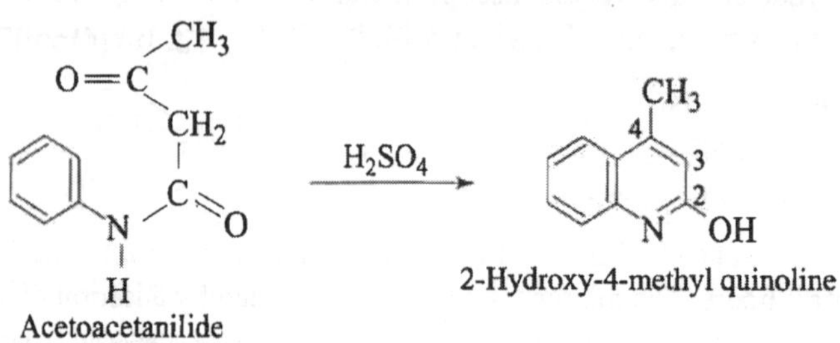

Acetoacetanilide : 8.9 g
Conc. sulphuric acid : 10 ml

Concentrated sulphuric acid (10 ml) is taken in a three necked flask (100 ml capacity) fitted with a mechanical stirred and a thermometer. The acid is heated to 70°C, the source of heat removed and acetoacetanilide (8.9 g, 0.05 mole) is added in small portions (25–30 minutes). The temperature of the reaction mixture is maintained at 70–75°C by intermittant cooling till all the acetoacitanilide has been added. The reaction mixture is heated to 90–95°C for 30 minutes, cooled, and water (250 ml) is added to it with vigorous stirring. The separated product is filtered, washed with water (3 × 50 ml) and cold methanol (2 × 15 ml). It is crystallised from methanol. The yield is 7 g (87%); m.p. is 223–224°C.

54.7 Barbituric Acid

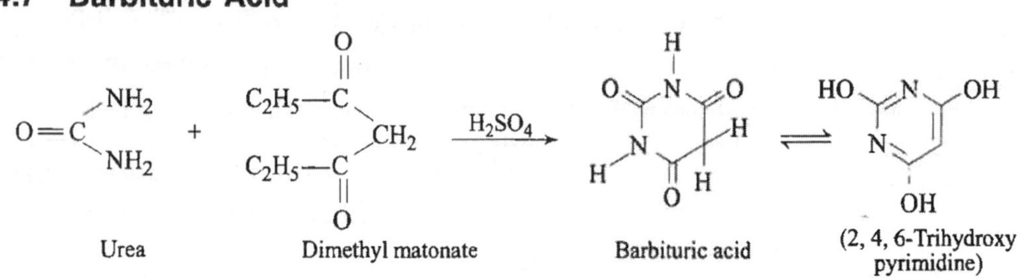

Sodium : 5.75 g
Absolute alcohol : 250 ml
Diethylmalonate : 40 g (38 ml)
Urea : 15 g

Absolute alcohol (125 ml) is added to sodium (5.75 g, 0.25 mol) contained in a R.B. flask (1 litre capacity) fitted with a reflux condenser, which is protected by a calcium chloride guard tube. In case the reaction becomes too vigorous, it is cooled for a minute (ice-cold water bath). When all the sodium has reacted, diethyl malonate (38 ml, 0.25 mol) and a solution of dry urea (15 g) in absolute alcohol (125 ml) is added. The mixture is shaken well and refluxed for 7 hours at 110°C (oil bath). To the reaction mixture (in which a white solid separates) is added hot water (225 ml, 50°C) and concentrated hydrochloric acid (about 25 ml) until the solution is acidic. The resultant solution is cooled (overnight in a refrigerator) and the separated barbituric acid is filtered and washed with cold water (15 ml). The yield is 25 g (78%); m.p. is 245°C (decom).

54.8 2,3-Diphenylquinoxalines (Quinoxalines)

Quinoxaline or benzopyrazines are obtained by the condensation of *o*-phenylene diamines and α-diketones. Thus,

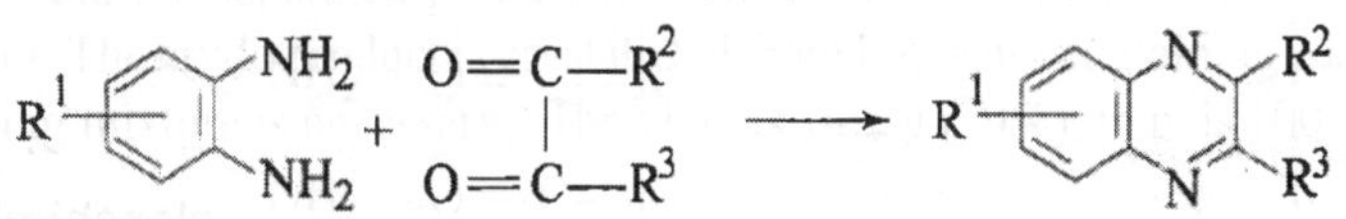

54.8.1 2,3-Diphenylquinoxaline

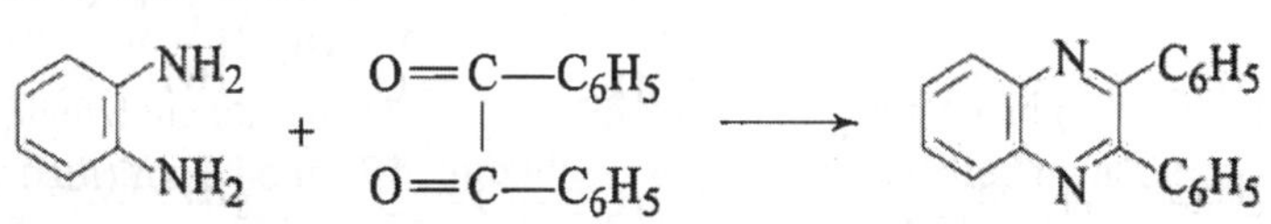

o-Phenylenediamine : 1.65 g
Benzil : 3.15 g

A solution of *o*-phenylenediamine (1.65 g, 0.015 mol) in rectified spirit (12 ml) is added to a warm solution of benzil (3.15 g, 0.015 mol) in rectified spirit (12 ml). The mixture is warmed for 30 minutes on a water bath. Water is added dropwise until slight cloudiness persists. The solution is cooled and the separated product filtered and crystallised from alcohol. The yield is 2.75 g (51%); m.p. is 125–126°C.

54.9 Piperazine-2,5-dione (Diketopiperazine)

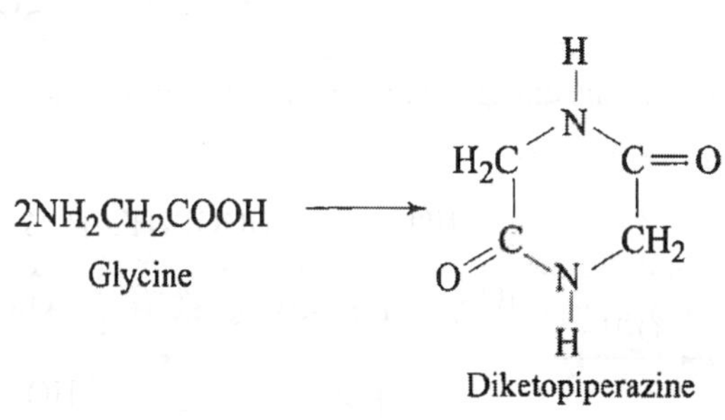

Glycine : 25 g
Ethylene glycol: 125 ml

A mixture of glycine (25 g, 0.44 mol) and ethylene glycol (125 ml) is stirred in a three necked R.B. flask (250 ml capacity) fitted with a sealed mechanical stirrer, a reflux condenser and a thermometer (bulb dipping in the liquid). The mixture is stirred and heated to 175°C for one hour. It is cooled and left in a refrigerator for 24 hours. The separated product is filtered by decantation. It is crystallised from hot water (decolourising carbon). The yield is 8.5 g (45%); m.p. is 310–312°C (decomp.)

54.10 Furans

A general procedure for the synthesis of a five-membered oxygen-containing heterocyclic system (furans) consists in the cyclodehydration of 1, 4-diketones by acidic reagents, e.g., sulphuric acid, zinc chloride etc. Thus,

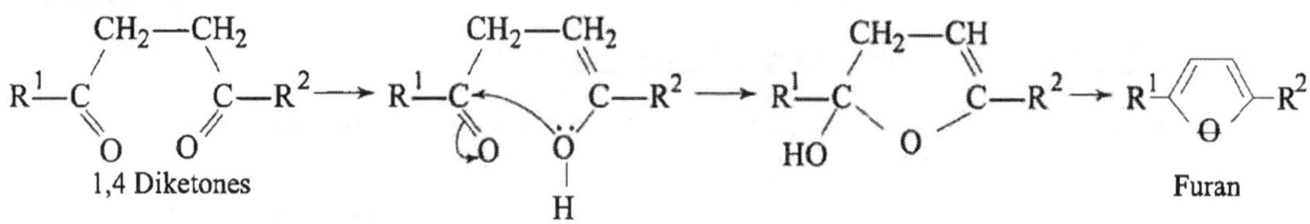

54.10.1 2, 5-Dimethylfuran

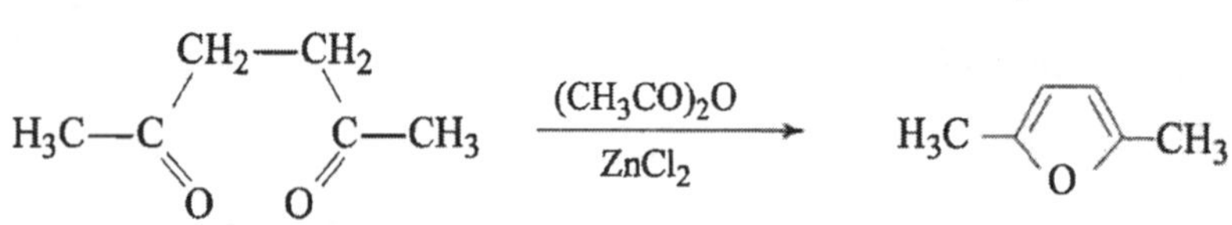

Hexane-2,5-dione : 8.55 g (8.8 ml)
Zinc chloride (fused): 0.5 g.
Acetic anhydride : 8.4 g (7.75 ml)

Anhydrous and freshly fused zinc chloride (0.5 g) is dissolved in acetic anhydride (8.4 g, 7.75 ml, 0.082 mol) in a R.B. flask (50 ml capacity) fitted with a reflux condenser. Hexane-2,5-dione (8.55 g, 8.8 ml, 0.075 mol see page 174) is added and the mixture warmed under anhydrous conditions. As soon as the vigorous reaction starts the source of heat is removed. The reaction mixture is finally refluxed gently for 2 hours. It is cooled and basified with aqueous sodium hydroxide (6 M, about 25 ml). The mixture is then steam distilled until no more oily drops appear in the distillate. It is extracted with ether (25 ml), the extract dried (anhydrous calcium chloride) and the ether evaporated. 2,5-Dimethyl furan is collected by distillation. The yield is 4.7 g (62%); b.p. is 93–95°C.

54.10.2 Benzofuran (Coumarone)

It is obtained from salicylaldehyde by conversion into the corresponding aryloxyacetic acid by reaction with sodium chloroacetate in the presence of alkali followed by cyclisation with a mixture of acetic anhydride, acetic acid and sodium acetate. The reaction proceeds by an internal Perkin's reaction followed by decarboxylative dehydration. Thus,

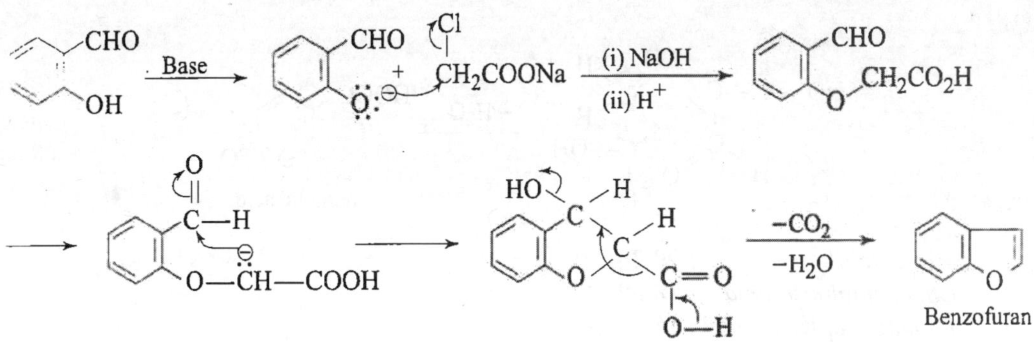

Step 1 o-Formylphenoxyaceticacid

Salicylaldehyde : 20 g (17.5 ml)
Chloroacetic acid : 15.75 g
Sodium hydroxide : 13.35 g

Sodium hydroxide solution (13.35 g, 0.33 mol) in water (350 ml) is added to a stirred mixture of salicylaldehyde (20 g, 17.5 ml, 0.165 mol) and chloroacetic acid (15.75 g, 0.165 mol). The mixture is refluxed for 3 hours, cooled and acidified with concentrated hydrochloric acid (30 ml). It is steam distilled to remove unreacted salicylaldehyde (about 12–13 g). The remaining mixture is cooled. The separated product is filtered and crystallised from water. The yield is 13.5 g (17%); m.p. is 132–133°C.

Step 2 Benzofuran

o-Formylphenoxyacetic acid : 10 g
Sodium acetate (anhyd.) : 20 g
Acetic anhydride : 50 ml
Acetic acid (glacial) : 50 ml

A mixture of *o*-formylphenoxy acetic acid (10 g, 0.05 mol), freshly fused and powdered sodium acetate (20 g), acetic anhydride (50 ml) and glacial acetic acid (50 ml) is refluxed for 8 hours. The reaction mixture is poured into cold water (300 ml) and stirred. It is extracted with ether (3 × 75 ml) and the combined ether extract washed with aqueous sodium hydroxide (5%) (until the aqueous layer is basic) and then with water. It is dried (anhydrous calcium chloride) and the ether evaporated. Benzofuran is obtained by distilling. The yield is 4.75 g (91%); b.p. is 170–172°C.

54.11 Pyrones

54.11.1 Coumalic acid

It is obtained by heating malic acid with concentrated sulphuric acid and fuming sulphuric acid. Formyl acetic acid is obtained as an intermediate by decarbonylation and dehydration of malic acid, which then undergoes self-condensation, Thus,

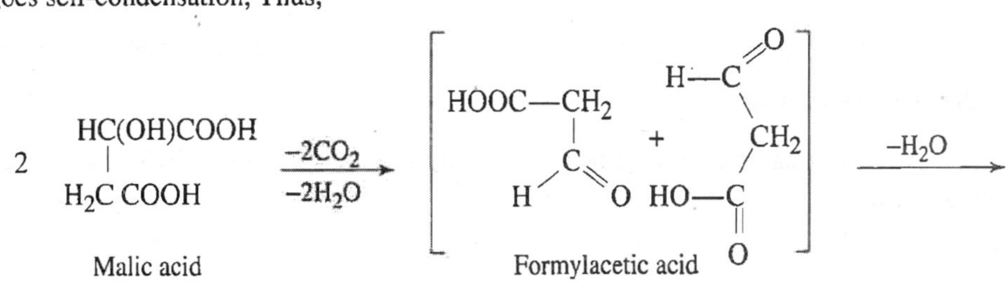

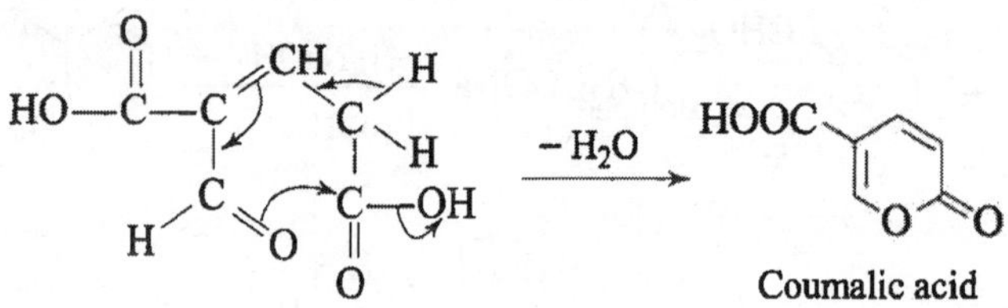

Malic acid	: 20.2 g
Conc. sulphuric acid	: 18 ml
Fuming sulphuric acid (25% SO_3)	: 15 ml

Concentrated sulphuric acid (18 ml) is added to finely powdered malic acid (20.2 g, 0.15 mol) contained in a R.B. flask (250 ml capacity). Fuming sulphuric acid (25% SO_3) is added in three lots (3 × 5 ml) at intervals of 40 minutes. An exothermic reaction with a steady evolution of gas sets in. When the reaction becomes slow, the mixture is heated for 2 hours (water bath), cooled and poured into cold water (400 ml) with stirring. The mixture is cooled in a refrigerator for 24 hours. The separated product is filtered, washed with cold water and recrystallised from methanol. The yield is 6.5 g (62%); m.p. is 206–208°C.

54.12 Benzo pyrones

54.12.1 Coumarin—See Perkins reaction and Pechmann's reaction

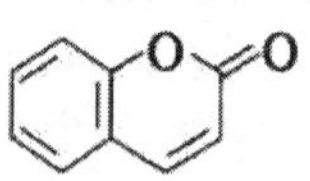

54.12.2 4-Methyl-7-hydroxy coumarin—See Pechmann's reaction

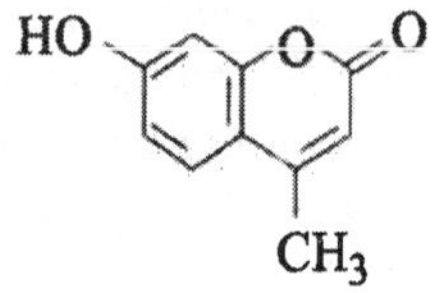

54.12.3 2-Methyl-7-hydroxychromone—See Kostanechki acylation

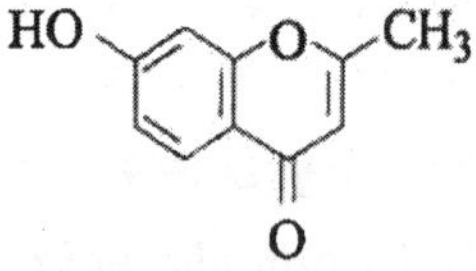

54.12.4 7-Hydroxyflavone—See Baker-Venkataraman reaction

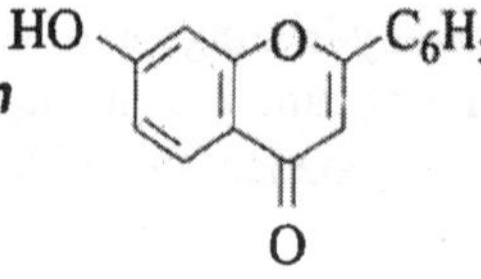

54.13 Anthraquinone

It is obtained by cyclisation of *o*-benzoyl benzoic acid by heating with concentrated sulphuric acid or polyphosphoric acid.

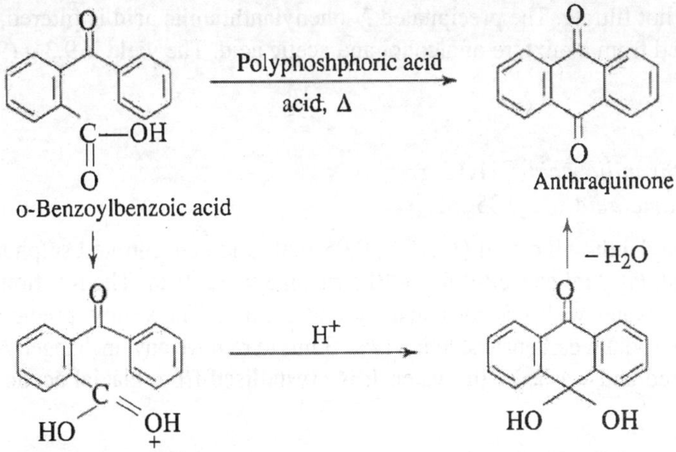

o-Benzoylbenzoic acid : 2.5 g
Polyphosphoric acid : 25 ml

A mixture of *o*-benzoylbenzoic acid (2.5 g, 0.01 mole) and polyphosphoric acid (25 ml) is heated in a boiling water bath for 2 hours (anhydrous conditions.) The reaction mixture is cooled and poured with stirring into ice-water (40 ml). The separated product is filtered, boiled with concentrated ammonia solution (10 ml) for 5 minutes to remove any unchanged acid. It is crystallised from glacial acetic acid (decolorizing carbon). The yield is 1.8 g (78.3%); m.p. is 285–286°C.

Note: Polyphosphoric acid is obtained by dissolving 18 g of phosphorus pentoxide in 10 g of commercial orthophosphoric acid (d. 1.75).

54.14 Acridone

It is obtained by the cyclisation of *N*-phenylanthranilic acid with concentrated sulphuric acid

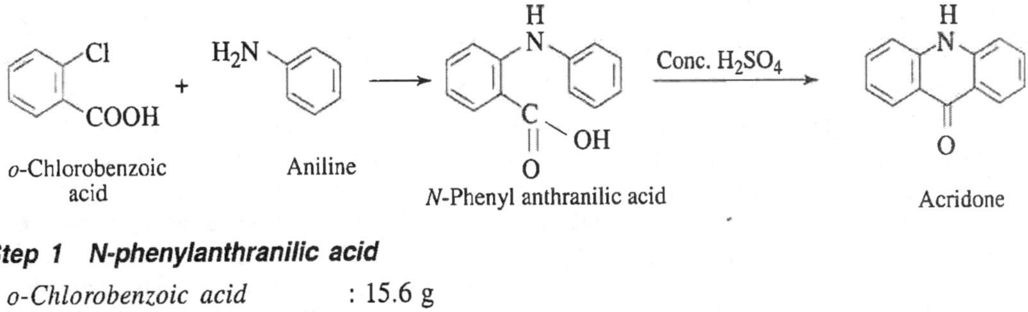

Step 1 N-phenylanthranilic acid

o-Chlorobenzoic acid : 15.6 g
Aniline : 40 ml
Potassium carbonate (anhyd.) : 10 g
Cupric oxide : 0.5 g

A mixture of *o*-chlorobenzoic acid (15.6 g, 0.1 mol), aniline (40 ml), anhydous potassium carbonate (10 g) and cupric oxide (0.5 g) is heated on a wire guaze with a free flame in a R.B. flask (100 ml capacity) fitted with a short air condenser, so that some water escapes from the top of the condenser. Gentle refluxing is done for 3 hours, the unreacted aniline removed by the steam distillation and the remaining solution filtered in hot. It is again boiled with norit and filtered. Concentrated hydrochloric acid is added to the hot filtrate. The precipitated *N*-phenylanthranilic acid is filtered, washed with cold water and crystallised from a mixture of alcohol and acetic acid. The yield is 9.3 g (90%); m.p. is 183–184°C.

Step 2 Acridone

N-phenylanthranilic acid : 11.15 g
Conc. sulphuric acid : 35 ml

A mixture of *N*-phenylanthranilic acid (11.15 g, 0.05 mol) and concentrated sulphuric acid (35 ml) is heated in a R.B. flask (100 ml capacity) for 3-4 hours on a water bath. The reaction mixture is cooled and poured into cold water with vigorous shaking and cooling. The yellow precipitate of acridone is filtered, warmed with sodium carbonate solution (2%, 3 ml) to remove any unchanged *N*-phenylanthranilic acid and again filtered and washed with water. It is crystallised from glacial acetic acid. The yield is 8 g (90%).

54.15 2,5-Dialkylthiophens

These are prepared by the reaction of 1,4-dicarbonyl compounds with phosphorus pentasulphide.

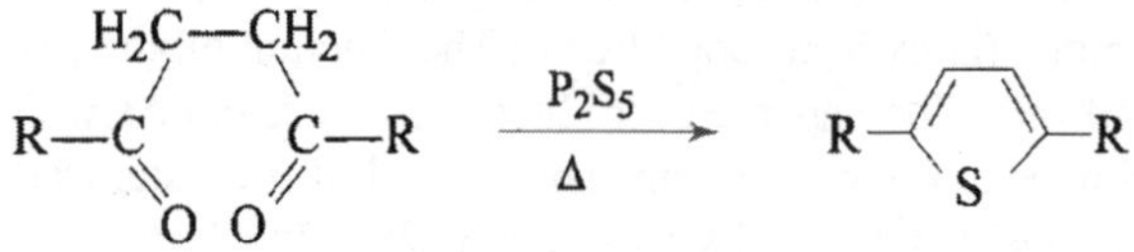

54.15.1 2,5-Dimethylthiophen

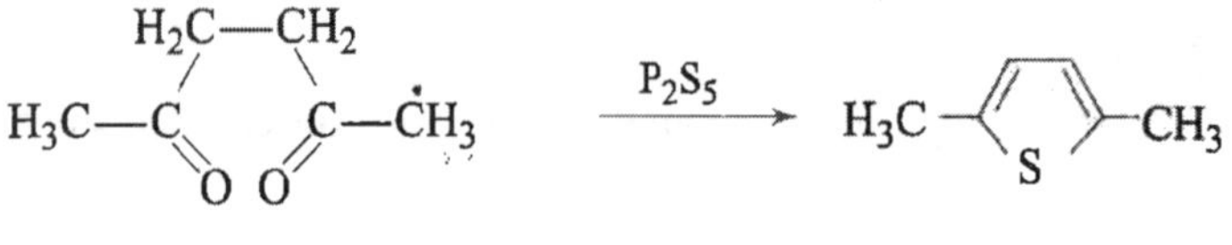

Hexane-2,5-dione (acetonyl acetone) : 29.5 ml
Phosphorus pentasulphide : 16.25 g

Phosphorus pentasulphide (16.25 g, 0.062 mol) is taken in a three necked R.B. flask fitted with a reflux condenser, a dropping funnel and a stopper. Acetonyl acetone (6 ml, see page 174) is added (dropping funnel) and the reaction mixture heated with a small luminous flame till the exothermic reaction starts. The flame is removed and acetonyl acetone (23.5 ml, total 0.25 mol) is added during 30-40 minutes so that the mixture refluxes gently. After the addition is complete and the reaction subsides, the reaction mixture is refluxed for one hour, cooled and poured into ice water (50 ml). It is extracted with ether (4 × 15 ml). The combined ether extract is washed with aqueous sodium carbonate and water, dried (magnesium sulphate) and the ether removed by distillation. 2,5-Dimethylthiophen is obtained by redistillation of the residual liquid. The yield is 13 g (46%); b.p is 135–136°C.

54.16 Benzothiophens

The benzothiophen system is obtained by the reaction of an appropriate thiophenol with α-chloroacetone under basic conditions, followed by cyclisation of the resultant ketone with polyphosphoric acid.

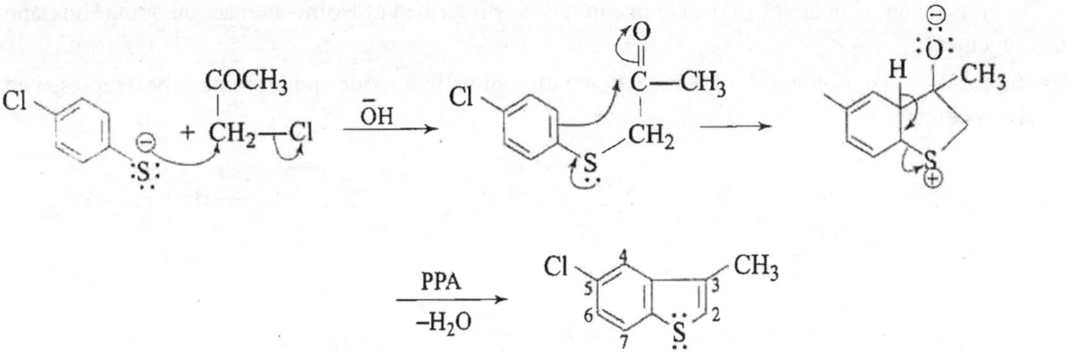

5-Chloro-3-methyl benzothiophen

54.16.1 5-Chloro-3-methylbenzothiophen

Step 1 (p-Chlorophenylthio)acetone

p-Chlorothiophenol : 10.1 g
Chloroacetone : 6.5 g
Sodium hydroxide : 2.8 g

p-Chlorothiophenol (10.1 g, 0.07 mol) is added to sodium hydroxide solution (2.8 g, 0.7 mol in 100 ml water). To the reaction mixture, Chloroacetone (6.5 g, 0.07 mol) is added with stirring. The mixture is stirred for 40 minutes at room temperature and extracted with ether. The extract is washed with water, dried over magnesium sulphate and distilled. (*p*-Chlorophenylthio) acetone is obtained by distillation of the residual product under reduced pressure. The yield is 5.75 g (82%); b.p. is 180–182°C/16 mm.

Step 2 5-Chloro-3-methylbenzothiophen

(p-Chlorophenylthio) acetone : 10 g
Polyphosphoric acid : 100 g

Polyphosphoric acid is first obtained by dissolving phosphorus pentoxide (35 g) in orthophosphoric acid (63 g). (*p*-Chlorophenylthio) acetone (10 g, 0.05 mole) is added and the mixture heated in an oil bath at 120-140°C. The reaction mixture is gently shaken during heating. The heating is continued for 2 hours and the reaction mixture poured into water (100 ml). It is extracted with ether. The ether extract is washed with water, dried over magnesium sulphate and distilled. The yield is 4.7 g (52%); b.p is 98–100°C/0.1 mm.

55. HOFMANN REACTION (HOFMANN REARRANGEMENT)[1]

Treatment of an amide with sodium hypobromite or sodium hypochlorite (or with sodium hydroxide and bromine), results in the formation of a primary amine with one carbon atom less than the starting amide.

$$R{-}\underset{\underset{O}{\|}}{C}{-}NH_2 + NaOBr \longrightarrow R{-}N{=}C{=}O \xrightarrow[-CO_2]{\text{Hydrolysis}} RNH_2$$

1. A.W. Hofmann, *Ber.*, 1881, **14**, 2725.

The conversion of an amide into an amine in this way is termed as Hofmann reaction or the Hofmann rearrangement.

Mechanism: The mechanism of the reaction with sodium hydroxide and bromine is best represented as shown below:

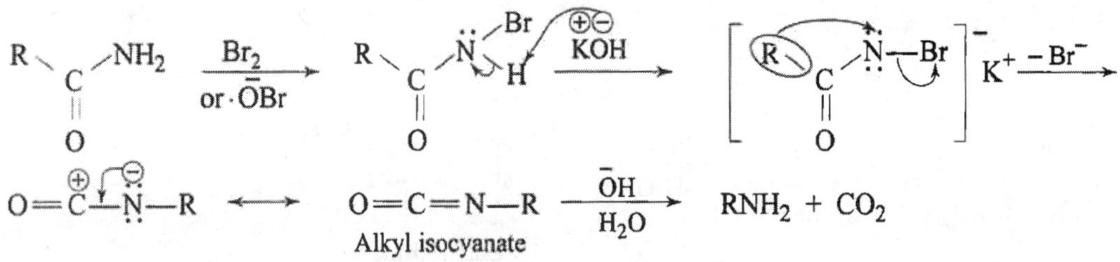

It was earlier believed that an acyl nitrene, an electronically deficient nitrogen fragment, is obtained, which rearranges to the isocyanate.

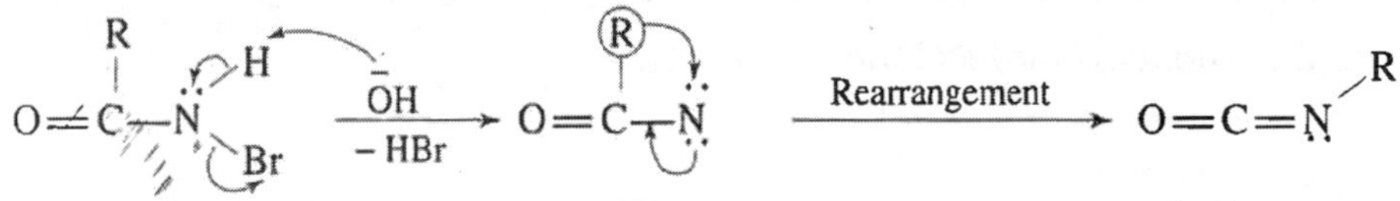

N-Bromoamide Nitrene Alkyl isocyanate

This pathway has not been established beyond doubt.

55.1 Methylamine Hydrochloride

$$C_3HCONH_2 + Br_2 + 2NaOH \longrightarrow C_3HNH_2 + 2NaBr + CO_2 + H_2O$$

Acetamide : 12.5 g
Bromine : 11.5 ml
Sodium hydroxide solution : 10%, 80 ml

Bromine (11.5 ml, 0.22 mole) is added cautiously to pure acetamide (12.5g, 0.21 mole) placed in a conical flask (150 ml capacity). The flask is shaken gently until acetamide dissolves to give a deep red solution. The mixture is cooled and sodium hydroxide solution (10%, 80 ml) is added with vigorous shaking in small portions until the solution becomes pale yellow. In case any solid separates, more water is added to get a clear solution of bromoacetamide.

Sodium hydroxide solution (30 g sodium hydroxide dissolved in 75 ml water) is taken in a three necked R.B. flask (250 ml capacity) fitted with a dropping funnel, a thermometer (the stem of the thermometer dipping below the surface of alkali solution) and a distillation arrangement. The distillate is passed into dilute hydrochloric acid (1:1, 50 ml) with the help of an inverted funnel connected to the condenser. The bromoacetamide solution obtained above, is added (dropping funnel) slowly at such a rate that the temperature does not rise above 70°C. In case the temperature rises above 75°C, the flask is cooled by the cold-water bath. When all the bromoacetamide solution has been added, the reaction mixture is maintained at 65–70°C for 10–15 minutes. A clear solution is obtained. The solution is boiled gently to drive off the methylamine vapour into dilute hydrochloric acid (40–60 minutes).

The methylamine hydrochloride solution is concentrated (wire guaze) to about 15 ml, final evaporation is done on a water bath till a solid product is obtained. It is heated with absolute ethanol (60 ml), only the methylamine hydrochloride dissolves in absolute alcohol. The clear alcohol solution is cooled. The separated methylamine hydrochloride is filtered and the product is dried in a vacuum desiccator. Extraction with absolute alcohol is repeated 3–4 times to get all the product. Total yield is 10 g (70.4%).

55.2 Anthranilic Acid

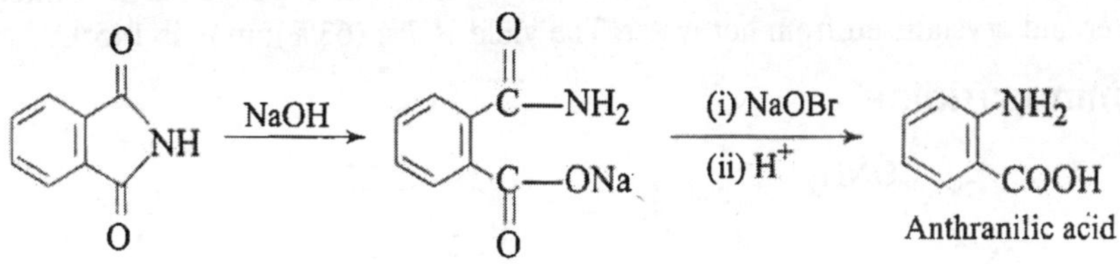

Phthalimide

Step 1 Phthalimide

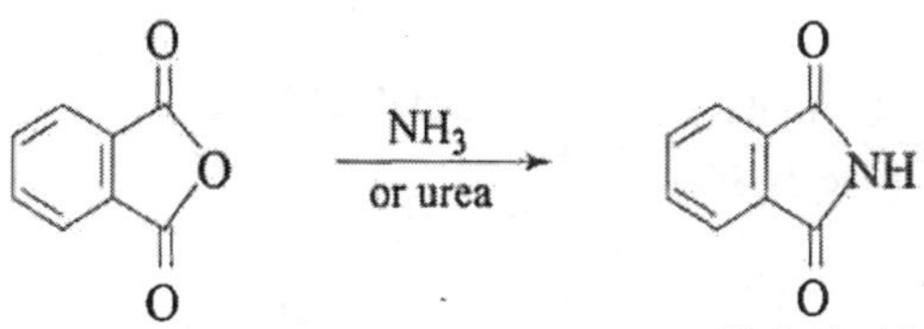

Phthalic anhydride Phthalimide

Phthalic anhydride : 25 g
Conc. ammonia (d : 0.88) : 30 ml
or Urea : 5.5 g.

A mixture of phthalic anhydride (25 g, 0.169 mole) and concentrated ammonia solution (d. 0.88, 30 ml) is heated in a R.B. flask fitted with an air condenser, initially over a wire guaze and subsequently on a free flame until the mixture becomes a homogenous melt (temperature approximately 300°C) in 1.5–2 hours. All the water is evaporated during the first hour. The mixture is poured while still hot into a porcelain dish. The phthalimide then obtained is quite pure. The yield is 23 g (92.77); m.p. is 233–234°C. It can, however, be crystallised from alcohol.

Alternatively, an intimate mixture of pure phthalic anhydride (25 g, 0.169 mole) and urea (5.5 g) is heated in a long necked R.B. flask (250 ml capacity) at 130–135°C (oil bath). The mixture melts, effervescence commences and frothing takes place (10–20 minutes). The temperature of the mixture rises to 150–160°C and becomes solid. It is cooled and water (40 ml) is added to disintegrate the solid product. It is filtered, washed with water and dried at 100°C. The yield is 22 g (88.7%); m.p. is 233–234°C.

Step 2 Anthranilic acid

Phthalimide : 12 g
Sodium hydroxide : 26 g
Bromine : 4.2 ml

Bromine (4.2 ml, 0.082 mole) is added in one lot to a cooled solution (< 0°C) of sodium hydroxide (15 g) in water (15 ml). The mixture is shaken till all the bromine has reacted. The temperature is maintained below 0°C, and finely powdered phthalimide (12 g, 0.082 mole) is added in one lot. The mixture is shaken vigorously and a cold solution of sodium hydroxide (11 g) in water (40 ml) added. The reaction mixture is shaken and the temperature rises to about 70°C. The mixture is warmed to 80°C for 5 minutes. The clear solution (filter if necessary) is cooled (ice bath) and concentrated hydrochloric acid is added slowly (about 30 ml) with stirring until the solution is just neutral. Anthanilic acid is precipitated by the gradual addition of glacial acetic acid (10–12 ml). The separated acid is filtered, washed with cold water and crystallised from hot water. The yield is 7 g (63%); m.p. is 145°C.

55.3 3-Aminopyridine

Nicotinamide 3-aminopyridine

Step 1 Ethyl nicotinate

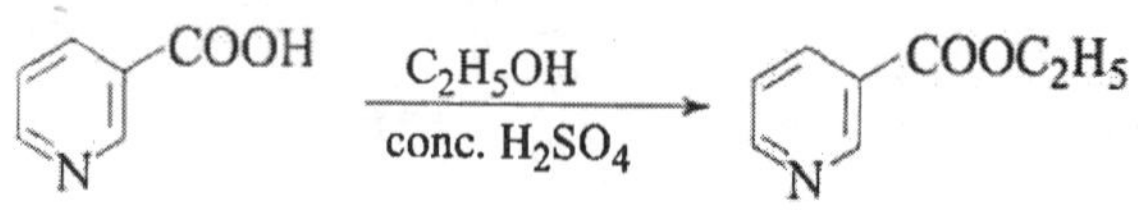

Nicotinic acid : 18.5 g
Absolute ethanol : 60 ml
Conc. sulphuric acid : 25 ml

A mixture of nicotinic acid (18.5 g, 0.15 mole), absolute ethanol (60 ml) and concentrated sulphuric acid (25 ml) is refluxed (steam bath) for 4 hours. The cooled reaction mixture is poured with stirring on to crushed ice (100 g). The solution is rendered strongly alkaline by adding a sufficient amount of ammonia solution. The mixture is extracted with ether (5 × 20 ml), the combined ether extract dried (anhydrous magnesium sulphate) and the solvent removed (water bath). The residual oily ethyl nicotinate is distilled under reduced pressure. The yield is 16 g (70.5%); b.p. is 117–118°C/16 mm; b.p. is 222–224°C/760 mm.

Step 2 Nicotinamide

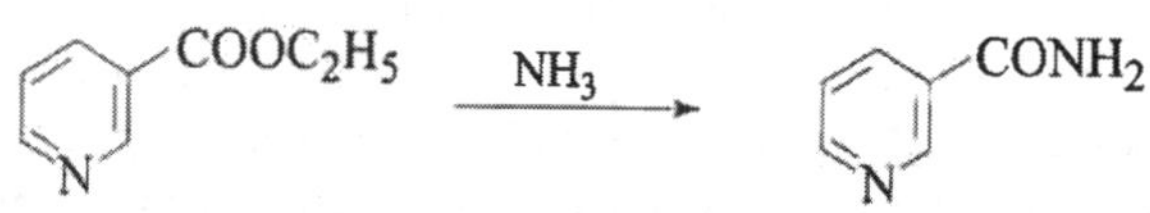

Ethyl nicotinate Nicotinamide

Ethyl nicotinate : 25 g
Conc. ammonia solution : 40 ml

A mixture of ethyl nicotinate (25 g, 0.165 mole) and cold concentrated ammonia solution (40 ml) is saturated with ammonia gas at 0°C. The reaction mixture is kept loosely stoppered for 18 hours. The mixture is shaken and again saturated with ammonia gas and left for 4 hours at room temperature. The mixture is evaporated to dryness in a dish (steam bath) and dried at 120°C. The yield is 8 g (89%); m.p. is 130°C.

Step 3 3-Aminopyridine

Nicotinamide : 15 g
Sodium hydroxide : 20 g
Bromine : 8 ml

A cold solution of sodium hypobromite is prepared from sodium hydroxide (20 g) and bromine (8 ml, 0.155 mole) in water (120 ml) (as in the case of anthranilic acid). Finely powdered nicotinamide (15 g, 0.123 mole) is added to the above solution and the mixture stirred vigorously for 15 minutes. The solution is warmed to 75°C (water bath) for 40–45 minutes. The cooled solution is saturated with common salt and extracted with ether (3×25 ml). The combined ether extract is dried (potassium hydroxide pellets) and ether removed by distillation (water bath). The 3-aminopyridine thus obtained is crystallised from benzene-petroleum ether (b.p. 60–80°C) mixture (4:1). The yield is 7.5 g (65%). m.p. is 63°C.

56. HOUBEN-HOESCH REACTION[1, 2]

Synthesis of acylphenols from phenols or phenolic ethers, by the action of organic nitrilies, in the presence of hydrogen chloride and lewis acid catalyst (eg, zinc chloride, aluminium trichloride) is known as Houben-Hoesch reaction or simply Hoesch reaction.

Mechanism: The mechanism of the reaction is not certain. However, the accepted view is that it involves the attack of the conjugate acid of nitrile on the reactive aromatic substrate to form an imine salt, which on hydrolysis gives the acyl phenol.

$$R—C\equiv N + \overset{+}{H} \xrightleftharpoons{ZnCl_2/HCl} R—\overset{+}{C}=NH\cdot HCl$$

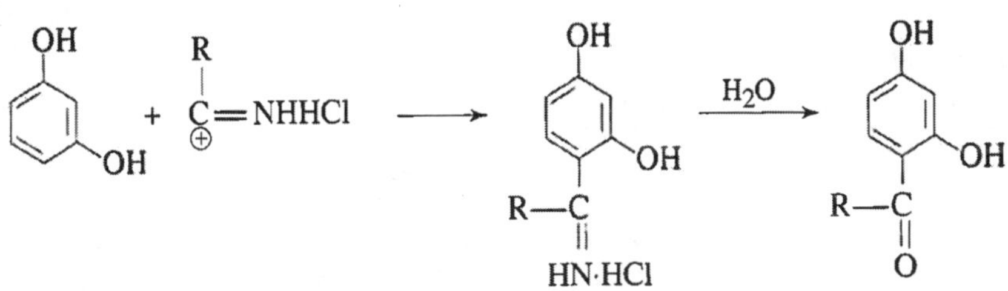

Imine salt

56.1 2,4-Dihydroxyacetophenone (β-Resacetophenone)

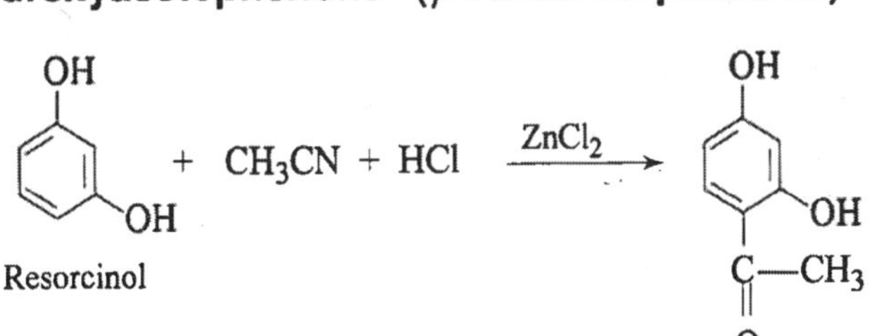

Resorcinol

β-Resacetophenone

Resorcinol (dry)	: 18 g
Acetonitrile	: 12.6 ml

1. K. Hoesch, *Ber.*, 1915, **48**, 1122.
2. J. Houben, *Ber.*, 1926, **59**, 2878.

Zinc chloride (fused) : 8.5 g
Ether (dry) : 90 ml

Dry hydrogen chloride gas is passed into a cold mixture (ice bath) of dry resorcinol (18 g, 0.163 mole), acetonitrile (12.6 ml, 0.24 mole), fused and powdered zinc chloride (8.5 g) and dry ether (90 ml) contained in a conical flask with a side arm (anhydrous conditions). After one hour the reaction mixture is allowed to stand overnight (after stoppering the flask). The ether solution is decanted and the separated imine salt washed with dry ether (2 × 25 ml). It is refluxed with water (30 ml) for about 30 minutes (water bath). The solution is cooled and the separated 2,4-dihydroxyacetophenone filtered and crystallised from boiling water. The yield is 20 g (80.4%); m.p. is 145°C.

Note: β-Resacetophenone can also be obtained by the Nencki reaction (page 150).

56.2 2,4,6-Trihydroxyacetophenone (Phloroacetophenone)

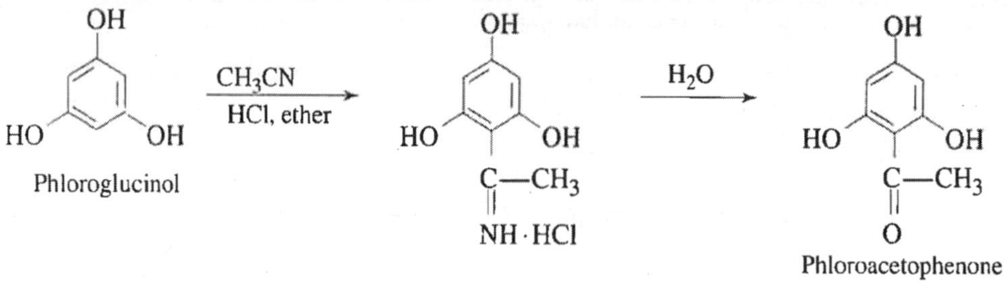

Phloroglucinol (dry) : 20 g
Acetonitrile : 15 ml
Ether (Anhyd.) : 80 ml
Zinc chloride (fused) : 4 g

Dry phloroglucinol (20 g, 0.123 mole), anhydrous ether (80 ml), freshly distilled acetonitrile (15 ml) and freshly fused and powdered zinc chloride (4 g) are taken in a Buchner flask (500 ml capacity) fitted with a wide gas inlet tube and the side arm is protected with a calcium chloride guard tube. The mixture is cooled to 0°C (ice-salt mixture) and a rapid steam of dry hydrogen chloride gas is passed through the solution for 2 hours with occasional shaking. The flask is allowed to stand in an ice-chest for 24 hours and hydrogen chloride gas passed for two more hours into the cooled solution (0°C). The flask is stoppered and left in a refrigerator for 24 hours. The ether is decanted and the solid ketimine hydrochloride is washed with dry ether (2 × 25 ml). The solid product is refluxed with water (1 litre) in a R.B. flask (free flame) for 2 hours. The solution is cooled and the separated product filtered. It is recrystallised from boiling water (100 ml). The yield is 22 g (95.6%); m.p. is 218–219°C.

57. INDICATORS

57.1 Congo Red

Benzidine : 3.7 g
Sodium nitrite : 3 g
Sodium naphthionate : 16 g
Sodium acetate : 16 g

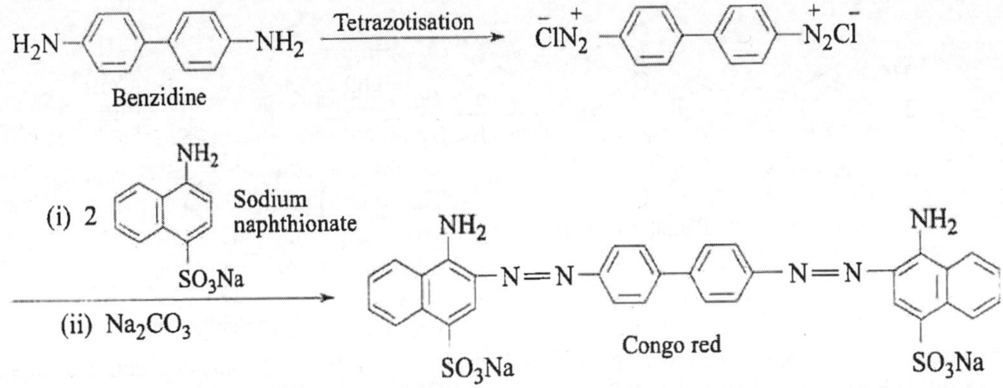

Benzidine (3.7 g, 0.02 mole) is dissolved in hot dilute hydrochloric acid (10 ml concentrated hydrochloric acid and 80 ml water). The solution is cooled to 0–5°C (ice bath) and diazotized (tetrazotized) by the addition of sodium nitrite solution (3 g in 20 ml water) within one minute. The solution is left for 5 minutes and then added to a solution of sodium napthionate (16 g, 0.065 mole) and sodium acetate (16 g) in water (200 ml). The mixture is stirred well and left for one hour. The coupling is complete when a test solution on warming with hydrochloric acid does not evolve nitrogen. The blue-black precipitate of the dye is dissolved by the gradual addition of sodium carbonate (about 8–10 g), followed by stirring and warming to 90°C. The hot solution is filtered and the filtrate just saturated with sodium chloride and allowed to cool to room temperature and then cooled in an ice bath. The separated congo red is filtered and washed with a saturated solution of sodium chloride and dried in air. The yield is 10 g (71.4%).

In presence of dilute mineral acids congo red gives blue colour. This blue colour may be attributed to resonance among the following charged canonical structures.

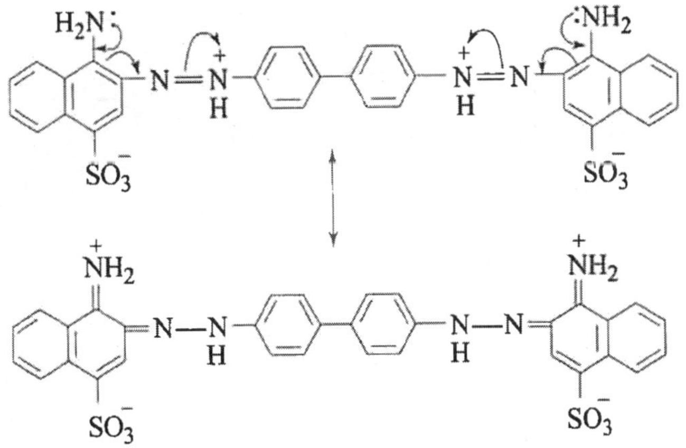

57.2 Methyl Orange

See coupling reaction. Diazo compounds (p. 66)

57.3 Phenolphthalein

Phenol : 18.8 g
Phthalic anhydride : 14.8 g
Conc. sulphuric acid: 8 ml

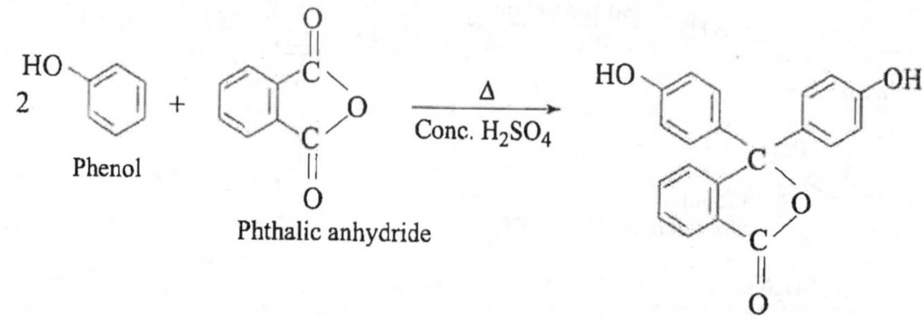

Phenol

Phthalic anhydride

Phenolphthalein

A mixture of phenol (18.8 g, 0.2 mole), phthalic anhydride (14.8 g, 0.1 mole) and concentrated sulphuric acid (8 ml) is heated in a R.B. flask (250 ml capacity) for 4 hours at 120°C (oil bath). The reddish oily product is poured into water (500 ml). Excess phenol is removed by steam distillation. The residual product is filtered and washed with water. It is dissolved in aqueous sodium hydroxide solution (10%) and the clear solution acidified with acetic acid and a few drops of hydrochloric acid. After allowing to stand overnight, the separated product is filtered and dried between folds of a filter paper.

The crude product is purified by refluxing with excess alcohol (5 ml absolute alcohol for each gm of the substance) and decolourising charcoal (0.5 g). The clear filtrate is concentrated to half the bulk. It is diluted with water (8 times the volume of alcohol solution), filtered and the solution evaporated on a water bath till phenolphthalein crystallise. It is obtained as a grannular white powder. The yield is 6 g (19%); m.p. is 261°C.

In an acidic medium, phenolphthalein (I) is colourless. In presence of dilute alkali, the lactone ring of phenolphthalein opens up to yield a triphenylcarbinol structure (II). This undergoes typical loss of water to produce a red resonating ion (III). However, if phenolphthalein is treated with concentrated alcoholic alkali, the red colour first produced disappears due to the formation of (IV). In which the quinonoid structure is destroyed. Thus,

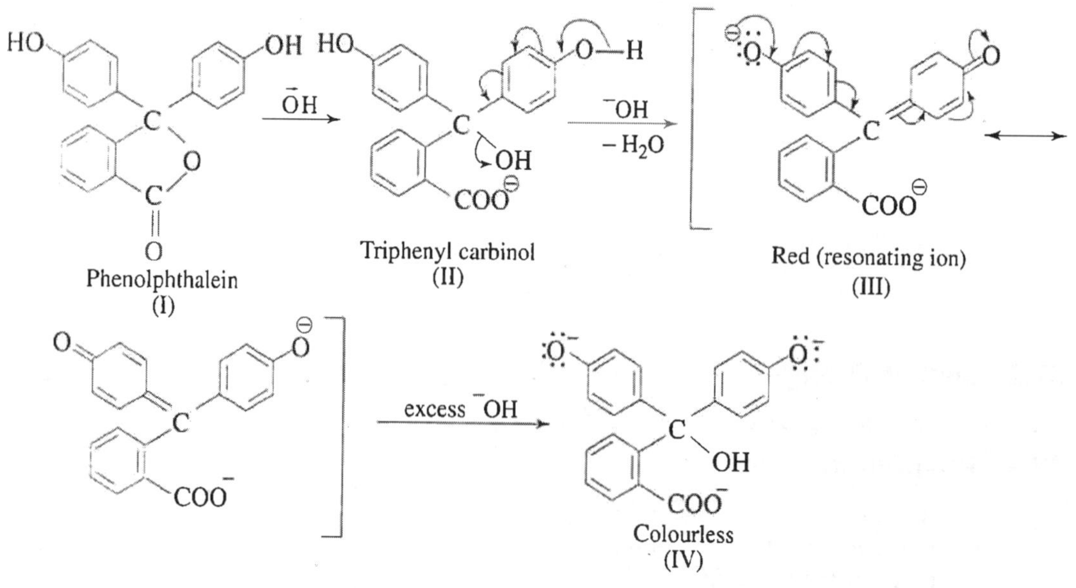

Phenolphthalein
(I)

Triphenyl carbinol
(II)

Red (resonating ion)
(III)

Colourless
(IV)

58. JONES OXIDATION[1]

The oxidation of primary and secondary alcohols to aldehydes and ketones, by the addition of calculated amount of chromic anhydride in dilute sulphuric acid solution to a solution of the alcohol in acetone, is known as Jones oxidation. This procedure does not attack triple bonds and also doesn't shift double bonds into conjugation with the ketone formed in the oxidation.

See also Sarett oxidation (p 206)

58.1 Cinnamaldehyde from Cinnamyl Alcohol

Chromium trioxide : 1 g
Conc. sulphuric acid : 0.9 ml
Cinnamyl alcohol : 1 g
Acetone : 10 ml

A solution of chromium trixiode (1 g) in water (10 ml) and concentrated sulphuric acid (0.9 ml) is added during 2 hours to a solution of cinnamyl alcohol (1 g, 0.007 mole) in acetone (10 ml) at 5–10°C with stirring. The stirring is continued for additional 30 minutes. Then the mixture is diluted with water and extracted with ether. The ether extract is dried over magnesium sulphate and concentrated. Cinnamaldehyde is collected on distillation at 248°C. The yield is 0.68 g (70%).

59. KNOEVENAGEL CONDENSATION[2, 3]

The condensation or aldehydes or ketones with active methylene compounds (especially malonic ester) in presence of very weak bases like ammonia or amines is known as Knoevenagel condensation.

When Knoevenagel condensation takes place with malonic ester/acid or cyano acetic acid in the presence of pyridine as a base, decarboxylation usually occurs. This is known as Doebner modification.

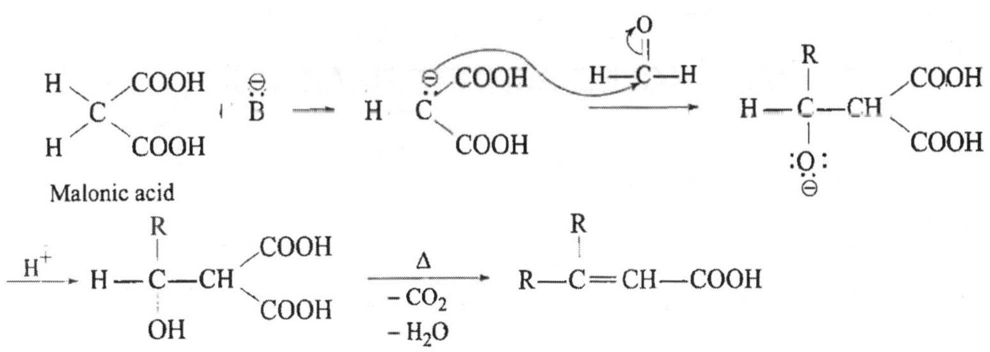

59.1 Cinnamic Acid

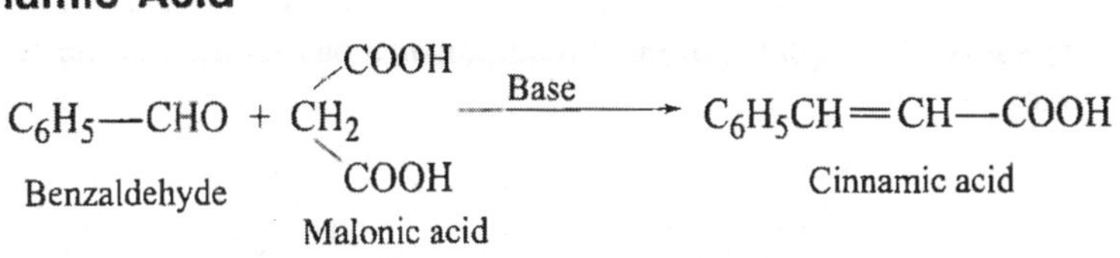

1. K. Bowden, I. M. Heilbron, E.R.H. Jones and B.C.L. Weodon, *J. Chem. Soc.*, 1946, 39.
2. E. Knoevenagel, *Ber.*, 1898, **31**, 2596.
3. O. Doebner, *Ber.*, 1900, **33**, 2140.

Benzaldehyde : 10.6 g
Malonic acid : 10.4 g
Ethanolic ammonia solution (8%) : 50 ml

A mixture of benzaldehyde (10.6 g, 0.01 mol), malonic acid (10.4 g, 0.01 mol) and ethanolic ammonia solution (8%, 50 ml) is heated on a water bath till a clear solution is obtained. Excess of alcohol is distilled off and the residue is heated at 150°C until carbon dioxide ceases to evolve. The residual product is dissolved in water and the solution acidified with hydrochloric acid. The solution is cooled and the separated cinnamic acid is filtered, washed with cold water and crystallised from hot water. The yield is 11.1 g (75%); m.p. is 134°C.

59.2 3,4-Methylenedioxycinnamic Acid (Doebner modification)

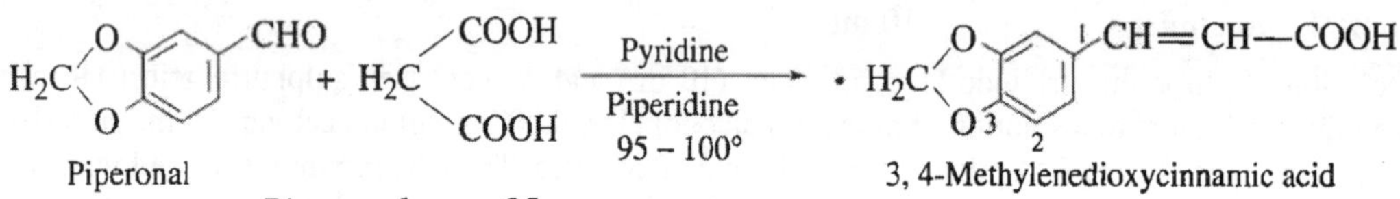

Piperonal : 25 g
Malonic acid : 37.5 g
Pyridine : 75 ml
Piperidine : 1.5 ml

A mixture of piperonal (25 g, 0.165 mol), malonic acid (37.5 g, 0.36 mol), dry pyridine (75 ml) and piperidine (1.5 ml) is refluxed for one hour. A rapid evolution of carbon dioxide takes place. The cooled reaction mixture is poured into water containing enough hydrochloric acid to dissolve pyridine. The precipitated acid is filtered and washed with water crystallised from acetic acid. The yield is 32 g (98%); m.p. is 238°C.

60. KOLBE-SCHMITT REACTION

Formation of aromatic hydroxy acids by carboxylation of phenolates, mostly in the ortho position, by carbon dioxide is known as Kolbe-Schmitt reaction.

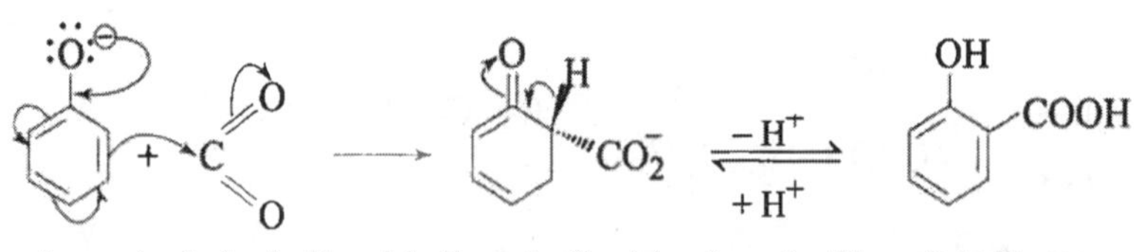

This reaction is particularly facile with di- and trihydric phenols. Thus, 2,4-dihydroxybenzoic acid is obtained readily by passing carbon dioxide through a boiling aqueous solution of the potassium salt or sodium salt of resorcinol.

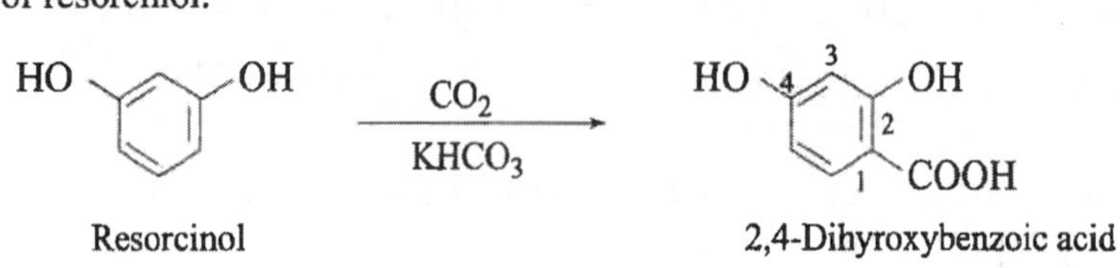

60.1 Salicylic Acid

Sodium	: 5 g
Ethanol (absolute)	: 60 ml
Phenol	: 20 g

Clean sodium (5 g, 0.217 mol), cut into small pieces is added to absolute alcohol in a three necked R.B. flask fitted with a reflux condenser (anhydrous conditions). In case the reaction becomes too vigorous it is cooled by using a cold water bath. When all the sodium has reacted, pure phenol (20 g, 0.212 mol) is added and most of the alcohol distilled off (water bath). The last traces of alcohol are removed on a water bath under reduced pressure. The solid sodium salt of phenol is powdered in the flask only with the help of a glass rod. The flask is heated to 110°C (oil bath) and a current of dry carbon dioxide is passed into the flask for one hour (the inlet tube for carbon dioxide should be almost touching the bottom). The temperature is gradually raised to 190°C and finally maintained at 200°C for 1.5 hours. Carbon dioxide is passed during heating and the contents of the flask stirred frequently with a glass rod in order to expose a fresh portion of the solid to the gas. The flask is cooled and water (50 ml) added. The clear solution is acidified with hydrochloric acid and the mixture cooled in an ice bath. The separated salicylic acid is filtered and crystallised from hot water (decolourising carbon). The yield is 8 g (27%); m.p. is 158–159°C.

60.2 2,4-Dihydroxybenzoic acid (β-Resorcyclic acid)

Resorcinol	: 30 g
Potassium bicarbonate	: 150 g

A mixture of resorcinol (30 g, 0.273 mol), potassium bicarbonate (150 g) and water (300 ml) is heated on a steam bath in a three nacked R.B. flask (500 ml capacity) fitted with a reflux condenser and a gas inlet tube. The mixture is heated gently on a steam bath for 3.5 hours and then refluxed for 30 minutes while passing a rapid stream of carbon dioxide gas through the solution. The solution is acidified while hot by adding concentrated hydrochloric acid (90 ml) from a separatory funnel with a long tube delivering the acid to the bottom of the flask. The flask is allowed to cool to room temperature and then cooled in an ice bath. The separated β-resorcylic acid is filtered and crystallised from hot water (decolourising carbon). The yield is 18 g (64%); m.p. is 216–217°C.

61. KOSTANECKI-ROBINSON ACYLATION[1]

o-Hydroxyacetophenone on heating with sodium acetate and acetic anhydride gives the corresponding 2-methyl-3-acyl chromones; these on hydrolysis with aqueous alkali undergo deacylation to afford the corresponding 2-methyl chromones.

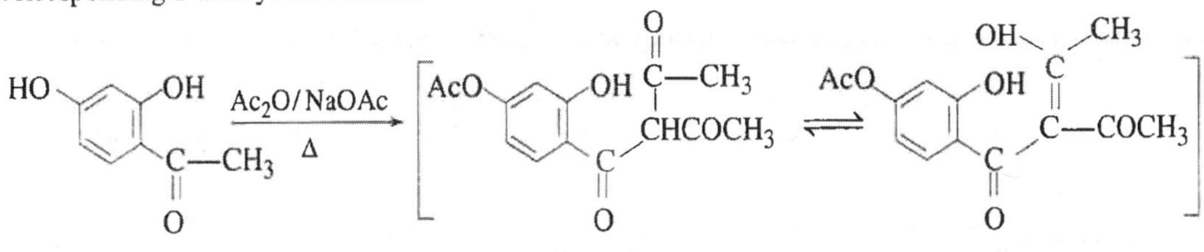

1. S. Von Kostanecki and A. Rozycki, *Ber.*, 1901, 34, 102.

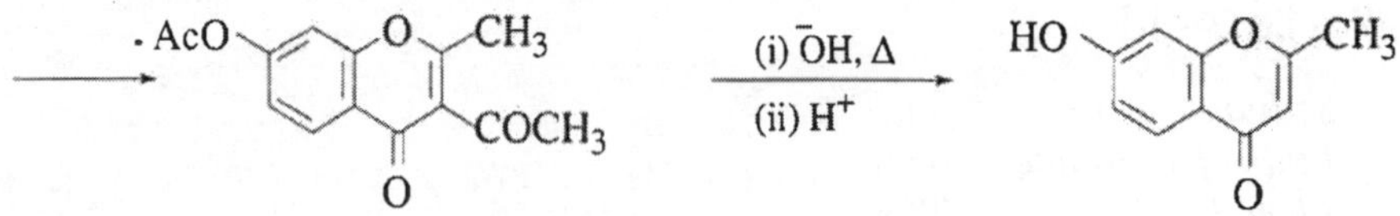

61.1 7-Hydroxy-2-methylchromone

Step 1 7-Acetoxy-3-acetyl-2-methyl chromone

β-Resacetophenone : 10 g
Fused sodium acetate: 10 g
Acetic anhydride : 20 ml

β-Resactophenone (page 151) (10 g, 0.065 mol) is heated with fused sodium acetate (10 g) in acetic anhydride (20 ml) for 3 hours at 150–160°C (oil bath) under anhydrous conditions. The reaction mixture is poured over crushed ice, stirred and left overnight. The separated product is filtered and washed with water. It is crystallised from alcohol. The yield is 10 g (58%); m.p. is 127°C.

Step 2 7-Hydroxy-2-methylchromone

The above 7-acetoxy-3-acetyl-2-methylchormone (5 g) is refluxed (2 hours) with aqueous sodium carbonate solution (10%, 250 ml). The clear solution is then acidified with hydrochloric acid (1:1). The separated product is filtered, washed with water and crystallised from alcohol. The yield is 2.5 g (73.5%); m.p. is 249–250°C.

62. MANNICH REACTION[1]

It consists in the condensation of the active methyl group (eg, in acetophenone) with formaldehyde and dimethylamine (in the form of its hydrochloride). Thus,

$$\underset{\text{Acetophenone}}{C_6H_5-\overset{\overset{\displaystyle O}{\|}}{C}-CH_3} + H-\overset{\overset{\displaystyle O}{\|}}{C}-H + (CH_3)_2\overset{+}{N}H_2Cl^- \longrightarrow \underset{\beta\text{-Dimethyl aminopropiophenone hydrochloride}}{C_6H_5-\overset{\overset{\displaystyle O}{\|}}{C}-CH_2-CH_2-N(CH_3)_2HCl} + H_2O$$

Mechanism: It is believed that the reaction may involve the formation of the intermediate, methylamine, which in presence of acids is converted into a reactive methylene-ammonium salt. The latter condenses either with the enol form of the ketone or with the carbanion derived from the ketone. Thus,

$$(CH_3)_2\overset{+}{N}H_2\overset{-}{Cl} + H_2C{=}O \rightleftharpoons (CH_3)_2\ddot{N}H + H_2C{=}\overset{+}{O}H + \overset{-}{Cl}$$

$$(CH_3)_2\ddot{N}H + H_2C{=}\overset{+}{O}H \rightleftharpoons (CH_3)_2\overset{\oplus}{\underset{H}{N}}-CH_2-OH \xrightarrow{-H_2O} (CH_3)_2\overset{+}{N}{=}CH_2$$

1. C. Mannich and W. Krosche, *Arch. Pharm.*, 1912, **250**, 647.

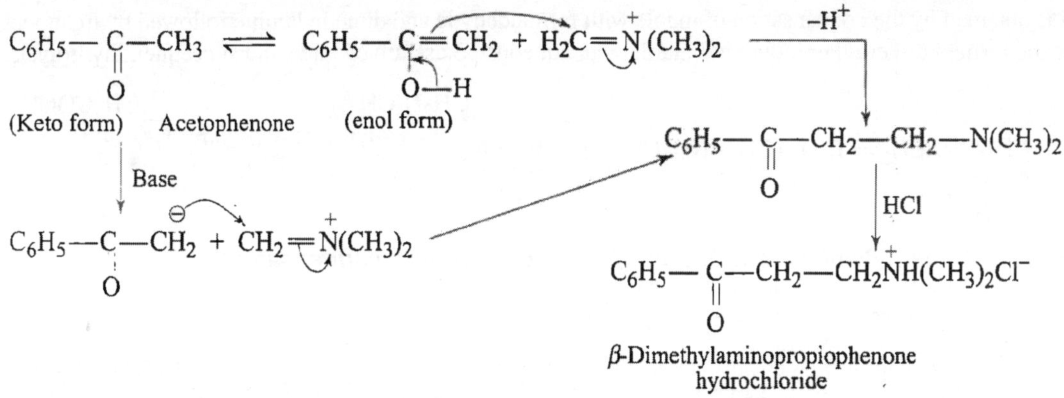

The Mannich base (quarternary salt) produced by the mannich reaction are of synthetic utility. Thus, on heating, the mannich bases eliminate an amine hydrochloride to yield α, β-unsaturated ketone. Thus,

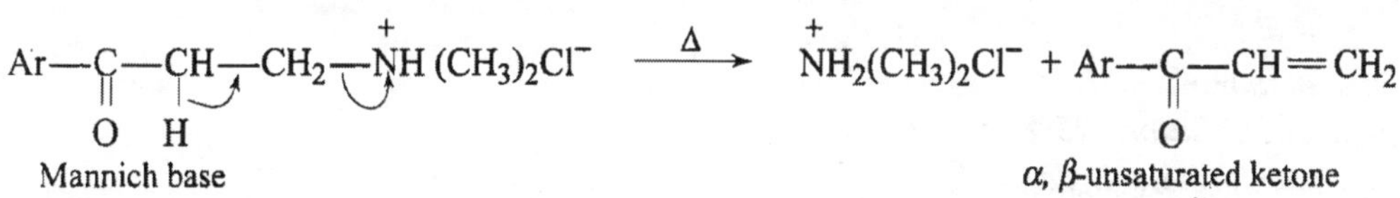

This procedure is used for in situ preparation of α, β-unsaturated carbonyl compounds.

Another usefulness of mannich bases is the replacement of the dimethylamino group by the nitrile group on heating with aqueous potassium cyanide.

$$Ar{-}\underset{\underset{O}{\|}}{C}{-}CH_2CH_2{-}\overset{+}{N}H(CH_3)_2\ Cl^- \xrightarrow[\Delta]{KCN} Ar{-}\underset{\underset{O}{\|}}{C}CH_2CH_2CN$$

62.1 Dimethylaminopropiophenone Hydrochloride

$$C_6H_5\underset{\underset{O}{\|}}{C}CH_3 + CH_2O + (CH_3)_2\overset{+}{N}H_2Cl^- \longrightarrow C_6H_5\underset{\underset{O}{\|}}{C}{-}CH_2{-}CH_2{-}\overset{+}{N}H(CH_3)_2Cl^- + H_2O$$

Dimethylamine hydrochloride	: 13.25 g
Paraformaldehyde	: 5 g
Acetophenone	: 15 g(14.65 ml)

A mixture of dimethylamine hydrochloride (13.25 g, 0.163 mol), paraformaldehyde (5 g, 0.165 mol), acetophenone (15 g, 14.65 ml, 0.125 mol), ethanol (20 ml) and concentrated hydrochloride acid (0.25 ml) is refluxed on a water bath for 2 hours. The solution is filtered. To the clear solution is added acetone (100 ml) and the mixture allowed to cool to room temperature and left overnight in a refrigerator. The separated product is filtered, washed with acetone (5 ml) and dried at 40–50°C. The yield is 19 g (71%); m.p. is 152–154°C. It can be purified by dissolving in hot rectified spirit (25 ml) and slowly adding to acetone (115 ml). The separated product is filtered. The m.p. is 155–156°C.

62.2 3-Indolylacetic Acid

It is obtained by the condensation of indole with formaldehyde and dimethylamine followed by treatment of the formed dimethylaminomethyl indole with aqueous potassium cyanide and subsequent hydrolysis.

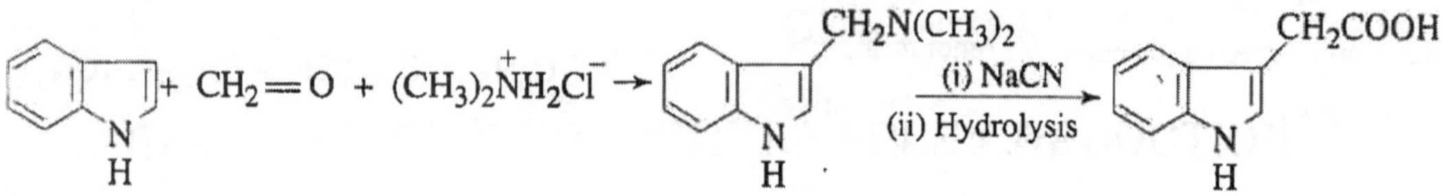

Step 1 Dimethylaminomethylindole (Gramine) (Mannich condensation)

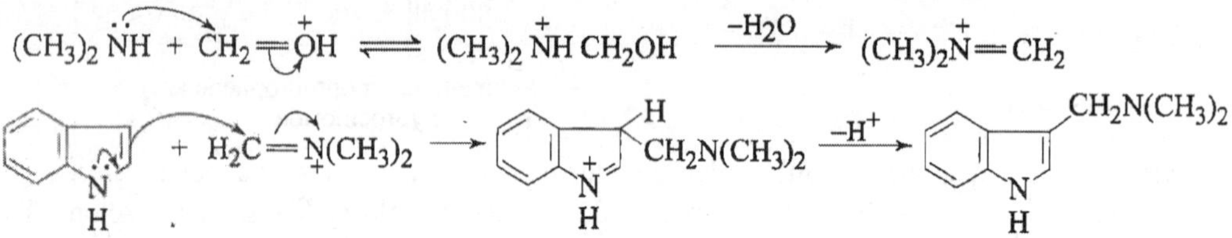

Indole	: 11.7 g
Dimethylamine	
(aqueous solution) 25% v/v	: 21.25 ml
Aqueous formaldehyde	
(formalin) 37%	: 8.6 g
Acetic acid	: 15 ml

To the cooled aqueous dimethylamine solution (25%, v/v, 21.25 ml, 0.118 mol) is added cold acetic acid (15 ml). The mixture is cooled (ice bath) and cold aqueous formaldehyde solution (37%, 8.6 g, 0.105 mol) is added. The solution is poured on indole (11.7 g, 0.1 mol). The mixture is allowed to warm to room temperature with occasional shaking. A clear solution is obtained. It is kept overnight at 30–40°C and then poured with stirring into a solution of potassium hydroxide (20 g) in water (150 ml). The solution is cooled (ice bath). The separated product is filtered, washed with cold water and dried. The yield is 17 g (47.5%). It is crystallised from acetone-hexane. The m.p. is 133–134°C.

Step 2 3-Indolylacetic acid

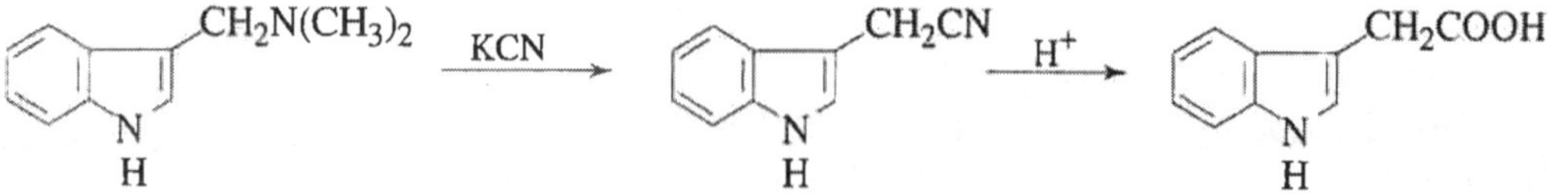

Dimethylaminomethylindole	: 12.5 g
(gramine)	
Sodium cyanide	: 17.6 g
Ethanol (95%)	: 150 ml

To a solution of dimethylaminomethylindole (12.5 g, 0.072 mol) in ethanol (150 ml) is added a solution of sodium cyanide (17.6 g) in water (35 ml). The mixture is refluxed for 80 hours. It is cooled, diluted with water (150 ml), shaken with activated charcoal and filtered. The filtrate is concentrated under reduced pressure to about 150 ml to remove most of the alcohol. The solution is cooled (5°C), the separated product (3-indolylacetamide, m.p. 150–151°C, 2 g) is filtered. To the filtrate is added dropwise concentrated hydrochloric acid with vigorous stirring (Caution—use fume cupboard as hydrogen cyanide is evolved) to pH 1-2 (about 25 ml). The precipitated 3-indolyl acetic acid is filtered and crystallised from 1, 2-dichloroethane containing a small amount of alcohol. The yield is 8.75 g (70%); m.p. is 167–168°C.

63. MADELUNG INDOLE SYNTHESIS[1]

It consists in the formation of indole derivative by intramolecular cyclisation of an *N*-acyl-*o*-toluidine with a strong bases at high temperatures.

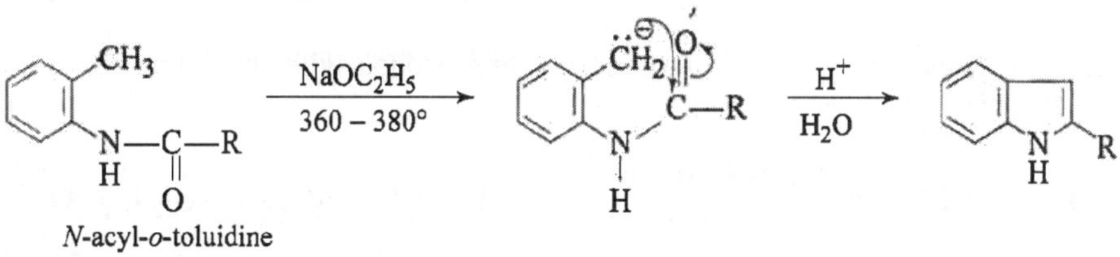

63.1 Indole

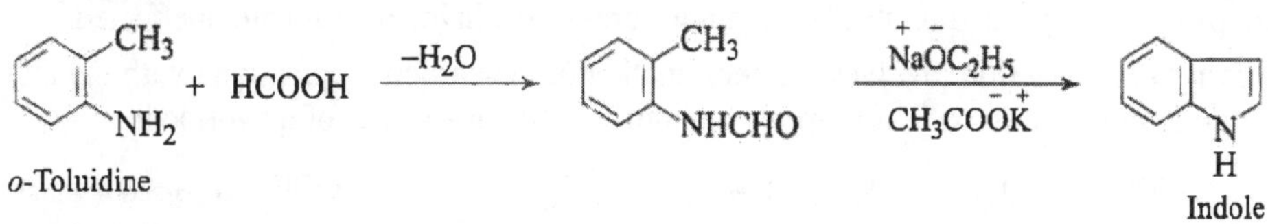

Step 1 N-formyl-o-toluidine

o-Toluidine	: 20.6 ml
Formic acid (90% w/w)	: 10.5 g

A mixture of *o*-toluidine (20.6 ml, 0.2 mole) and formic acid (90%, 10.5 g, 8.75 ml, 0.2 mol) is heated in a R.B. flask fitted with a reflux condenser. The reaction mixture is distilled under reduced pressure. *N*-formyl-*o*-toluidine distils at 173–175°C (25 mm). It solidifies on cooling. The yield is 26 g, (80%); m.p. is 57–59°C. It can be crystallised from a mixture of benzene-petroleum ether, m.p. is 61°C.

Step 2 Indole

Sodium	: 2.85 g
Methanol (Anhyd.)	: 70 ml
Potassium acetate (fused)	: 25 g
N-formyl-o-toluidine	: 17 g

Sodium (2.85 g, 0.125 mol) is added to anhydrous methanol (70 ml) in a R.B. flask fitted with a reflux condenser and calcium chloride guard tube (anhydrous condition). When the sodium has dissolved, *N*-formyl-*o*-toluidine (17 g, 0.125 mol) is added followed by the addition of freshly fused and powdered potassium acetate (25 g, 0.255 mol). The mixture is refluxed till a clear solution is obtained. Excess methanol is removed under reduced pressure (rotary evaporator). The reaction mixture is heated at 300–350°C using a bath of molten wood's metal. The *o*-toluidine is allowed to distil away. The heating is continued (30 minutes) till no more *o*-toluidine distils. The last traces of *o*-toluidine are removed by applying partial vacuum. The reaction mixture is cooled, water (50 ml) added and steam distilled.

1. M. MADELUNG, *Ber.*, 1912, **45**, 1128.

Colourless plates of indole separate out on cooling the distillate. The yield is 2.5 g, (17%); m.p. is 48–50°C. It can be crystallised from petroleum ether, m.p. is 52°C.

Note: Fused potassium acetate is obtained by heating it in a procelain dish until it melts. It is allowed to stay in the molten state for 2–3 minutes and stirred. It is allowed to cool in a desiccator and finally powdered.

64. MEERWEIN-PONNDORF-VERLEY REDUCTION[1-3]

The reduction of aldehydes or ketones to the corresponding alcohols with aluminium alkoxides is known as Meerwein-Ponndorf-Verley reduction.

$$R{-}\underset{\underset{O}{\|}}{C}{-}R' \xrightarrow[(CH_3)_2CHOH]{Al\,[OCH(CH_3)_2]_3} R{-}\underset{\underset{OH}{|}}{CH}{-}R' + (CH_3)_2\,C{=}O$$

It is a mild method of reducing carbonyl compounds in good yield and is especially valuable since other groups eg, a conjugated double bond, a nitro group or a halogen atom are uneffected.

Mechanism: The reaction involves the coordination of the carbonyl compound with the aluminium atom in aluminium isopropoxide followed by an intramolecular transfer of a hydride ion.

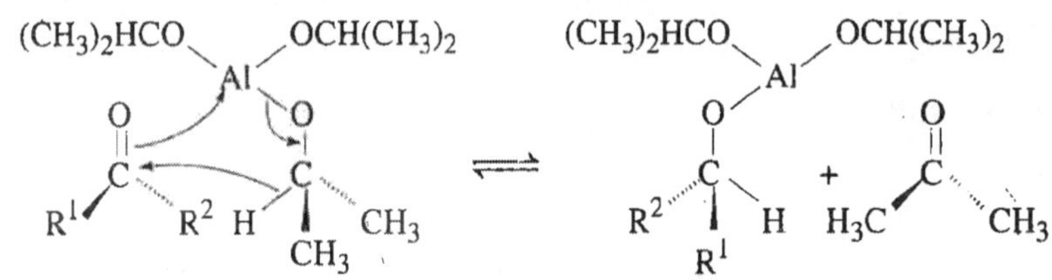

$$[(CH_3)_2CHO]_2AlOCHR^1R^2 \xrightarrow{2R^1{-}\overset{\overset{O}{\|}}{C}{-}R^2} [R^1R^2CHO]_3Al + 2(CH_3)_2C{=}O$$

$$[R^1R^2CHO]_3Al \xrightarrow{H_3O^+} 3R^1R^2CHOH + Al^{3+}$$

64.1 But-2-en-1-ol(Crotyl Alcohol)

It is obtained by the reduction of croton aldehyde with aluminium isopropoxide.

$$CH_3CH{=}CH{-}\underset{\underset{O}{\|}}{CH} \xrightarrow[\text{Isopropanol}]{\text{Aluminium isopropoxide}} CH_3CH{=}CH{-}\underset{\underset{OH}{|}}{CH_2} + (CH_3)_2C{=}O$$

Croton aldehyde : 21 g
Aluminium isopropoxide : prepared as given below

A mixture of clean aluminium foil (4.7 g, 0.166 mol), dry isopropyl alcohol (50 ml, 0.65 mol), and mercuric chloride (0.5 g) is refluxed in a R.B. flask (250 ml capacity) fitted with a reflux condenser carrying a calcium chloride guard tube. When the liquid is boiling, carbon tetrachloride (2 ml, a catalyst for reaction between aluminium and dry alcohols) is added through the condenser. The heating is continued

1. H. MEERWEIN, R. SCHMIDT, *Ann.*, 1925, **444**, 221.
2. W. PONNDORF, *Angew. Chem.*, 1926, **39**, 138.
3. A. VERLAY, *Bull. Soc. Chim., France*, 1925, **37**, 537, 871.

till all the aluminium has reacted. The mixture turns grey and within few minutes a vigorous reaction of hydrogen commences. If the reaction becomes too vigorous, the flask has to be cooled in ice-water. Finally, the reaction mixture is refluxed for 6 hours. The solution (there may be suspended particles present) is used as such.

To the above solution of aluminium isopropoxide is added redistilled crotonaldehyde (b.p. 102–103°C, 21 g) and dry isopropyl alcohol (100 ml). An efficient fractionating column is attached on to the flask with arrangement for distillation. The reaction mixture is heated at 110°C (oil bath) and the acetone formed is allowed to distil away. When no more acetone distills (as indicated by a negative 2,4-dinitrophenyl hydrazine test), the remaining isopropyl alcohol is removed under reduced pressure. The reaction mixture is cooled and sulphuric acid (3M, 150 ml from 24.2 ml of concentrated sulphuric acid and 133 ml water) is added. The upper oily layer is separated (separatory funnel), washed with water and distilled at 100°C/20 mm to get crotyl alcohol. The combined aqueous layer obtained above is distilled until the distillate does not give test for unsaturation (bromine in carbon tetrachloride solution). The aqueous distillate is saturated with potassium carbonate and the oily layer separated. The oily layer is combined with the crotyl alcohol obtained above. It is dried (anhydrous potassium carbonate) and distilled using a fractionating column. Crotyl alcohol distills at 119–121°C. The yield is 11 g (50%).

65. METHYLATION

Phenols can be conveniently methylated to give the methyl ethers, which are useful intermediates for organic synthesis. The nature of the methylation procedure to be used depends on the nature of the compound to be mthylated. The following procedures are normally used.

Alcohols (e.g, methanol, ethanol) in presence of catalysts (concentrated sulphuric acid, p-toluene sulphonic acid). This procedure is generally used for the esterification of carboxylic acids.

Diazomethane: This is used for the esterification of carboxylic acids or methylation of phenolic hydroxy groups. It is used for methylation of very small quantity of the compound.

Dimethyl sulphate: This procedure is the most versatile and is commonly use for methylation of phenols and acids. Methylation can be affected either in an alkaline medium (sodium hydroxide solution) or in anhydrous conditions (anhydrous acetone and freshly ignited potassium carbonate). Sometimes the hydroxy compounds (especially those can be easily oxidised) are converted into acetate and then methylated with dimethyl sulphate and alkali.

Following examples illustrate the utility of dimethyl sulphate.

65.1 β-Naphthyl methyl ether (Nerolin)

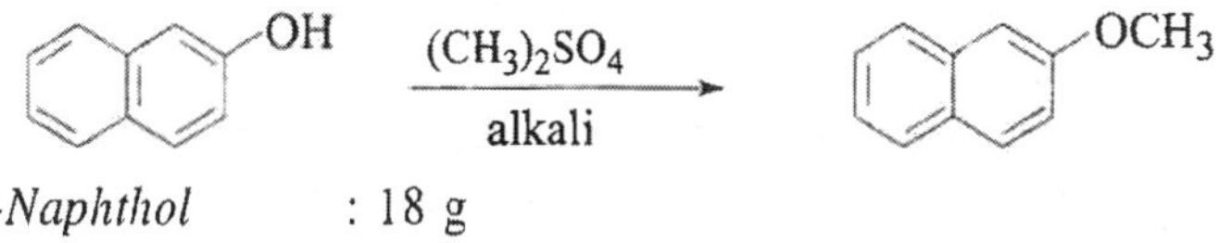

β-Naphthol : 18 g
Sodium hydroxide : 5.25 g
Dimethyl sulphate : 15.75 g, (11.75 ml)

Dimethyl sulphate (15.75 g, 11.75 ml, 0.125 mol) is added dropwise to a stirred and cooled solution (10–15°C) of β-naphthol (18 g, 0.25 mole) in sodium hydroxide solution (5.25 g, 0.20 mol in 75 ml water). After the addition is over, the mixture is warmed for one hour at 70-80°C and then cooled. The separated product is filtered, washed with sodium hydroxide solution (10%) and then with water, dried and crystallised from alcohol. The yield is 16.5 g (84%); m.p. 72°C.

65.2 Resorcinol dimethyl ether (1,3-Dimethoxybenzene)

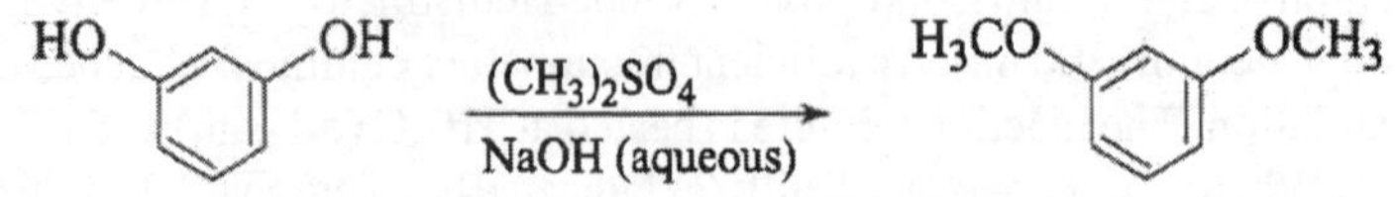

Resorcinol	: 25 g
Aqueous sodium hydroxide	: 40%, 87.5 ml
Sodium hydrosulphite	: 10 g
Dimethyl sulphate	: 55 ml
Alcohol	: 75 ml

Aqueous sodium hydroxide (40%, 87.5 ml) is added dropwise during 3–4 hours to a well stirred mixture of resorcinol (25 g, 0.25 mol), sodium hydrosulphite (10 g, 0.057 mol), dimethyl sulphate (55 ml, 0.57 mol) and alcohol (75 ml) at 18–20°C. Ice (approx. 500 g) is added to the reaction mixture and left overnight. It is extracted with ether (2 × 100 ml). The ether extract is washed with water, dried (anhydrous sodium sulphate) and distilled. The oily residue is distilled and the fraction distilling between 216–218°C is collected. The yield is 26.37 g, 25 ml (84.1%).

65.3 1,2,4-Trimethoxybenzene

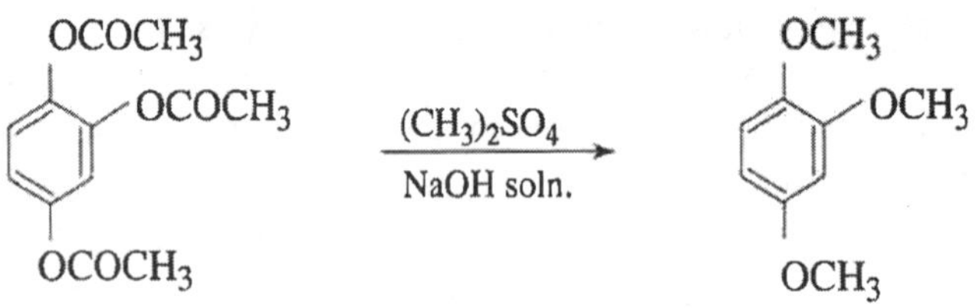

1,2,4-Triacetoxybenzene	: 12.5 g
(Hydroxy hydroquinone triacetate)	
Methanol	: 50 ml
Dimethyl sulphate	: 45 ml
Sodium hydroxide solution	: 37.5 g in 37.5 ml water

Aqueous sodium hydroxide solution (37.5 g in 37.5 ml water) is added dropwise during 2 hours to a cooled and well stirred solution of 1,2,4-triacetoxybenzene (12.5 g, 0.049 mol, See page 7) in methanol (50 ml) and dimethyl sulphate (45 ml, 0.475 mol). The temperature of the reaction mixture is kept between 25–30°C throughout the addition. The reaction mixture is stirred for 30 minutes more and left overnight. It is extracted with ether (2 × 70 ml). The ether extract is washed with water, dried (sodium sulphate) and distilled. The yield is 6 g (72.3%); b.p. is 247°C.

65.4 3,4-Dimethoxybenzaldehyde (Veratraldehyde)

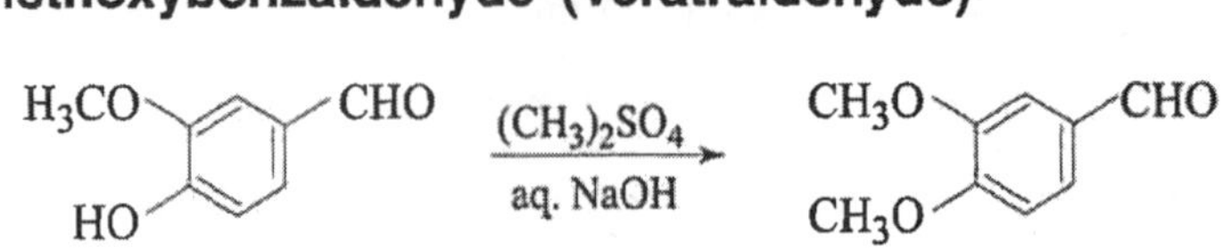

Vanillin : 19 g
Potassium hydroxide : 11.2 g
Dimethyl sulphate : 15 ml

Vanillin (19 g, 0.125 mol) is melted by warming on a steam bath in a three necked flask (250 ml capacity) fitted with a mechanical stirrer and two separatory funnels. A solution of potassium hydroxide (11.2 g, 0.375 mole) in water (75 ml) is added dropwise through a separatory funnel. Simultaneously, dimethyl sulphate (20 g, 15 ml, 0.157 mol) is added dropwise through the second separatory funnel to the vigorously stirred mixture. The external heating is stopped after a few minutes and the mixture continues to reflux from the heat of the reaction. The addition of both the reagents is completed in 20–25 minutes. It is ascertained that the reaction mixture remains alkaline throughout the reaction. It is allowed to stand overnight. The crystalline mass of veratraldehyde is removed, ground in a mortar with ice-cold water (100 ml) and filtered. It is dried and crystallised from petroleum ether as colourless needles. The yield is 19.8 g (95%); m.p. is 42–43°C.

65.5 4-Methyl-7-methoxycoumarin

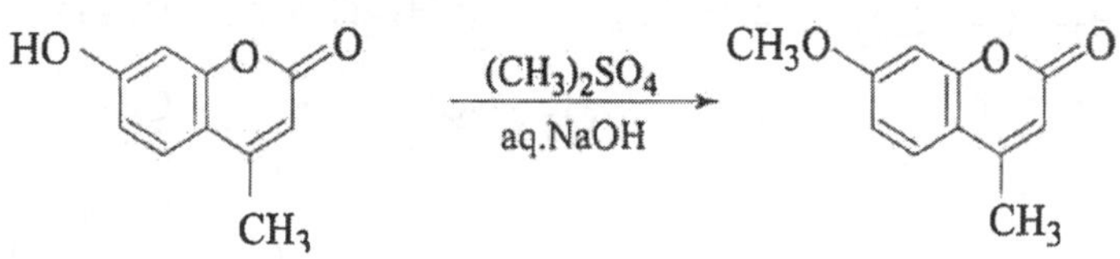

4-Methyl-7-hydroxycoumarin : 5.53 g
(4-methyl umbelliferone)
Acetone (anhyd.) : 75 ml
Dimethyl sulphate : 4.62 g, 3.55 ml
Ignited potassium carbonate : 12 g

4-Methyl-7-hydroxycoumarin (5.53 g, 0.33 mol, See page 177) is dissolved in dry acetone (75 ml) and refluxed with dimethyl sulphate (3.55 ml, 0.34 mol) and ignited potassium carbonate (12 g) for 4 hours under anhydrous conditions. The acetone solution is filtered and the inorganic salts washed with hot acetone (25 ml.). The combined acetone solution is distilled and the oily residue is macerated with ice. The solid product is filtered, washed with water and crystallised from methanol. The yield is 5.6 g (90%); b.p. is 216–218°C.

66. MICHAEL CONDENSATION[1]

It is the addition reaction between an α, β-unsaturated carbonyl compound and a compound with an active methylene group (eg, malonic ester, acetoacetic ester etc) in presence of a base (eg, sodium ethoxide or secondary amine).

Mechanism: The mechanism of the reaction using malonic ester and mesityl oxide to give dimedone is represented as given below:

$$CH_2(COOC_2H_5)_2 + :\overset{\ominus}{\ddot{O}}:C_2H_5 \rightleftharpoons :\overset{\ominus}{C}H(COOC_2H_5)_2 + C_2H_5OH$$

1. A. Michael, *J. Prakt. Chem.*, 1887, **35**, 349.

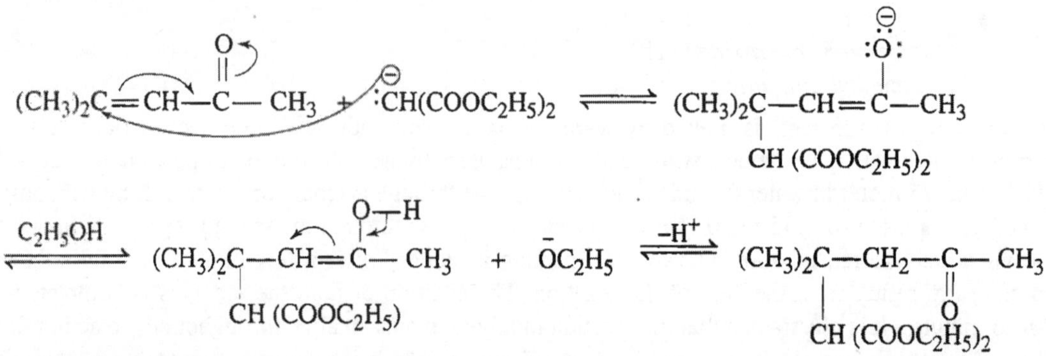

In this case, the Michael condensation is followed by an internal claisen condensation leading to the formation of dimedone. Thus,

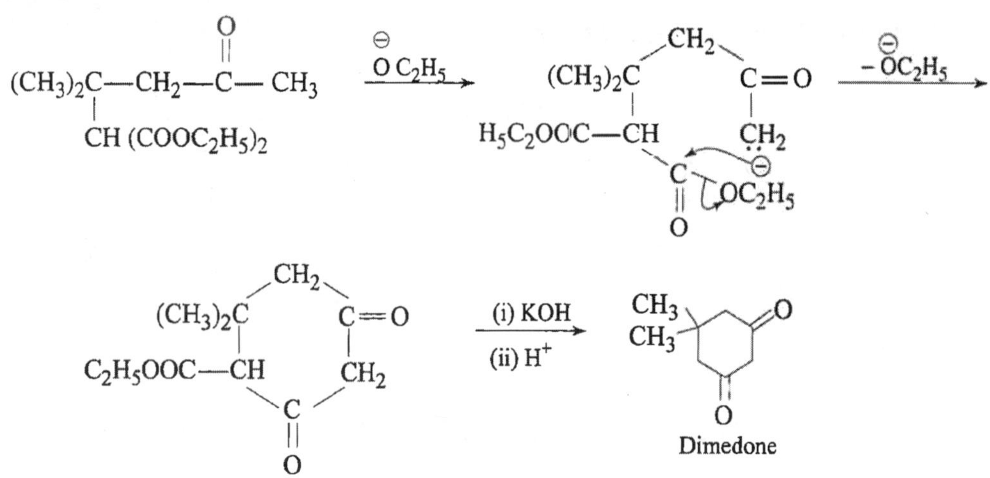

66.1 Dimedone

Sodium : 2.3 g
Absolute alcohol : 40 ml
Mesityl oxide : 10 g (11.6 ml)
Diethyl malonate : 17 g (16.1 ml)

Clean sodium (2.3 g, 0.1 mole) cut into small pieces is added to absolute alcohol (40 ml) contained in a three necked R.B. flask (250 ml capacity) fitted with a reflux condenser, a mercury-sealed stirrer and a dropping funnel (anhydrous conditions). If the reaction becomes too vigorous, it is cooled in an ice-water bath. When all the sodium has reacted, diethyl malonate (17 g, 16.1 ml, 0.1 mol) is added through a dropping funnel. Mesityl oxide (10 g, 11.6 ml, 0.1 mol) is subsequently added through the dropping funnel. The reaction mixture is kept stirred for 2 hours with refluxing. Potassium hydroxide solution (12.5 g in 60 ml water) is added and the mixture refluxed for 2 hours more on a water bath. The reaction mixture is acidified (dilute hydrochloric acid 1:2 by volume 55 ml). As much alcohol as possible is distilled (water bath). The residual solution is further acidified with dilute hydrochloric acid until it is just acidic to methyl orange. The mixture is cooled and the separated dimedone is filtered and washed with ice-cold water. It is dried and crystallised from acetone-petroleum ether. The yield is 12 g (86%); m.p. is 147–148°C.

67. NATURAL PRODUCTS—ISOLATION

67.1 Isolation of Caffeine from Tea Leaves

Tea leaves contain 1–4% caffeine, which is soluble in hot water. Its structure is 1,3,7-trimethyl xanthine,

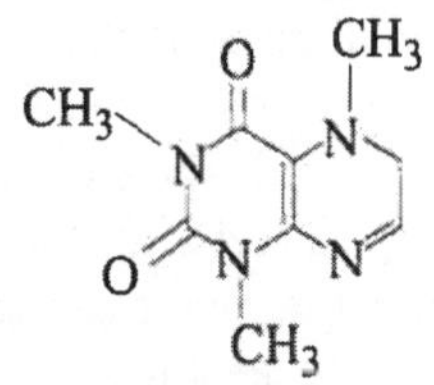

Procedure: The tea leaves (25 g) are boiled with water (250 ml) for 15–20 minutes in a beaker. The aqueous solution is filtered while hot. The remaining tea leaves are boiled again with water (100 ml) for 15–20 minutes and the solution filtered while hot. To the combined aqueous solution is added lead acetate solution (10%, 55 ml) until no more precipitate is formed. The mixture is filtered while hot. The precipitated material consists of naturally occurring tannins (polyphenols). To the clear solution (filtrate) is added dilute sulphuric acid until the precipitation of lead as lead sulphat is complete. The precipitate of lead sulphate is filtered, the filtrate boiled with a little activated charcoal, filtered and concentrated on a sand bath to about 50 ml. The solution is cooled to room temperature and extracted with chloroform (3×30 ml). The combined chloroform solution is dried (sodium sulphate) and distilled (water bath). To the residue is added petroleum ether (30 ml) and the mixture is stirred. The solid product of crude caffeine is filtered. The yield is 0.25 g. It is crystallised from a minimum quantity of water. The m.p. is 237–238°C.

67.2 Isolation of Casein and Lactose from Milk

Casein is a mixture of related phosphoproteins occurring in milk and cheese. It is present to the extent of 3% in milk along with lactose (4–5%), fats (3–4%) and minerals (1%), the remainder being water. Casein gets precipitated from milk at pH 4.5.

Lactose is a reducing disaccharide and consists of *D*-glucose and *D*-galactose units. The latter is linked glycosidically (through its reducing end) to C-4 of the glucose unit by *β*-linkage. It is designated as 4-O-*β*-D-galactopyranosyl-*D*-glucose and is called milk sugar.

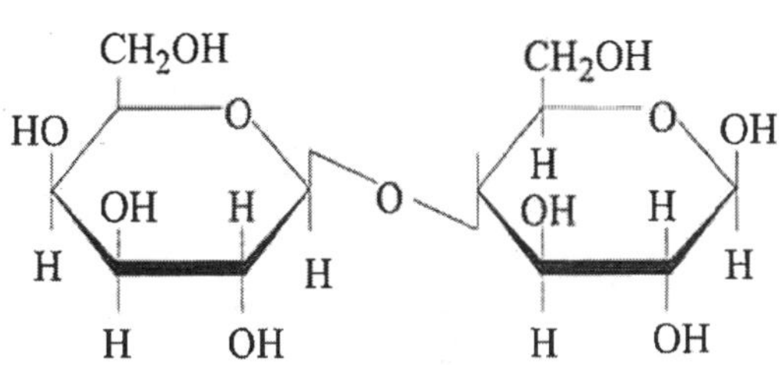

Lactose

Isolation of Casein. Milk (200 ml) is diluted with water (100 ml) in a beaker and warmed to 40–45°C. Acetic acid (10%) is added dropwise to the stirred solution till the precipitation of casein is complete. The mixture is stirred (2–3 minutes) and the casein is filtered through a muslin cloth. It is washed with water (2–3 times) and then with alcohol (the combined filtrate and washings is used for isolation of lactose). The casein obtained is titurated with ether to remove fat. The fat free protein is dissolved in minimum quantity of sodium hydroxide solution (5%) and the clear alkaline solution (filter if necessary) is acidified with acetic acid (10%). The separated casein is filtered, washed with water and then with ethanol. It is dried in a desiccator.

Isolation of lactose. The combined filtrate and washings obtained above is neutralised with dilute sodium hydroxide solution (5%). The solution is boiled to precipitate the water-soluble proteins (albumins). It is filtered and the clear filtrate is concentrated over a steam bath to a small volume. It is boiled with a little activated charcoal and filtered. The clear filtrate is concentrated on a boiling water bath to a syrup. Ethanol is added to the syrup and set aside for crystallisation. Lactose can be crystallised by dissolving in minimum volume of water, filtering if necessary and adding ethanol dropwise until a permanent turbidity persists in the solution. It is kept in a refrigerator for some time for crystallisation. It is filtered and dried in a desiccator. The m.p. is 201–202°C.

67.3 Isolation of Piperine from Black Pepper

Piperine is an alkaloid present in black pepper to the extent of 8–10%, along with volatile oils (1–2%), starch (20–40%) and water (8–10%). Its structure is represented as:

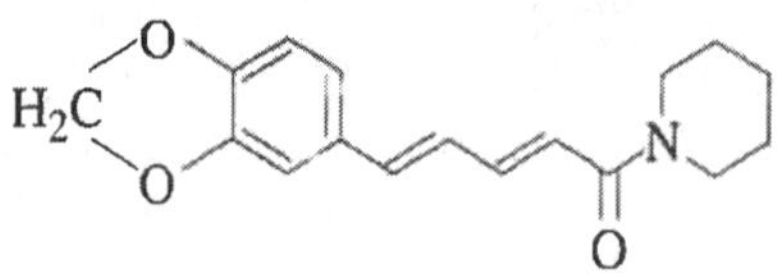

Procedure: Finely ground black pepper (25g) is taken in a R.B. flask (500 ml capacity). To this is added ethanol (95%, 300 ml) and the mixture is refluxed. The clear filtrate is concentrated to about 25 ml by distillation on a water bath. To the concentrated residue is added alcoholic potassium hydroxide solution (2N, 25 ml). The solution is warmed and filtered. The clear filtrate is heated on a steam bath and water added dropwise till a turbidity appears. The solution is cooled overnight in a refrigerator. It is filtered and crystallised from acetone as fine yellow needles. The m.p. is 129–130°C.

68. NENCKI REACTION[1]

The ring acylation of phenols with acids in the presence of zinc chloride is known as Nencki reaction. Thus,

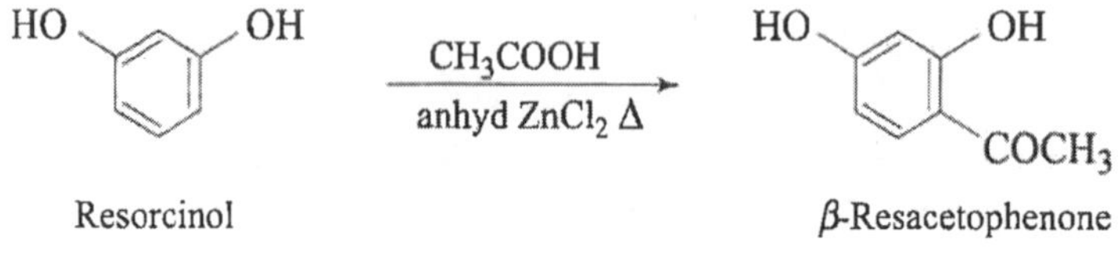

1. M. NENCKI and N. SIEBER, *J. Prakt. Chem*, 1881, **23**, 147.

68.1 β-Resacetophenone (2,4-Dihydroxyacetophenone)

Resorcinol : 11 g
Zinc chloride (anhydrous) : 16.5 g
Acetic acid (glacial) : 16 ml

Freshly fused and powdered zinc chloride (16.5 g, 0.27 mol) is dissolved in glacial acetic acid (16 ml) by heating in a beaker on a sand bath. Dry resorcinol (11 g, 0.1 mol) is added with stirring to the mixture at 140°C. The solution is heated until it just begins to boil and kept for 20 minutes at 150°C. Dilute hydrochloric acid (1:1, 50 ml) is added to the mixture and the solution cooled (5°C). The separated product is filtered, washed with dilute hydrochloric acid (1:3) and crystallised from hot water containing a little hydrochloric acid, the yield is 14 g (93%); m.p. is 142–144°C.

Note: β-Resacetophenone can also be obtained by the Houben-Hoesch reaction (page 133).

69. NITRATION

Aromatic compounds (eg, hydrocarbons, phenols, carboxylic acids etc) can be conveniently nitrated by the use of nitrating mixture, which is normally a mixture of concentrated nitric acid and concentrated sulphuric acid. The nature of the product obtained by nitration depends on the nature of the compound, nature of the nitrating agent and reaction temperature. Following procedures are generally used:

(i) concentrated nitric acid and concentrated sulphuric acid,
(ii) fuming nitric acid and concentrated sulphuric acid,
(iii) sodium nitrite and dilute sulphuric acid,
(iv) tetranitromethane, and
(v) BF_3–nitrate.

Aromatic hydrocarbons can be nitrated with concentrated nitric acid in the presence of concentrated sulphuric acid. Thus,

$$Ar\text{–}H + HNO_3 + H_2SO_4 \longrightarrow Ar\text{–}NO_2$$

The function of sulphuric acid is to convert nitric acid into a highly reactive, electrophilic, nitronium ion, $^{\oplus}NO_2$, which is the effective nitrating agent.

$$HONO_2 + 2H_2SO_4 \longrightarrow \overset{\oplus}{N}O_2 + H_3\overset{+}{O} + 2HSO_4^-$$

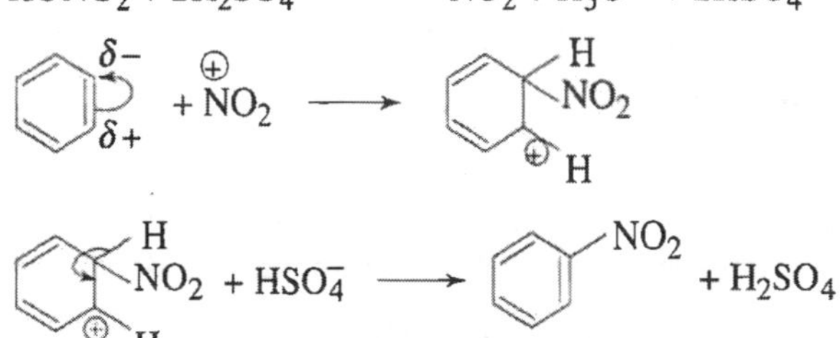

The nitration of aromatic hydrocarbons is usually carried out with the above reagent, viz. concentrated nitric acid and concentrated sulphuric acid at low temperatures (50°C approx.) and is used in the preparation of nitrobenzene, 1-nitronaphthalene and *p*-bromo nitrobenzene. Higher temperature is avoided, since polynitration is more likely and the oxidative breakdown of the ring system may occur.

Nitration of aromatic compounds containing electron releasing groups (viz. $-OH$, $-NH_2$, $-NHCOCH_3$) is carried out under milder condition, since the aromatic nucleus is activated towards the attack of the electrophilic species. This is illustrated by the nitration of phenol to give a mixture of *o*- and *p*-nitrophenols (the reagent used is dilute nitric acid or sodium nitrate and dilute sulphuric acid) and nitration of acetanilide to give *p*-nitroacetanilide (the nitration is carried out in acetic acid medium with concentrated nitric acid and concentrated sulphuric acid at 0°C).

In case, the aromatic compound contains an electron withdrawing group (eg, $-NO_2$, $-SO_3H$, $-CHO$, $-COOH$) the nitration requires forcing or drastic conditions, i.e., use of fuming nitric acid and concentrated sulphuric acid. This is illustrated by the nitration of nitro benzene to give *m*-dinitrobenzene, *p*-nitrotoluene to 2, 4-dinitrotoluene, and phenol to 2, 4, 6-trinitrophenol (picric acid).

69.1 Nitrobenzene

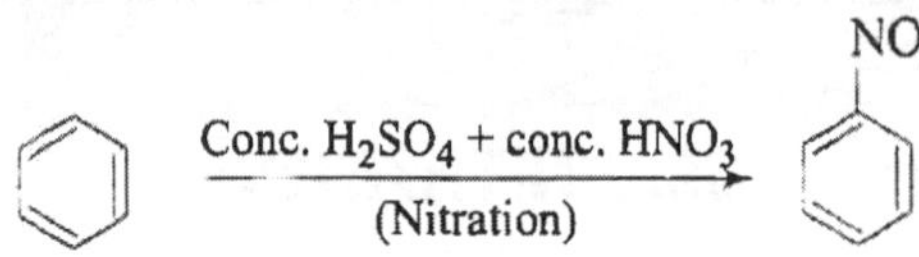

Benzene	:	15 ml
Conc. nitric acid	:	17.5 ml
Conc. sulphuric acid	:	18.4 ml

Concentrated sulphuric acid (33 g, 18.4 ml, 0.33 mole) is added in portions with shaking to concentrated nitric acid (25 g, 17.5 ml, 0.25 mol) contained in a R.B. flask (100 ml capacity). The reaction mixture is kept cool by immersing the flask in a cold water bath. Benzene (13 g, 15 ml, 0.165 mol) is added in portions of 1–2 ml. The mixture is shaken after each addition. The temperature is kept between 50–55°C by the use of a cold water bath if necessary. After the addition is over, the reaction mixture is heated for 10-15 minutes at 60°C (water bath), and the flask is shaken occasionally. The contents are poured into cold water (250 ml) and the mixture stirred. The lower layer of nitrobenzene is separated (separatory funnel), washed with cold water (2 × 25 ml), dried anhydrous calcium chloride) and distilled, Nitrobenzene is collected at 210°C. The yield is 17 g (85%).

Note: (1) During distillation adequate precaution should be taken to see that the material is not distilled to dryness; and explosion may result if the temperature is allowed to raise appreciably due to formation of *m*-dinitrobenzene and higher nitro compounds. (2) Nitrobenzene is toxic. Its vapours should not be allowed to escape in the laboratory. It is a skin poison and if comes in contact with skin, it should be immediately washed with rectified spirit, followed by a soap and water.

69.2 1-Nitronaphthalene

It is obtained by nitration of naphthalene with concentrated nitric acid and concentrated sulphuric acid. The nitro group enters position 1, since the intermediate carbonium ion formed is more stable (resonance stabilisation) if the attack occurs at position 1 than at position 2.

$$2H_2SO_4 + HONO_2 \longrightarrow \overset{+}{N}O_2 + 2HSO_4^- + H_3\overset{+}{O}$$

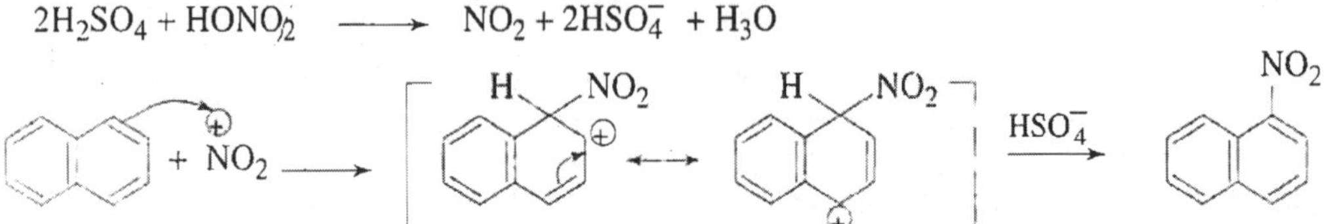

2-Nitronaphthalene

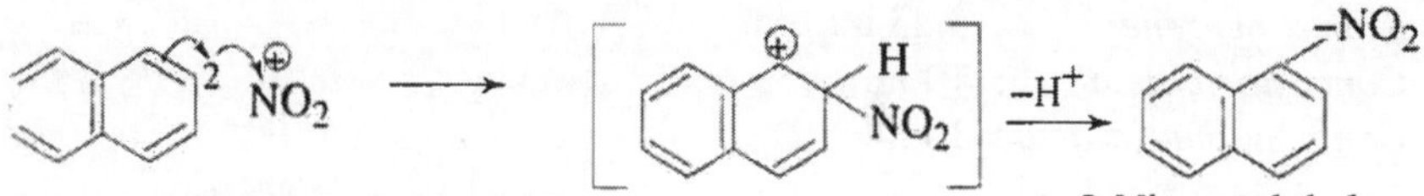

2-Nitronaphthalene

Naphthalene	: 5 g
Conc. nitric acid	: 10 ml
Conc. sulphuric acid	: 10 ml

Finely powdered naphthalene (5 g, 0.039 mol) is added in small quantities to the nitrating mixture prepared from conc. nitric acid (10 ml) and conc. sulphuric acid (10 ml) as in the case of nitrobenzene, (see page 152). The reaction mixture is vigorously shaken during the addition and the temperature is maintained at 45–50°C (cold water bath is used if necessary). When all the naphthalene has been added, the reaction mixture is heated at 55–60°C for 30–40 minutes until the odour of naphthalene has disappeared. The reaction mixture is poured into cold water (50 ml). The aqueous layer is decanted and the solid product (nitronaphthalene) boiled with water (50 ml) for 15 minutes and the aqueous layer decanted. The remaining product is steam distilled to remove unreacted naphthalene. The solid product is filtered, washed with water and crystallised from dilute alcohol. The yield is 6 g (80%); m.p. is 61°C.

69.3 *p*-Bromonitrobenzene

It is obtained by the nitration of bromobenzene with concentrated nitric acid in presence of concentrated sulphuric acid; a small amount of *o*-isomer is also formed.

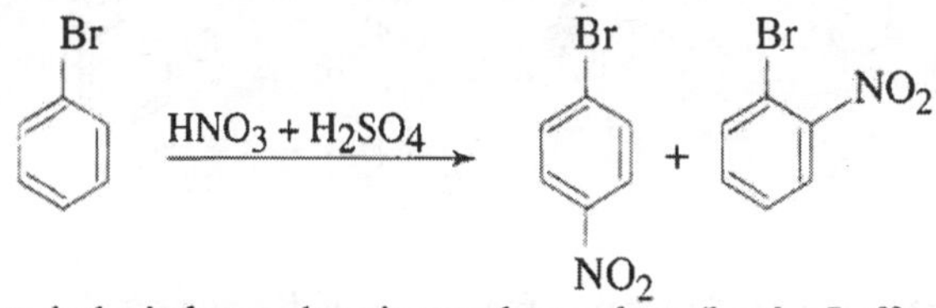

In the above case, though the halogen deactivates the nucleus (by the-I effect), it directs the incoming nitronium ion to the *ortho*- and *para*-positions as a result of the mesomeric interaction of the halogen lone pair of electrons with the charge developed in the corresponding intermediates

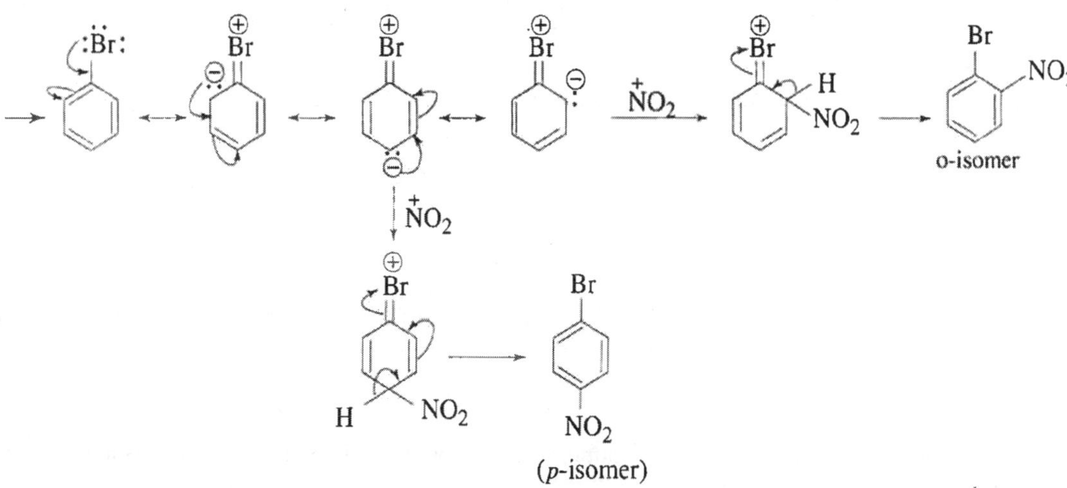

o-isomer

(*p*-isomer)

Bromo benzene : 5.25 ml
Conc. nitric acid : 10 ml
Conc. sulphuric acid: 10 ml

To the nitrating mixture (prepared from 10 ml each of concentrated nitric acid and concentrated sulphuric acid as in the case of nitrobenzene, page 152) is added at room temperature bromo benzene (8 g, 5.25 ml, 0.05 mol) in small lots (1–2 ml each time) during 15 minutes. The reaction mixture is shaken vigorously during the addition and temperature is not allowed to rise above 50–60°C (if necessary the reaction flask is cooled in ice water). Finally, the reaction mixture is heated on a boiling water bath for 30 minutes. The cooled reaction mixture is poured , with stirring, into cold water (100 ml). The separated product is filtered, washed with cold water and crystallised from rectified spirit (50–60 ml). The yield is 7 g, (70%); m.p. is 124–125°C.

69.4 *o*-and *p*-Nitrophenols

Phenol on nitration with dilute nitric acid gives a mixture of *o*- and *p*-nitrophenols. However, the yield of the *o*-isomer increases if a mixture of sodium nitrate and dilute sulphuric acid is employed. The separation of *o*- and *p*-isomer is affected by steam distillation—the *o*-isomer is steam volatile.

Mechanism: Phenol is resonance hybrid of five forms as represent below:

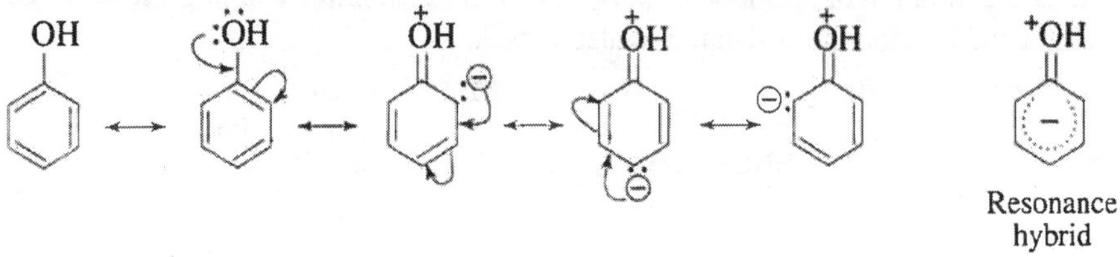

The electrophilic substitution by $\overset{\oplus}{NO_2}$ takes place in *o*-and *p*-positions.

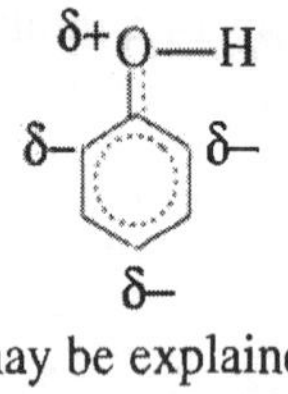

The formation of *o*- and *p*-nitro phenol may be explained as given below:

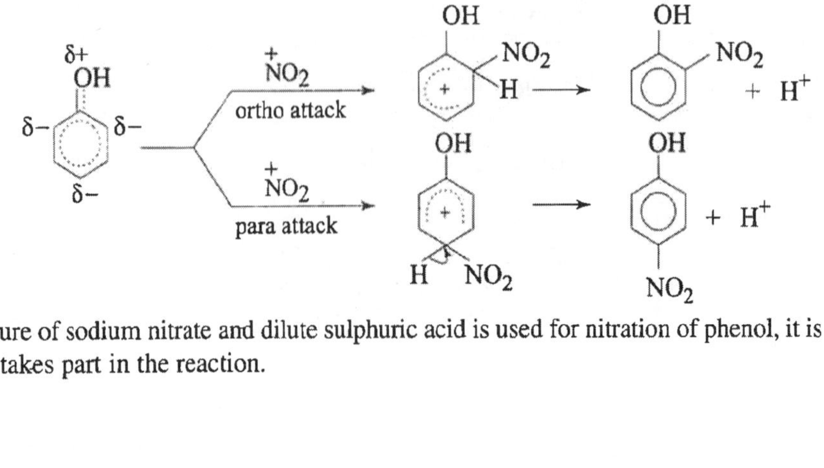

If a mixture of sodium nitrate and dilute sulphuric acid is used for nitration of phenol, it is nitrosonium ion, which takes part in the reaction.

$$2NaNO_3 + H_2SO_4 \longrightarrow 2HNO_2 + Na_2SO_4 + H_2O + O$$

$$HONO + H_2SO_4 \rightleftharpoons NO^+ + HSO_4^- + H_2O$$

Even with dilute nitric acid the nitrosonium ion, $\overset{\oplus}{N}O$, is the electrophile. Some nitrous acid is always present in solution and this catalyses the reaction.

$$HONO + 2HNO_3 \rightleftharpoons \overset{+}{N}O + 2NO_3^- + H_3O^+$$

Thus, in the nitration of phenol, initial nitrosation takes place to form a mixture of *o*- and *p*-niotroso phenol, which gets oxidised to *o*- and *p*-nitro phenol

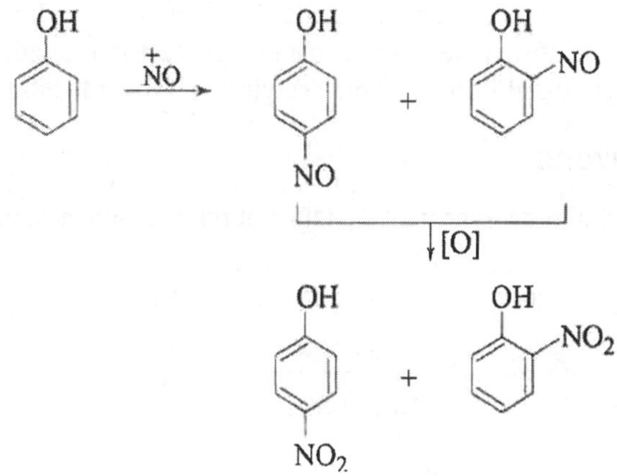

Phenol	:	9.4 g
Conc. sulphuric acid	:	13.6 ml
Sodium nitrate	:	15 g

Sodium nitrate (15 g) is added to dilute sulphuric acid (13.6 ml concentrated sulphuric acid and 40 ml water prepared cautiously). The solution is cooled in an ice bath and a solution of phenol in water (9.4 g, 0.1 mole of phenol is mixed with 2 ml water) is added dropwise from a dropping funnel to the stirred solution. The temperature is not allowed to rise above 20°C. After the addition of phenol is complete, the reaction mixture is allowed to stand for one hour with frequent shaking. The supernatant liquid is poured out and the residue washed (2–3 times) with water to remove any residual acid. The mixture is steam distilled until no more *o*-nitrophenol passes over. The separated *o*-nitrophenol in the steam distillate, is filtered and dried. The yield is 3 g; (43%); m.p. is 46°C.

The residual product in the flask is filtered. It is boiled with dilute hydrochloric acid (2%) together with decolourizing carbon (1–2g) for 5–10 min and filtered through a hot water funnel. The clear solution is cooled and the separated *p*-nitrophenol is filtered. The yield is 2 g (28.8%); m.p. is 112°C.

69.5 *p*-Nitroacetanilide

It is obtained by nitration of acetanilide.

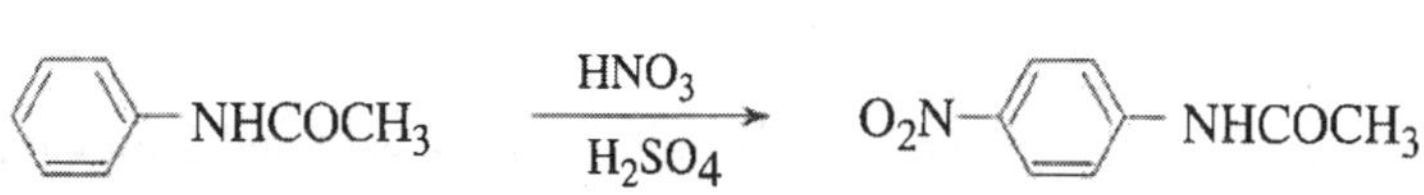

p-Nitro acetanilide is used to prepare p-nitroaniline, which cannot be obtained by direct nitration of aniline.

Acetanilide	: 8.4 g
Acetic acid (glacial)	: 10 ml
Conc. sulphuric acid	: 19.4 ml
Conc. nitric acid	: 3.7 ml

Concentrated sulphuric acid (31 g, 17 ml) is added to a well stirred solution of acetanilide (8.4 g, 0.002 mole) in glacial acetic acid (10 ml). The clear solution is cooled (freezing mixture of ice-salt), stirred, and a mixture of concentrated nitric acid (3.7 ml) and concentrated sulphuric acid (2.4 ml) is added slowly through a dropping funnel. The temperature of the reaction mixture is maintained below 10°C. The mixture is allowed to attain room temperature and left for 30 minutes. It is poured on to crushed ice (about 80 g) and the mixture stirred well. The separated product is filtered, washed with cold water (until free of acid) and crystallised from alcohol. The yield is 4.8 g; (43%), m.p. is 213–14°C.

69.6 *m*-Dinitrobenzene

It is prepared by nitration of nitrobenzene with a mixture of fuming nitric acid and concentrated sulphuric acid.

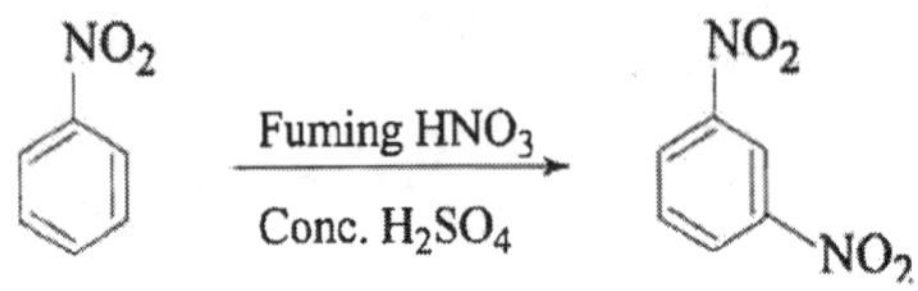

Nitrobenzene	: 5 ml
Fuming nitric acid (sp. gr. 1.5)	: 10 ml
Conc. sulphuric acid	: 10 ml

Nitrobenzene (5 ml, 0.049 mole) is added slowly in small lots to the nitrating mixture [prepared by the careful addition of 10 ml concentrated sulphuric acid to 10 ml fuming nitric acid (sp. gr. 1.5)] contained in a conical flask (100 ml capacity). The mixture is shaken frequently during the addition. It is heated with shaking on a boiling water bath for 15 minutes. The cooled mixture is poured cautiously into water (150 ml). The separated dinitrobenzene is filtered and washed with cold water (till free of acid). It is crystallised from alcohol. The yield is 5.5 g (67.4%); m.p. is 89–90°C.

69.7 2,4-Dinitrochlorobenzene

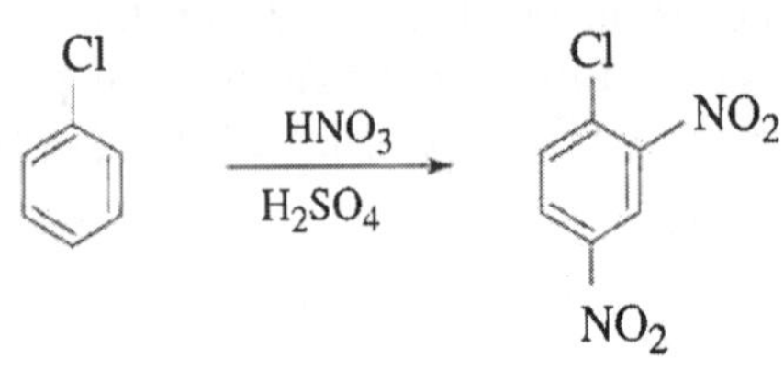

Chlorobenzene	: 14 g (12.5 ml)
Conc. sulphuric acid	: 30.5 ml
Nitric acid	: 20 ml (12 ml fuming nitric acid and 8 ml Conc. nitric acid)

Chlorobenzene (12.5 ml, 0.123 mole) is added to concentrated sulphuric acid (15.5 ml) contained in an R.B. flask (250 ml capacity). The mixture is shaken and warmed on a water bath. The mixture thus obtained, is added dropwise (dropping funnel) to a mixture of nitric acid (20 ml, 12 ml fuming nitric acid and 8 ml concentrated nitric acid) and concentrated sulphuric acid (15 ml). The temperature of the mixture should not rise above 100°C. When all the mixture has been added, the reaction mixture is heated with frequent shaking, for one hour on a boiling water bath and then for 1.5 hours on a sand or oil bath at 120°C, until a drop of the reaction mixture solidifies on addition to water. The cooled reaction mixture is poured into ice-water. The separated product is filtered, washed with cold water (until free of acid) and crystallised from alcohol. The yield is 15 g (60.2%); m.p. is 52–53°C.

69.8 Picric Acid (2,4,6-Trinitrophenol)

It is prepared by nitration of phenol with concentrated nitric acid in presence of concentrated sulphuric acid.

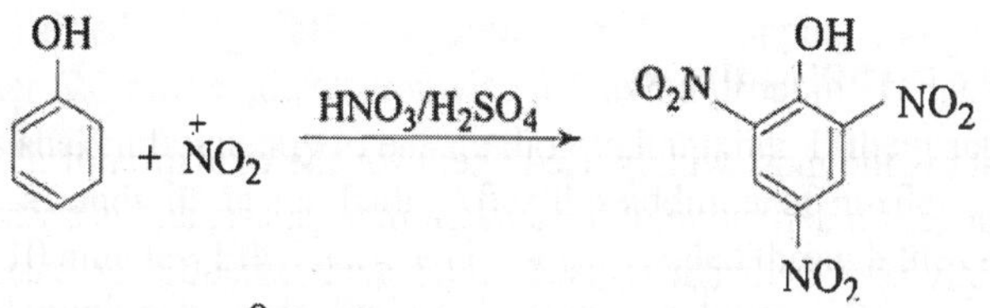

Phenol : 8 g
Conc. sulphuric acid : 20 ml
Conc. nitric acid : 20 ml

A mixture of phenol (8 g, 0.085 mole) and concentrated sulphuric acid (20 ml) is shaken and heated for 30 minutes on a boiling water bath until a clear solution of phenol sulphonic acid is obtained. The solution is cooled (ice-water) and concentrated nitric acid (20 ml) added in small lots while shaking the reaction mixture. A vigorous reaction sets in, copious red fumes of oxides of nitrogen are evolved in 1–2 minutes and the reaction mixture becomes deep red in colour. After the reaction has subsided, it is heated for 1.5 hours in a boiling water bath with occasional shaking. The cooled reaction mixture is poured into cold water (70 ml) and cooled. The separated picric acid is filtered, washed well with cold water and crystallised from dilute alcohol. The yield is 11 g, (56.4%); m.p. is 121–122°C.

70. OXIDATION

Oxidation is an important step for the structural elucidation of organic compounds, especially the natural products. It is also used in the laboratory and industry for the preparation of variety of organic compounds.

Generally speaking, oxidation is either the addition of oxygen to the substrate (as in the conversion of ethylene into ethylene oxide) or the removal of hydrogen (as in the oxidation of ethanol to acetaldehyde). Oxidation refers to the loss of electrons, at an atom or ion or an increase in the oxidation number. A wide variety of agents are available for the oxidation of organic compounds. Following oxidising agents are generally used:

(i) potassium permanganate,
(ii) chromium trioxide,
(iii) potassium dichromate,

(iv) concentrated nitric acid,
(v) hydrogen peroxide,
(vi) selenium dioxide,
(vii) yellow mercuric oxide,
(viii) sodium or potassium bromate,
(ix) sodium hypobromate,
(x) potash fusion,
(xi) potassium persulphate,
(xii) peracids,
(xiii) iodine-pyridine, and
(xiv) chromyl-chloride.

The choice of an oxidising agent to be used depends on the nature of the compound to be oxidised and the product to be obtained.

70.1 Potassium Permanganate

Potassium permanganate is the most widely used agent for the oxidation of a wide variety of organic compounds. Permanganate is a derivative of hexavalent manganese, and is a very powerful oxidant. Its reactivity depends mainly on whether it is used under acidic, neutral or basic conditions. In acidic solution it is reduced to the divalent manganese II ion, Mn^{2+}, with net transfer of five electrons (Mn VII $\longrightarrow$ Mn II), while in neutral or basic media manganese dioxide, MnO_2, is usually formed, corresponding to a three electron change (Mn VII $\longrightarrow$ Mn IV). Permanganate is generally used in aqueous solution and this restricts its usefulness since not many organic compounds are sufficiently soluble in water and only a few organic compounds are resistant to the oxidising action of the reagent. Solutions in acetic acid, *t*-butanol or dry acetone or pyridine can sometimes be employed. Alternatively, oxidation with aqueous solutions of permanganate can be effected in the presence of crown ethers or phase transfer catalysts.

Following oxidations are easily carried out: Veratraldehyde to veratric acid (neutral $KMnO_4$); *o*-chlorotoluene to *o*-chlorobenzoic acid (neutral $KMnO_4$); *p*-xylene to terephthalic acid (alkaline $KMnO_4$); toluene to benzoic acid (alkaline $KMnO_4$).

70.1.1 Veratric acid (3,4-dimethoxybenzoic acid)

It is obtained by oxidation of veratraldehyde with potassium permanganate under neutral conditions.

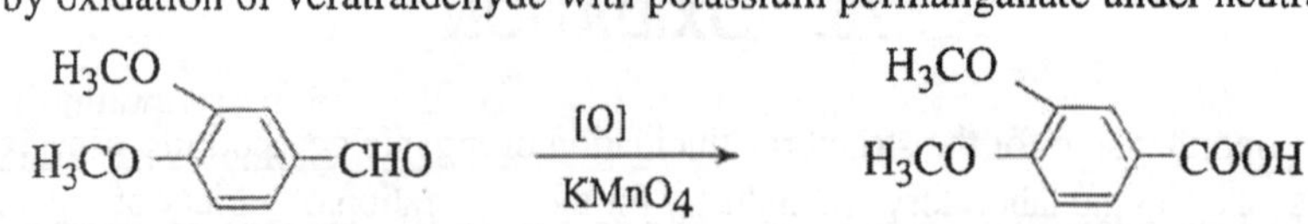

3,4-Dimethoxybenzaldehyde
(veratraldehyde) : 25 g
Potassium permanganate : 35.5 g

Veratraldehyde (25 g, 0.15 mol) is suspended in water (150 ml) and the suspension warmed to 50–60°C (water bath). A solution of potassium permanganate (35.5 g, 0.225 mol in 350 ml water) is then added dropwise during 2.5–3 hours to the stirred suspension. The mixture is stirred for one hour more and the precipitated manganese dioxide is destroyed by passing a current of sulphur dioxide gas. The separated crystalline product is filtered and dried. The yield is 25 g (91.5%); m.p. is 179–180°C.

70.1.2 o-Chlorobenzoic acid

It is obtained by oxidation of *o*-chlorotoluene with potassium permanganate under neutral conditions:

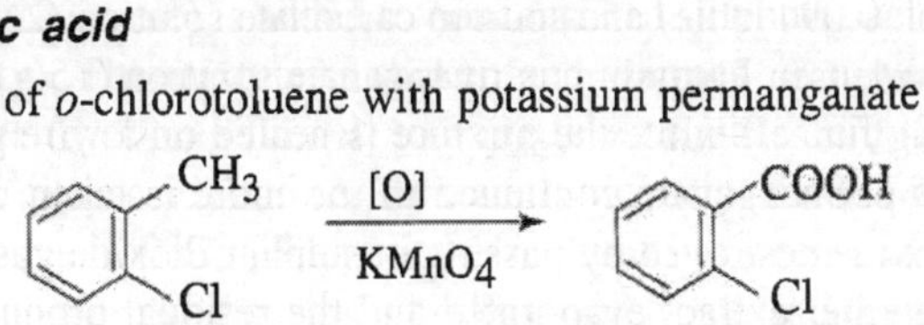

o-Chlorotoluene : 25 g
Potassium permanganate : 75 g

A mixture of potassium permanganate (37. 5 g), *o*-chlorotoluene (25 g, 0.2 mol) and water (600 ml) is stirred and refluxed in a three necked flask (1 litre capacity) fitted with a sealed stirrer and a reflux condenser. The mixture is refluxed with stirring till the permanganate colour has disappeared (1.5–2 hours). A second lot of potassium permanganate (18.75 g) is added and the mixture refluxed with stirring until the permanganate colour disappears (1.5–2 hours). Finally, a third lot of potassium permanganate (18.75 g) is added and refluxing continued until the permanganate colour has disappeared (2–2.5 hours). The precipitated manganese dioxide is destroyed by passing a current of sulphur dioxide gas. The separated *o*-chlorobenzoic acid is filtered and washed with cold water. It is purified by dissolving in sodium bicarbonate solution and acidification of the clear alkaline solution. It is finally crystallised from hot water. The yield is 21 g (68%); m.p. is 139–140°C.

70.1.3 Terephthalic acid

It is obtained by the oxidation of *p*-xylene with alkaline potassium permanagate.

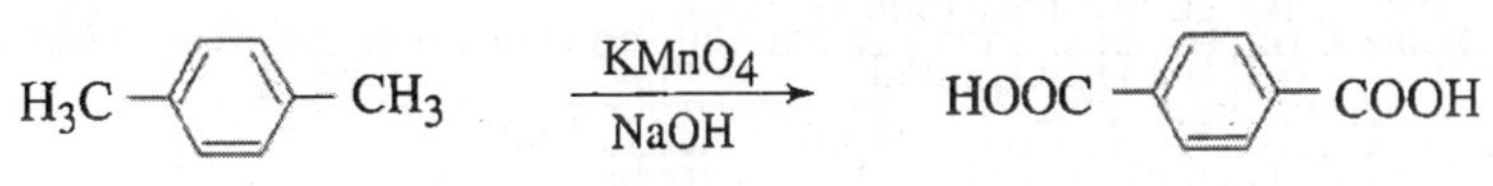

p-Xylene : 15 g
Potassium permanganate : 75 g
Sodium hydroxide solution (2N) : 7.5 ml

A mixture of *p* xylene (15 g, 0.14 mole), potassium permanganate (30 g), water (300 ml) and sodium hydroxide (2N, 7.5 ml) is refluxed in a R.B. flask till the permanganate colour disappears (about 1–1.5 hours). More potassium permanganate is added in lots (total 75g) and refluxing continued till the permanganate colour is discharged (total refluxing 3–4 hours). The mixture is cooled, sulphur dioxide gas passed and the precipitated acid is filtered and washed with cold water. The yield is 6 g (25.6%). It sublimes without melting at 300°C.

Note: Terephthalic acid is also obtained by oxidation of *p*-xylene with sodium dichromate and dilute sulphuric acid (p 164).

70.1.4 Benzoic acid

It is obtained by the oxidation of toluene with alkaline potassium permanganate.

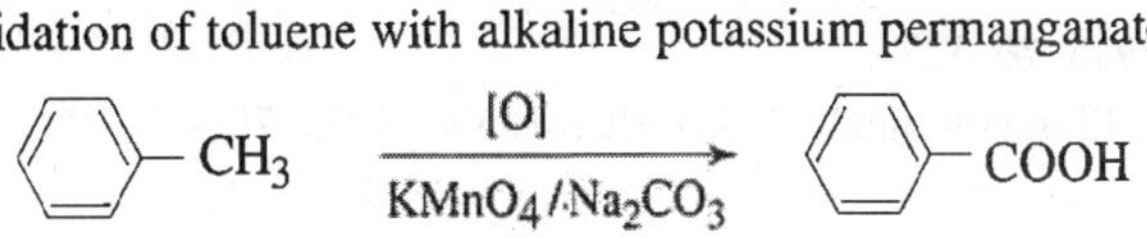

Toluene : 10 ml
Potassium permanganate: 15 g
Sodium carbonate (2N) : 7.5 ml

A mixture of toluene (10 ml, 0.094 mole) and sodium carbonate solution (2N, 7.5 ml) is taken in a R.B. flask fitted with a reflux condenser. Potassium permanganate solution (15 g dissolved in 150 ml water) is added dropwise (dropping funnel) while the mixture is heated on a wire guaze. The mixture is kept boiling and the addition of permanganate continued till no more is taken up. The mixture is cooled, precipitated manganese dioxide destroyed by passing in sulphur dioxide gas. The solution is extracted with ether (2 × 50 ml), the ether extract evaporated and the residual product crystallised from water. The yield is 4.5 g (39%); m.p. is 121°C.

70.2 Chromium Trioxide

Chromic acid, a derivative of hexavalent chormium is one of the most versatile of the available oxidising agent. It reacts with almost all types of oxidisable groups. The reactions can be controlled to yield mainly one product. So it is a useful process in organic synthesis. In oxidation, chromium is reduced from the hexavalent to the trivalent state (Cr VI ⟶ Cr III) with generation of a chromium III salt. It is used in solution of acetic anhydride, *t*-butanol or in pyridine. In these solutions the reactive species present are chromyl acetate, *t*-butyl chromates VI and pyridine in chromium (VI) oxide complex.

Following oxidations are easily carried out: (i) Anthraquinone from anthracene, and (ii) *p*-Nitrobenzaldehyde from *p*-nitrotoluene.

For oxidations, chromium trioxide is used as:

(i) Fischer's reagent[1] (CrO_3-HOAc),
(ii) Sareth reagent[2] (CrO_3-pyridine),
(iii) Cornforth reagent[3] (CrO_3-pyridine—H_2O),
(iv) Thiele reagent[4] (CrO_3-acetic anhydride-H_2SO_4),
(v) Jone's reagent[5, 6] (CrO_3—H_2SO_4), and
(vi) Kiliani reagent[4] (H_2SO_4—H_2CrO_4—H_2O) in acetic acid.

70.2.1 Anthraquinone

It is obtained by the oxidation of anthracene with a solution of chromium trioxide in glacial acetic acid.

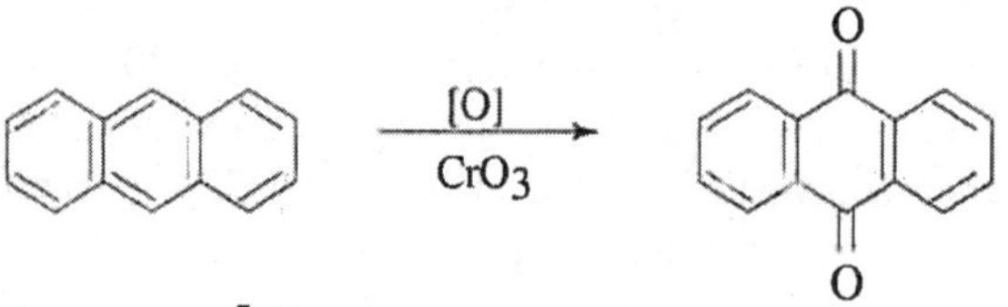

Anthracene	:	5 g
Chromium trioxide	:	10 g.
Acetic acid (glacial)	:	100 ml

A solution of chromium trioxide (10 g) in water (10 ml) and glacial acetic acid (25 ml) is added dropwise to a well stirred solution of anthracene (5 g, 0.028 mole) in glacial acetic acid (75 ml) (the

1. I. F. Fieser, *Am. Soc.*, 1948, **70**, 3237.
2. G.I. Poos, G.E. Arte, R.E. Beyler and L.H. Sareth, *Am. Soc.*, 1953, **75**, 422.
3. R.H. Cornforth, J.W. Cornforth and S. Popjak, Tetrahedron, 1962, **18**, 1351.
4. J. Threle and E. Winter, *Ann.*, 1900, **311**, 353.
5. E A. Brande and J.S. Fawcett, *Org. Synth. Coll. Vol.*, 1963, **4**, 698.
6. R. Wendland and J. Lalonde, *Org. Synth. Coll, Vol.*, 1963, **4**, 757.

anthracene solution is obtained by gentle warming). The reaction is carried out in a three necked R.B. flask fitted with sealed stirrer, a reflux condenser and a dropping funnel. The chromium trioxide solution is added at such a rate that the mixture continues to reflux (10–15 minutes). When all the solution has been added, the reaction mixture is refluxed for 10 minutes. The cooled mixture is poured into cold water (250 ml) and mixture stirred. The separated anthraquinone is filtered under gentle suction, washed with water, hot sodium hydroxide solution (N) and finally with cold water. It is crystallised from glacial acetic acid using decolorising carbon. The yield is 5.5 g (94.2%); m.p. is 285–286°C. Alternatively, it can be purified by sublimation.

Note: Anthraquinone is also obtained by the cyclisation of *o*-benzoyl benzoic acid (see page 126)

70.2.2 p-Nitrobenzaldehyde

It is obtained by the oxidation of *p*-nitrotoluene with chromium trioxide in acetic anhydride.

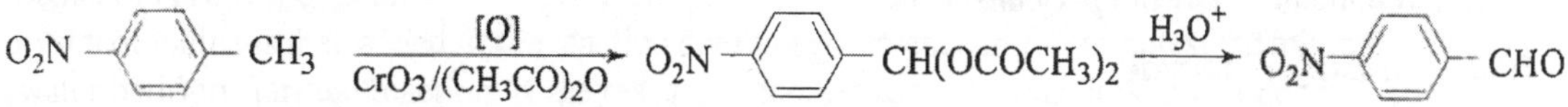

p-Nitrotoluene	: 12.5 g
Acetic anhydride	: 225 ml
Conc. sulphuric acid	: 20 ml
Chromium troxide	: 25 g

Acetic anhydride (100 g, 92.5 ml, 1 mol) and *p*-nitrotoluene (12.5 g, 0.09 mol) are taken in a three necked flask (500 ml capacity) fitted with a mechanical stirrer and a dropping funnel. The mixture is cooled to 0°C (ice-salt mixture) and concentrated sulphuric acid (20 ml) is added dropwise to the stirred solution. A solution of chromium trixoide in acetic anhydride (obtained by adding 25 g, 0.25 mol chromium oxide portionwise to well cooled 125 ml acetic anhydride) is added at such a rate that the temperature does not exceed 10°C. The mixture is stirred for 2 hours more and poured over crushed ice-water (approx. 1 kg). The separated *p*-nitrobenzylidene diacetate is filtered and washed with cold water until the washings are colourless. It is suspended in cold sodium carbonate solution (2%), stirred for 5 minutes, filtered and washed with cold water and finally with cold alcohol (5 ml). It is crystallised from alcohol. The yield is 15 g (65%); m.p. is 125–126°C.

The crude *p*-nitrobenzylidene diacetate is refluxed in a mixture of ethanol (35 ml), water (35 ml) and concentrated sulphuric acid (3.5 ml) for 30 minutes. The mixture is filtered and cooled. The separated *p*-nitrobenzaldehyde is filtered, washed with cold water and dried in a vacuum desiccator. The yield is 7.5 g (55%); m.p. is 106°C.

Note: This is convenient method for converting a CH_3 group to —CHO group. See also Jones oxidation (page 137) and Sarett oxidation (page 206).

70.3 Potassium Dichromate

Potassium or sodium dichromate in presence of dilute sulphuric acid, is a convenient oxidising agent for the controlled oxidation of primary alcohols to aldehydes. To avoid further oxidation to the corresponding acids, the aldehyde is removed as rapidly as possible by distillation through a fractionating column. Thus,

$$\underset{\substack{n\text{-butyl alcohol} \\ \text{b.p.}: 117^\circ}}{CH_3CH_2CH_2CH_2OH} \xrightarrow[K_2Cr_2O_7|H_2SO_4]{[O]} \underset{\substack{n\text{-butyraldehyde} \\ \text{b.p.}: 75^\circ}}{CH_3CH_2CH_2CHO} + H_2O$$

Sodium dichromate is also used for the oxidation of a side chain. The molecule should not contain any other groups (e.g. NH_2, OH) which are affected by the oxidising agent. Thus,

$$\underset{p\text{-nitrotoluene}}{p\text{-}NO_2C_6H_4CH_3} \xrightarrow[Na_2Cr_2O_7 + H_2SO_4]{[O]} \underset{p\text{-nitrobenzoic acid}}{p\text{-}NO_2C_6H_4COOH}$$

Following oxidations are easily carried out;

(i) *n*-butyraldehyde from *n*-butanol,
(ii) cyclohexanone from cyclohexanol,
(iii) adipic acid from cyclohexene,
(iv) *p*-nitrobenzoic acid from *p*-nitro toluene,
(v) phenanthraquinone from phenanthrene, and
(vi) terephthalic acid from *p*-xylene.

70.3.1 n-Butyraldehyde

It is obtained by the oxidation of *n*-butyl alcohol with sodium dichromate and dilute sulphuric acid

$$\underset{n\text{-butyl alcohol}}{CH_3CH_2CH_2CH_2OH} \xrightarrow[Na_2Cr_2O_7 + H_2SO_4]{[O]} \underset{n\text{-butyraldehyde}}{CH_3CH_2CH_2CHO} + H_2O$$

n-Butyl alcohol : 25.5 ml
Sodium dichromate : 28 g
Conc. sulphuric acid: 20 ml

n-Butyl alcohol (20.5 g, 25.5 ml, 0.279 mole) is taken in a two necked R.B. flask (250 ml capacity) fitted with a dropping funnel and a fractionating column (6″) with a distillation arrangement. A solution of sodium dichromate in dilute sulphuric acid (obtained by adding cautiously with stirring, 20 ml concentrated sulphuric acid to a suspension of 28 g sodium dichromate in 100 ml water) is added during 15 minutes to the warm *n*-butyl alcohol. The rate of addition should be such that the temperature does not rise above 80–85°C. Since the oxidation proceeds with evolution of heat the reaction flask has to be cooled in ice. However, the temperature of the reaction mixture should not be allowed to fall below 75°C (if necessary the flask has to be heated with a small flame). After the addition of dichromate solution is complete, the reaction mixture is heated with a small flame for 15 minutes. The fraction that passes over below 90°C is collected. It is dried (anhydrous magnesium sulphate) and distilled. The yield is 9 g (45%); b.p. is 74–75°C.

70.3.2 Cyclohexanone

It is obtained by oxidation of cyclohexanol with sodium dichromate and dilute sulphuric acid.

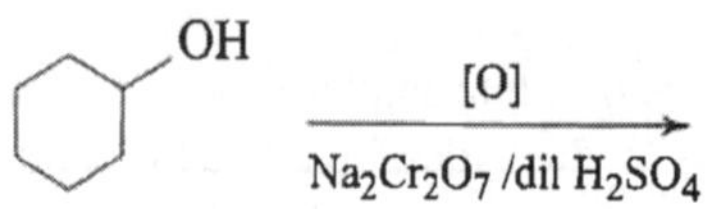

Cyclohexanol : 15.6 g
Sodium dichromate : 29.6 g
Conc. sulphuric acid: 15 ml

Sodium dichromate solution (obtained by cautiously adding 15 ml concentrated sulphuric acid to a suspension of 29.6 g sodium dichromate and 175 ml water) is added in one lot to cyclohexanol (15.6 g

0.156 mole) contained in an R.B. flask (250 ml capacity). The mixture is shaken. Considerable heat is evolved in the oxidation. The temperature of the reaction mixture is kept between 55–60°C (a cold water bath way be used if necessary). When the reaction is complete (as indicated by no further increase of temperature), water (125–150 ml) is added. The mixture is distilled and the distillate (about 100 ml) collected. It is saturated with salt (about 20–25 g) and the upper layer of cyclohexanone separated. The aqueous layer is extracted with ether (20 ml). The combined upper layer and ether extract is dried (anhydrous sodium sulphate) and ether evaporated and the residue distilled. The yield is 10.4 g (68%); b.p. is 154–155°C.

70.3.3 Adipic acid

It is obtained by the oxidation of cyclohexene with potassium dichromate and dilute sulphuric acid.

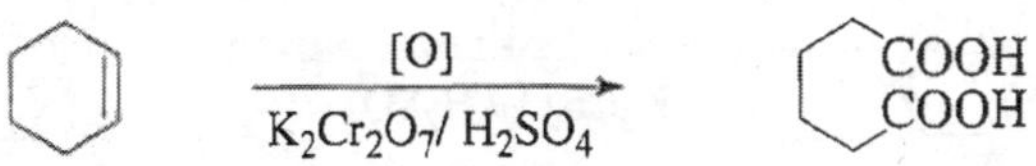

Cyclohexene Adipic acid

Cyclohexene : 12 g
Potassium dichromate : 10 g
Conc. sulphuric acid : 35 ml

Cyclohexene (12 g, 0.146 mole) is added to dilute sulphuric acid (obtained by cautiously adding 35 ml concentrated sulphuric acid to 35 ml water) at room temperature. The mixture is vigorously shaken and a solution of potassium dichromate (12 g) in water (15 ml) is added dropwise. The addition is done at such a rate so as to maintain the temperature of the reaction between 40–50°C. Finally, the reaction mixture is heated to 80–90°C for few minutes and then cooled to 5–10°C in an ice-salt bath and kept at this temperature for 10–15 minutes. The solution become dark green due to reduction of Cr^{6+} to Cr^{3+}. The separated product is filtered, washed with ice cold water and recrystallised from hot water. The yield is 10 g (46.3%); m.p. is 152°C.

Note: Adipic acid can also be obtained by the oxidation of cyclohexanol with nitric acid (page 165).

70.3.4 p-Nitrobenzoic acid

It is obtained by the oxidation of *p*-nitrotoluene with sodium dichromate and dilute sulphuric acid.

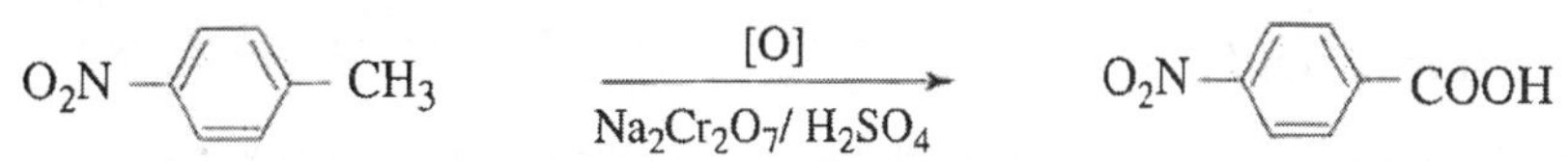

p-Nitrotoluene : 14 g
Sodium dichromate : 40 g
Conc. sulphuric acid : 49 ml

Concentrated sulphuric acid (49 ml) is added slowly (dropping funnel) to a stirred mixture of *p*-nitrotoluene (14 g, 0.1 mole), sodium dichromate (40 g) and water (90 ml). Considerable heat is evolved, the *p*-nitrotoluene melts and oxidation proceeds The reaction mixture is occasionally cooled in a cold water bath. After the addition of acid, the mixture is refluxed by gentle boiling on a wire guage for 15 minutes. It is cooled and poured into cold water (130 ml). The separated product is filtered and washed with water (60–70 ml). It is digested by heating on a water bath with dilute sulphuric acid (5%, 75 ml obtained by adding 2.7 ml concentrated sulphuric acid to 75 ml water) with stirring for 20 minutes (to remove the chromium salts). The mixture is cooled and filtered. The crude product is treated with dilute sodium hydroxide solution (5%, 100–120 ml) until the liquid is alkaline. The mixture is warmed to 50°C with decolourising carbon, stirred for 5 minutes and filtered. The clear alkaline filtrate is added to well stirred dilute sulphuric acid and (15%, 120 ml obtained by adding 15 ml concentrated sulphuric acid to 150 ml water). The precipitate of *p*-nitrobenzoic acid is filtered, washed with water and dried. It is crystallised from benzene or glacial acetic acid. The yield is 13,2 g (77.4%); m.p. is 238–239°C.

70.3.5 Phenanthraquinone

It is obtained by the oxidation of phenanthrene with potassium dichromate and dilute sulphuric acid.

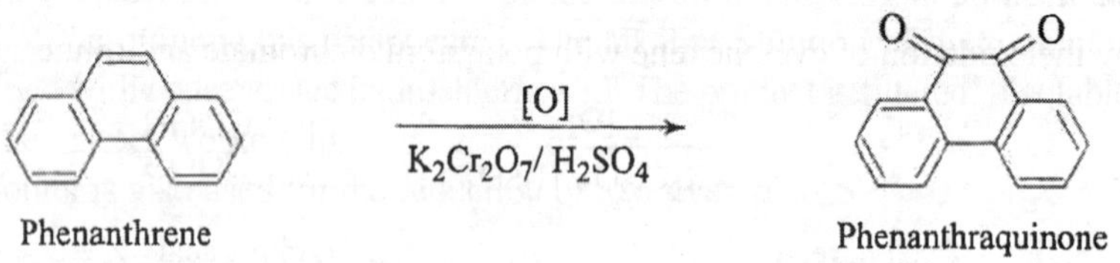

Phenanthrene : 3 g
Potassium dichromate : 18 g
Conc. sulphuric acid : 30 ml

To a suspension of phenanthrene (3 g, 0.017 mole) in dilute sulphuric acid (obtained by cautiously adding 30 ml concentrated sulphuric acid to 60 ml water with stirring) is added at 90–95°C (water bath), potassium dichromate (18 g) in small lots (0.5–1 g) until a vigorous reaction sets in. The external heating is removed and the temperature of the mixture is approximately 110–115°C. Addition of potassium dichromate is continued. The temperature of the reaction mixture is not allowed to fall below 85°C (hot water bath is used if necessary). Finally, the reaction mixture is heated on a boiling water bath for 30 minutes. It is cooled, water (200 ml) is added, and the crude product filtered and washed with water until it is free from chromium salts. It is purified by suspending in rectified spirit (30 ml) and stirring with saturated sodium bisulphite solution (30 ml). The mixture is stirred for 10 minutes, diluted with water (175 ml) and the clear solution (filter if necessary) containing the bisulphite addition product is treated with saturated sodium carbonate solution. The precipitated phenanthraquinone is filtered, washed with water and crystallised from glacial acetic acid. The yield is 2 g (57%); m.p. is 206–207°C.

70.3.6 Terephthalic acid

It is obtained by oxidation of *p*-xylene with sodium dichromate and dilute sulphuric acid.

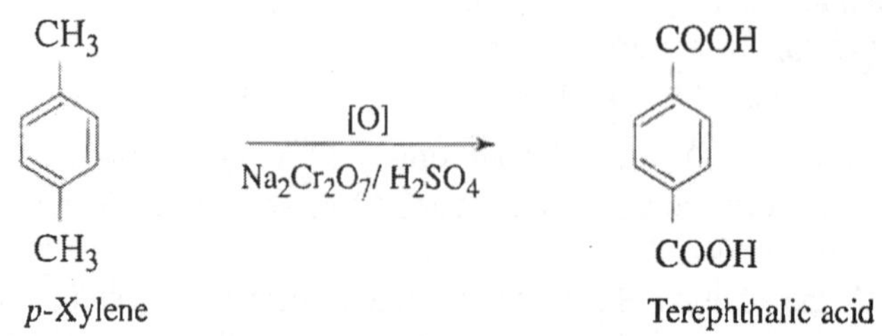

p-Xylene : 15 ml (12.5 g)
Sodium dichromate : 70 g
Conc. Sulphuric acid : 170 g (92.5 ml)

A mixtrure of *p*-xylene (12.5 g, 15 ml, 0.118 mole), sodium dichromte (70 g) and water (300 ml) is placed in a three necked one litre R.B. flask fitted with a sealed mechanical stirrer, a reflux condenser and a dropping funnel. Concentrated sulphuric acid (170 g, 92.5 ml) is added dropwise (dropping funnel) to the well stirred mixture (35–40 minutes). When all the acid has been added, the temperature begins to fall, and the mixture is gently refluxed for 30 minutes. The reaction mixture is poured into water (600 ml) and allowed to stand for one hour. The separated terephthalic acid is filtered, washed with cold water (30 ml) and then by ether (30 ml). It is purified by dissolving in sodium hydroxide solution (5%, 70 ml) and the clear filtrate (obtained by filtration) is added to well stirred dilute sulphuric acid (15%, 225 ml obtained by cautiously adding 20 ml concentrated sulphuric acid to 200 ml water). The yield is 9 g (46%). It sublimes at 300°C without melting.

Note: Terephthalic acid is also obtained by the oxidation of *p*-xylene with alkaline potassium permanganate (page 159).

70.4 Concentrated Nitric Acid

Nitric acid is a strong oxidising agent. Unless used under controlled conditions, it completely oxidises the organic compound into carbon dioxide and water. However, as has already been stated, it along with concentrated sulphuric acid is used for nitration of organic compounds.

Following examples illustrate the use of nitric acid as an oxidising agent: adipic acid from cyclohexanol, benzil from benzoin (see benzil-benzilic acid rearrangement, page 36), oxalic acid from cane sugar.

70.4.1 Adipic acid

It is obtained by the oxidation of cyclohexanol (or cyclohexanone) with nitric acid.

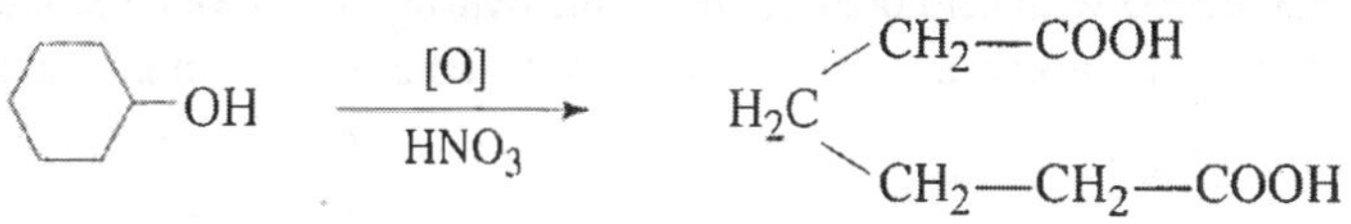

Adipic acid

Cyclohexanol : 30 g
Nitric acid (50%) : 126 g (96 ml)

Nitric acid (96 ml, 95%, sp. gr. 1.32) is taken in a two necked R.B. flask (500 ml capacity) fitted with a reflux condenser and a dropping funnel. Cyclohexanol (30 g, 0.3 mole) is added dropwise to the nitric acid, which has been previously heated just to the boiling point. A vigorous reaction with the evolution of nitrous oxide fumes sets in. Addition of cyclohexanol is continued dropwise (fresh drop of cyclohexanol is not added till the previous drop has reacted). The addition is so controlled so as to keep the reaction mixture at the boiling point. Under no circumstances is cyclohexanol allowed to accumulate in the flask, otherwise an explosive reaction may occur. The addition of cyclohexanol takes about 100 minutes. The reaction mixture is finally refluxed for 15 minutes. The separated adipic acid is filtered (sintered funnel) and washed with cold water (20 ml). It is crystallised from concentrated nitric acid (25–30 ml). The yield is 25 g (56.3%); m.p. is 152°C.

Note: Adipic acid is also obtained from cyclohexene by oxidation of potassium dichromate and dilute sulphuric acid (page 163).

70.4.2 Oxalic acid

It is obtained by oxidation of cane sugar with concentrated nitric acid.

$$\underset{\text{Cane sugar}}{C_{22}H_{12}O_{11}} + 18\,O \xrightarrow[\text{Oxidation}]{\text{Conc. } HNO_3} \underset{\text{Oxalic acid}}{6\,(COOH)_2} + 5H_2O$$

Cane sugar : 20 g
Conc. nitric acid : 100 g (70 ml, sp. gr 1.42)

Cane sugar (20 g, 0.058 mole) is added in one lot to concentrated nitric acid (100 g, 70 ml, sp. gr. 1.42) contained in a conical flask (500 ml capacity). The mixture is warmed in a water bath till the reaction starts. Subsequently, it is allowed to proceed with the application of external heat. In case the reaction becomes too violent with evolution of brown fumes of oxides of nitrogen, it is cooled in water. The mixture is finally kept for 24 hours at room temperature. The separated oxalic acid is filtered (sintered funnel). Further quantity is obtained by concentration of the mother liquor on a water bath and subsequent cooling. Total yield 5 g (16%). It is crystallised from hot water as a hydrate; m.p. is 101.5°C.

70.5 Hydrogen Peroxide

Hydrogen peroxide is a mild oxidising agent. It is used for the oxidation of phenolic aldehydes to a hydroxyl group. For details see Dakin's reaction.

Hydrogen peroxide is used for the preparation of peracids, which in turn are used to oxidise the aldehyde group to phenolic group. For details see Baeyer-Villiger oxidation.

It is also used for the oxidation of easily oxidisiable or reactive C=C bond in aromatic compound. Thus, diphenic acid is obtained when phenanthrene is oxidised with 30 per cent hydrogen peroxide in glacial acetic acid solution at 85°C. No phenanthraquinone is formed under these conditions (compare the oxidation of phenanthrene with acid dichromate). This hydrogen peroxide oxidation of phenanthrene is, essentially oxidation by peracetic acid, of the reactive 9–10 positions in phenanthrene.

70.5.1 Diphenic acid (biphenyl-2,2′-dicarboxylic acid)

It is obtained by the oxidation of phenanthrene with hydrogen peroxide in glacial acetic acid medium.

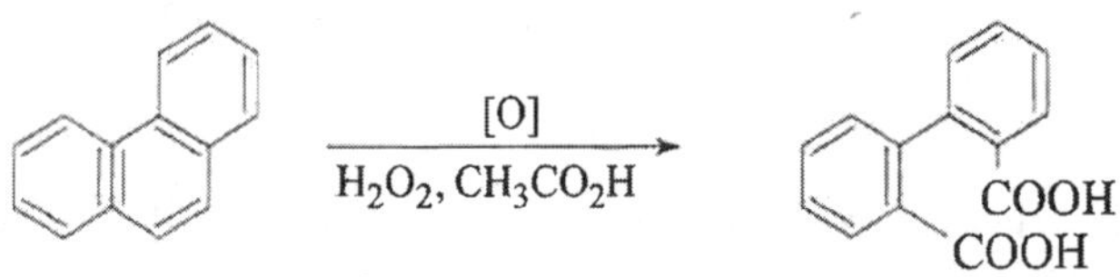

Phenanthrene : 17.8 g
Acetic acid (glacial) : 200 ml
Hydrogen peroxide (30%) : 71 ml

Hydrogen peroxide (30%, 71 ml, 2 mol) is added slowly during 40–45 minutes to a stirred solution of phenanthrene (17.8 g, 0.1 mol) in glacial acetic acid (200 ml, 85°C) contained in a three necked flask (1 litre capacity) fitted with a sealed stirrer, a reflux condenser and thermometer. After the addition is complete, the temperature falls to 80°C, the mixture is heated on a water bath with stirring for 3–4 hours. The solution is concentrated to approximately half by distillation under reduced pressure and then cooled. The precipitated diphenic acid is filtered, the filtrate is evaporated almost to dryness under reduced pressure. The residue is extracted with sodium carbonate solution (10%, 175 ml) by warming on a water bath. The alkaline extract is boiled with a little decolourising carbon, filtered and the filtrate acidified with dilute hydrochloric acid to pH 4.5. The solution is cooled to 0°C, any tarry material separated is filtered and the clear solution acidified with dilute hydrochloric acid. The separated diphenic acid is filtered and washed with cold water. Total yield is 16.6 g (69%). It is crystallised from glacial acetic acid, m.p. 229–230°C.

70.6 Selenium Dioxide

It is used for the oxidation of the CH_2 group adjacent to the carbonyl group of aldehydes and ketones to the CO group. Thus, aldehydes and ketones are oxidised to 1, 2-dicarbonyl compounds. Acetophenone, for example, is converted into phenyl glyoxal, C_6H_5—COCHO, in 70% yield by warming with selenium dioxide in aqueous dioxane. Similarly, cyclohexanone and phenylbenzyl ketone are converted into cyclohexane-1,2-dione and benzil, $C_6H_5COCOC_6H_5$ respectively.

Mechanism: The reaction is believed to proceed by way of a β-ketoseleninic acid, formed by electrophilic attack of seleninic acid on the enol. Various steps are:

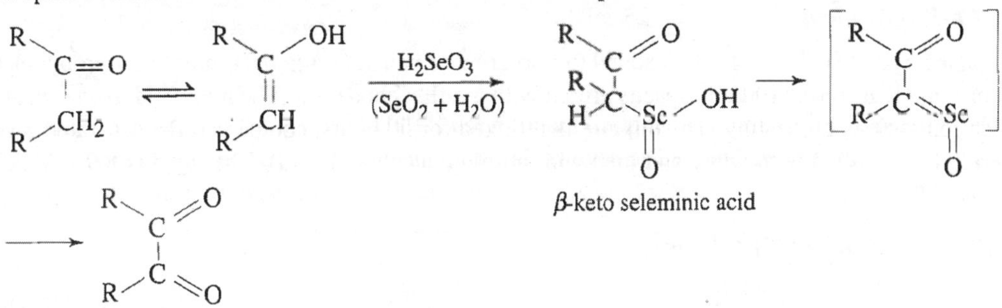

70.6.1 Benzil

It is obtained by the oxidation of phenylbenzyl ketone (desoxybenzoin) with selenium dioxide.

$$\underset{\text{Desoxybenzoin}}{C_6H_5\overset{O}{\overset{\|}{C}}CH_2C_6H_5} \xrightarrow[\text{[O]}]{SeO_2} \underset{\text{Benzil}}{C_6H_5-\overset{O}{\overset{\|}{C}}-\overset{O}{\overset{\|}{C}}-C_6H_5}$$

Desoxybenzoin	: 4 g
Selenium dioxide	: 3 g
Acetic acid (glacial)	: 30 ml

A mixture of desoxybenzoin (4 g, 0.02 mole) selenium dioxide (3 g, 0.027 mole), water (12 ml) and glacial acetic acid (30 ml) is heated at 90°C for 4–5 hours in a R.B. flask (100 ml capacity) fitted with a reflux condenser. Red selenium gradually separates. The mixture is cooled and extracted with ether (2 × 25 ml). The ether extract is washed with saturated potassium carbonate, water, and then dried (anhydrous magnesium sulphate). Ether is distilled and the residue crystallised from alcohol. The yield is 2.8 g (65.4%); m.p. is 95°C.

70.7 Mercuric Oxide

It is a mild oxidising agent for some specific use. Thus, oxidation of benzil dihydrazone with mercuric oxide gives diphenylethyne.

$$C_6H_5-\underset{N-NH_2}{\underset{\|}{C}}-\overset{N-NH_2}{\overset{\|}{C}}-C_6H_5 \xrightarrow[2HgO]{[O]} C_6H_5C\equiv CC_6H_5 + 2N_2 + 2H_2O + 2Hg$$

70.7.1 Tolan (Diphenylethyne)

Step 1 Benzil hydrazone

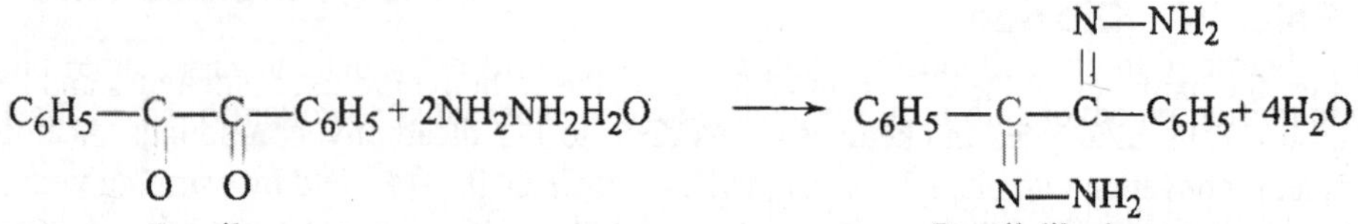

Benzil   Benzil dihydrazone

Benzil	: 8.4 g
Hydrazine hydrate (85%)	: 6.1 ml
Ethanol (absolute)	: 15 ml
n-Propyl alcohol	: 25 ml

Hydrazine hydrate (85%, 6.1 ml) is added to a solution of benzil (8.4 g, 0.04 mole) in *n*-propyl alcohol (25 ml) in a R.B. flask (100 ml capacity) fitted with a reflux condenser. The monohydrazone separates, which redissolves on heating. The mixture is refluxed for 50 hours, cooled and the separated product filtered. It is washed with water and then with absolute alcohol (15 ml). The yield is 8.9 g, (93.7%); m.p. is 150–151°C.

Step 2 Tolan (diphenylethyne)

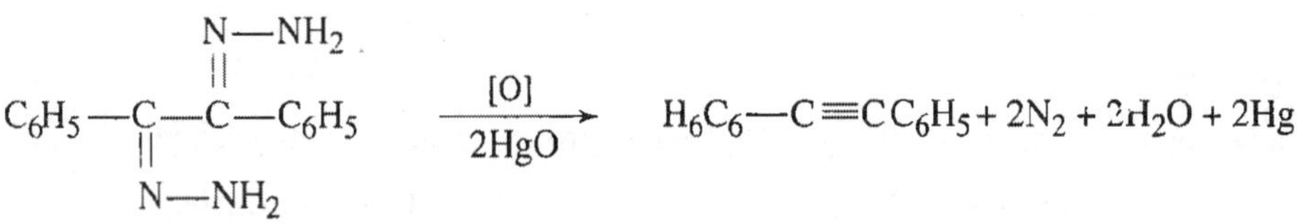

Benzildihydrazone	: 5 g
Benzene	: 30 ml
Yellow mercuric oxide	: 12 g

Yellow mercuric oxide (0.5 g) is added to a suspension of benzil-dihydrazone (5 g, 0.02 mole) in benzene (30 ml) in a three necked R.B. flask (100 ml capacity) fitted with a sealed stirrer and a reflux condenser. The reaction mixture is warmed on a water bath. Evolution of nitrogen takes place and the mixture turns grey. The remaining yellow mercuric oxide (11.5 g) is added in small lots so as to keep the reaction mixture gently refluxing. The reaction mixture is stirred for one hour, left overnight and filtered. The residual mercuric oxide is washed with benzene (5 ml). The combined benzene solution is dried (anhydrous sodium sulphate) and distilled. The residual product is distilled under reduced pressure (0.2–0.3 mm). The fraction between 95–105°C is collected. It solidifies on cooling and is crystallised from alcohol. The yield is 2.4 g (67.4%); m.p. is 60–61°C.

70.8 Potassium Bromate

Potassium or sodium bromate is used as a selective oxidising agent. It oxidises *p*-dihydroxy compounds to *p*-benzoquinone. Thus,

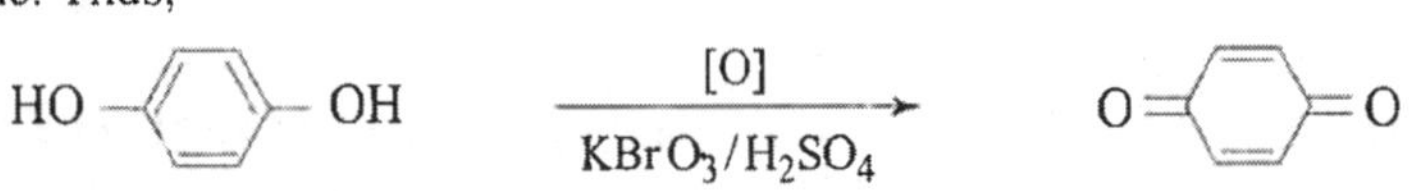

Hydroquinone   *p*-benzoquinone

It also oxidises the secondary alcoholic group to keto group. Thus, alkaline sodium bromate has been used for the conversion of benzoin into benzilic acid. This reaction occurs via the intermediate formation of benzil. Thus,

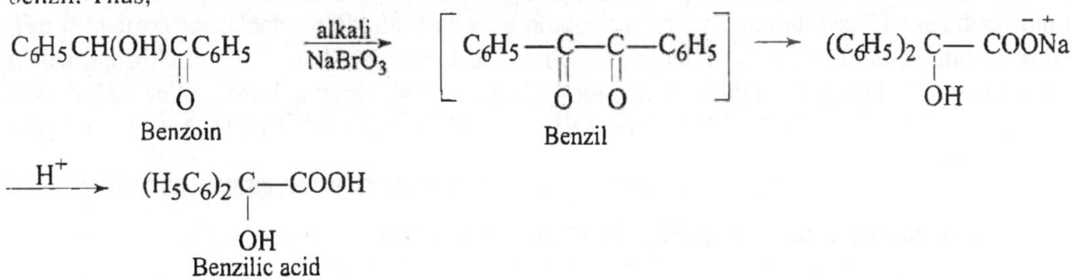

70.8.1 p-Benzoquinone

Hydroquinone : 10 g
Potassium bromate : 5.6 g
Sulphuric acid (1N) : 5 ml

A mixture of hydroquinone (10 g, 0.09 mole), potassium bromate (5.6 g, 0.33 mol), sulphuric acid (1N, 5 ml) and water (100 ml) is slowly heated to 60°C on a water bath. After a few minutes, the temperature is raised to 80°C and maintained for 10 minutes. The reaction mixture is cooled in an ice bath. The separated quinone is filtered, washed with cold water and dried in a vacuum desiccator over fused calcium chloride.The yield is 7 g (71.4%); m.p. is 115°C. It is crystallised from petroleum ether; m.p. is 117°C.

70.8.2 Benzilic acid

Benzoin : 4 g
Potassium bromate : 1 g
Sodium hydroxide : 3.7 g

Benzoin (4 g, 0.019 mole) is added in portions to a stirred solution of sodium hydroxide (3.7 g) and potassium bromate (1 g) (or sodium bromate 0.9 g) in water (7 ml) at 85–90°C in an evaporating dish. The temperature should not exceed 90°. A small quantity of water is added from time to time to prevent the mixture from becoming too thick. The heating and stirring is continued (1.5–2 hours) until a test portion is almost completely soluble in water. To the reaction mixture is added water (30 ml) and the mixture is cooled in ice-water. The separated benzhydrol (byproduct) if formed is filtered. A little of the filtrate (1–2 ml) is kept aside and the remaining filtrate is acidified with dilute sulphuric acid (1:3 by volume, 13 ml) added slowly with stirring to a point just short of liberation of bromine. If more acid has been added, then the filtrate kept separately is added and then sufficient sulphuric acid added to the end point. The separated benzilic acid is filtered and washed with water. The yield is 3.1 g, (72%); m.p. is 149–150°C.

Note: Benzilic acid is also obtained by the action of alcoholic potassium hydroxide on benzil (see page 36).

70.9 Sodium Hypobromite

Methyl ketones, R—CO CH_3, can be oxidised into a carboxylic acid by the use of the haloform reaction (see page 115). This involves the treatment of the methyl ketone with alkaline hypohalite reagent. A trihalomethyl ketone, which is initially formed undergoes hydrolysis under basic conditions used.

$$2H\bar{O} + Br_2 \rightleftharpoons :\ddot{\bar{O}}Br + :\bar{B}r + H_2O$$

$$RCOCH_3 + :\bar{O}Br \rightleftharpoons [RCO\ddot{C}H_2]^- + HOBr$$

$$RCO\ddot{C}H_2^{\ominus} + Br—OH \longrightarrow RCOCH_2Br + :\bar{O}H$$

$$RCOCH_2Br \longrightarrow RCOCBr_3 \xrightarrow{:\bar{O}H} RCOO^- + CHBr_3$$

70.9.1 2,2-Dimethylpropionic acid (trimethyl acetic acid)

It is obtained by the oxidation of pinacolone with sodium hypobromite.

$$\underset{\text{Pinacolone}}{(CH_3)_3C—\overset{\overset{O}{\|}}{C}—CH_3} \xrightarrow[\text{(ii) } H_3\overset{+}{O}]{\text{(i) NaOBr}} \underset{\text{Trimethyl acetic acid}}{(CH_3)_3C·CO_2H} + CHBr_3$$

Pinacolone : 40 g
Sodium hydroxide : 128 g
Bromine : 192 g (62.6 ml)

Bromine (62.6 ml, 1.2 mol) is added dropwise to a stirred and well cooled (0°C) solution of sodium hydroxide (128 g, 3.2 mol) in water (1220 ml) contained in a three necked R.B. flask fitted with a mechanical stirrer and a dropping funnel. During the addition of bromine, the solution is continuously stirred and the temperature is not allowed to rise above 10°C (15–20 minutes). The mixture is cooled again to 0°C (ice-salt bath) and pinacolone (40 g, 0.4 mol) is added keeping the temperature below 10°C. After the colour of bromine has disappeared (one hour) the mixture is stirred for 3 hours more at room temperature. The reaction mixture is steam distilled to remove any carbon tetrachloride and bromo form. The reaction mixture is heated with a bunsen flame during steam distillation. It is cooled and concentrated sulphuric acid (160 ml) added cautiously though a dropping funnel. The reaction mixture is heated again in the distillation assembly. Trimethyl acetic acid passes over with about 200 ml of water. Triethyl acetic acid is in the upper layer, when a substance heavier than water distils (brominated pinacolone), the distillation is stopped. The upper layer of trimethylacetic acid is separated, dried (anhydrous sodium sulphate) and distilled. The yield is 26.4 g (55%); b.p. is 75–80°C/20 mm; m.p. is 34–35°C.

70.9.2 3,3-Dimethylacrylic acid

It is obtained by oxidation of mesityl oxide with sodium hypochlorite.

$$\underset{\text{Mesityl oxide}}{(CH_3)_2C═CH—\overset{\overset{O}{\|}}{C}—CH_3} \xrightarrow[\text{(ii) } H_3O^+]{\text{(i) NaOCl}} \underset{\text{3,3-Dimethylacrylic acid}}{(CH_3)_2C═CH\,COOH} + CHCl_3$$

Mesityl oxide : 30 g (23.2 ml)
Dioxane : 60 ml
Sodium hypochlorite : (See note)

Sodium hypochlorite solution (see note) is added to a stirred and cooled (10°C) solution of mesityl oxide (30 g, 23.2 ml, 0.3 mol) in dioxan (60 ml) contained in a three necked flask (2 litre capacity) fitted with a sealed stirrer, a reflux condenser and a dropping funnel. Heat is evolved during the reaction.

In about 5–10 minutes chloroform commences to reflux. If the reaction becomes too vigorous, it is cooled (ice-water). The mixture is kept stirred after the addition of sodium hypochlorite for 2–3 hours till the mixture attains room temperature. Excess of hypochlorite is decomposed by the addition of sodium metabisulphite (about 1 g), ie, until a test portion does not liberate iodine from potassium iodide solution. Sulphuric acid (50% about 60 ml) is added to the cooled reaction mixture until it is acidic to congo red. It is extracted with ether (8 × 50 ml). The combined ether extract is dried (anhydrous calcium sulphate) and the ether evaporated. The residual product is distilled under reduced pressure (20 mm) and the fraction at 100–106°C is collected. The yield is 15.2 g (51%). It is crystallised from petroleum ether, m.p. is 68°C.

Note: Sodium hypochlorite can be prepared from sodium hydroxide and chlorine (F. Mallory, Org. Synth., 1963, coll vol **4**, 74). This reagent is commercially available as chlorox, a solution with minimum 5% chlorine. For the above experiment 600 ml of chlorox solution is diluted with 100 ml water and used.

70.9.3 2-Naphthoic acid

It is obtained by oxidation of 2-acetylnaphthalene with sodium hypochlorite.

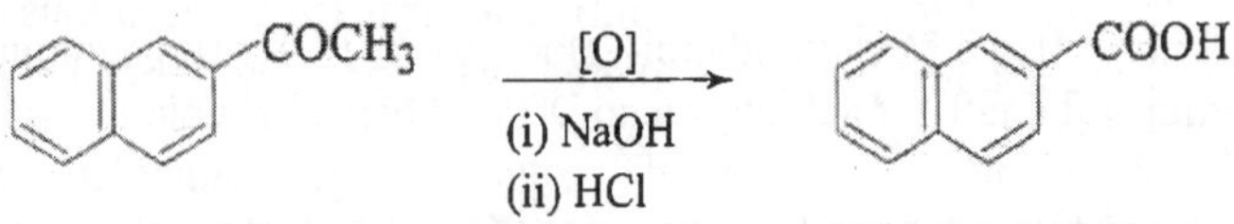

2-Acetylnaphthalene 2-Naphthoic acid

2-Acetylnaphthalene : 22.25 g
Sodium hypochlorite : 37.5 g (see note)

Sodium hypochloride solution (containing about 37.5 g, 0.5 mol) (see note) is taken in a three necked R.B. flask (2 litre capacity) fitted with a stirrer, a thermometer and a reflux condenser. 2-Acetyl naphthalene (22.25 g, 0.125 mol) is added to the stirred solution of sodium hypochlorite at 55°C. An exothermic reaction commences. The reaction mixture is maintained at 60–70°C by cooling in an ice bath if necessary (30 minutes). The mixture is stirred for 30 minutes more and excess hypochlorite destroyed by adding sodium metabisulphite solution (25%, 50 ml). It is to be ascertained that no hypochlorite remains, by testing a solution with acidified potassium iodide solution. The reaction mixture is acidified with concentrated hydrochloric acid (50 ml). The product obtained is filtered, washed with water and crystallised from alcohol. The yield is 18.75 g (87%); m.p. is 184–85°C.

Note: Sodium hypochlorite solution is obtained by diluting 600 ml of commercially available chlorox (contains about 5% chlorine) with 150 ml water.

70.10 Potash Fusion

On fusion of the compound with potassium hydroxide, the —CHO is oxidised to —COOH. Thus,

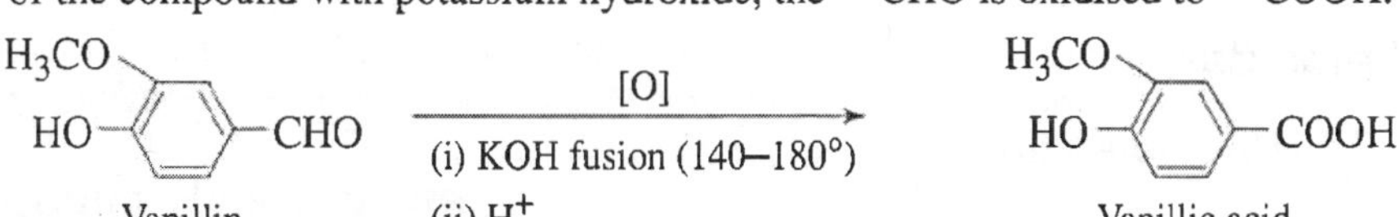

Vanillin Vanillic acid

However, on fusion with a mixture of sodium hydroxide and potassium hydroxide at a comparatively higher temperature (190–195°C), vanillin is converted into 3,4-dihydroxy benzoic acid (protocatechuic acid); in this case there is simultaneous oxidation of the —CHO group and demethylation of the methyoxyl group.

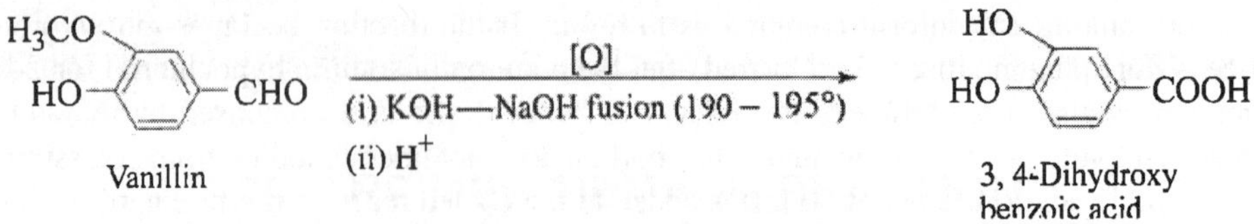

70.10.1 Vanillic acid (3-methoxy-4-hydroxybenzoic acid)

Vanillin : 20 g
Potassium hydroxide : 56 g

A mixture of potassium hydroxide (56 g, 1.0 mol) and water (8 ml) is placed in a nickel dish set in a deep sand bath and heated by the bunsen burner. The alkali is heated to 120°C with stirring. To the viscous mass is added vanillin (20 g, 0.13 mol) in small portions during 10 minutes. The reaction mixture is stirred after each addition. Considerable evolution of hydrogen takes place during the reaction. The temperature rises to 140°C and the mixture becomes fluid. After all the vanillin has been added, the pasty mixture is heated to 180°C and stirring continued for 10 minutes more at this temperature. The reaction mixture is cooled, dissolved in water (80 ml). The clear red solution thus obtained is acidified with hydrochloric acid. The separated product is filtered, washed with water and crystallised from dilute alcohol. The yield is 22g (99.5%); m.p. is 208–209°C.

70.10.2 3,4-Dihydroxybenzoic acid (protocatechuic acid)

Vanillin : 30.4 g
Sodium hydroxide : 16 g
Potassium hydroxide : 67.2 g

A mixture of sodium hydroxide (16 g, 0.4 mol), potassium hydroxide (67.2 g, 1.2 mole) and water (10 ml) is heated in a nickel dish at 160°C over a wire gauge and a bunsen burner. Vanillin (30.4 g, 0.2 mol) is added in portions during 3 minutes, at a rate sufficient to maintain the reaction temperature 190–195°C. The temperature is then raised to 240–245°C with constant stirring. The mixture is heated at this temperature for 5 minutes and then cooled with stirring. When the mixture has cooled to 150–160°C, water (200 ml) is added and the mixture stirred to get a clear solution. The clear solution is acidified with hydrochloric acid (6N, 300 ml). The separated protocatechuic acid is filtered and washed with water. Some more acid is obtained by ether extraction of the mother liquor. Total yield is 27 g (87%); m.p. is 198–199°C.

70.11 Potassium Persulphate

See Elbs persulphate oxidation (page 80).

70.12 Peracids

See Baeyer-Villiger oxidation (page 26).

70.13 Iodine-pyridine

Methyl ketones of the type Ar—$COCH_3$ react with iodine in presence of pyridine to give the pyridinium iodides in good yields (95%). The salts are cleaved by alkali to give the corresponding acids. Thus,

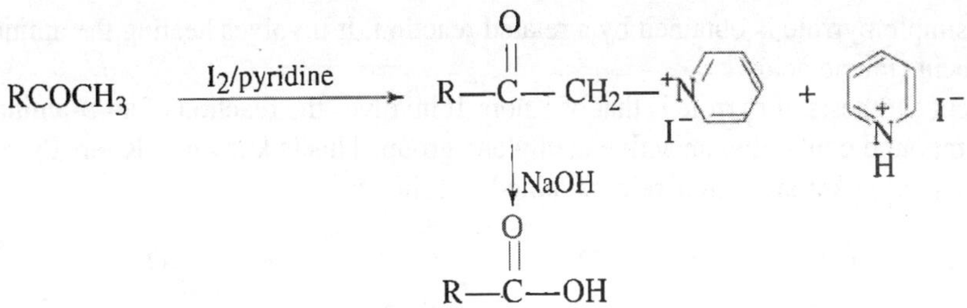

70.13.1 2-Hydroxy-4-Methoxybenzoic acid

It is obtained by the oxidation of 2-hydroxy-4-methoxyacetophenone with iodine-pyridine

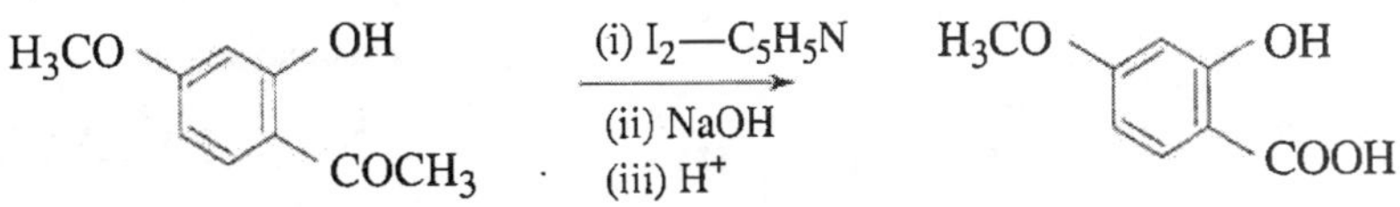

2-Hydroxy-4-methoxy acetophenone	: 2 g
Iodine	: 3 g
Pyridine	: 20 ml

2-Hydroxy-4-methoxyactophenone (2g, 0.012 mol), iodine (3 g, 0.012 mol) and dry pyridine (20 ml) is heated under anhydrous conditions on a steam bath for one hour. The solution is kept at 0°C for 24 hours. The separated pyridinium iodide adduct is filtered and washed with cold water. It is heated on a steam bath with aqueous potassium hydroxide (2%, 120 ml). The clear, cooled alkaline solution is acidified with hydrochloric acid. The separated product is filtered and purified by dissolving in sodium carbonate solution and acidifying the clear alkaline solution. It is crystallised from dilute alcohol. The yield is 1.2 g (60%); m.p. is 160°C–161°C.

70.14 Chromyl Chloride

See Etard's reaction (page 89).

71. PAAL-KNORR SYNTHESIS[1-5]

The heteroaromatic ring system, pyrrole is obtained by the cyclisation of a 1,4-dicarbonyl compound with ammonia (or ammonium carbonate) or a primary amine. This is known as Paal-Knorr synthesis and is illustrated by the preparation of 2,5-dimethyl pyrrole.

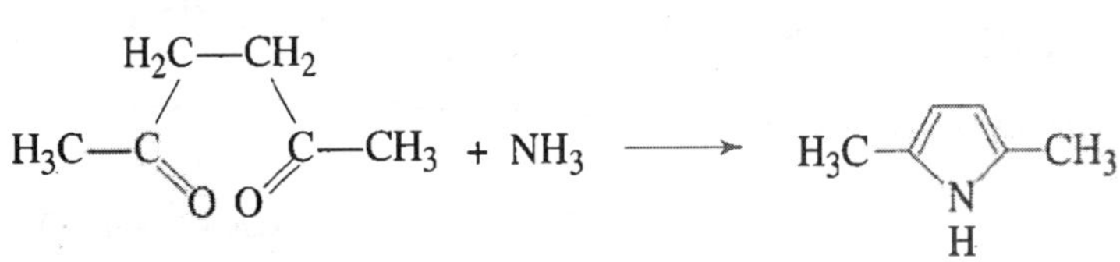

1. C. PALL, *Ber.*, 1885, **18**, 367.
2. L. KNOVV, *Ber*, 1885, **18**, 299.
3. L. KNORR, *Ber.*, 1884, **17**, 1635.
4. L. KNORR, *Ann.*, 1886, **236**, 290.
5. L. KNORR and H. LANSE, *Ber.*, 1902, **35**, 2998.

However, simple pyrrole is obtained by a related reaction. It involves heating the ammonium salt of *D*-galactaric acid (mucic acid).

A convenient synthesis of pyrrole is that of Knorr. It involves the reaction of an α-amino ketone with a carbonyl compound containing an active methylene group. This is known as Knorr Pyrrole synthesis. The required α-amino ketone is generally obtained as follows:

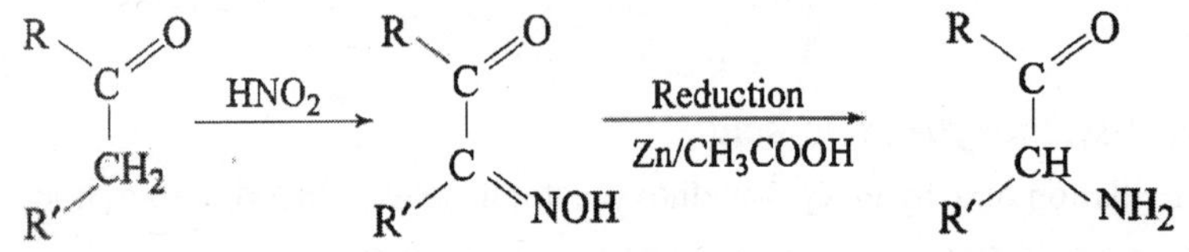

α-amino ketone

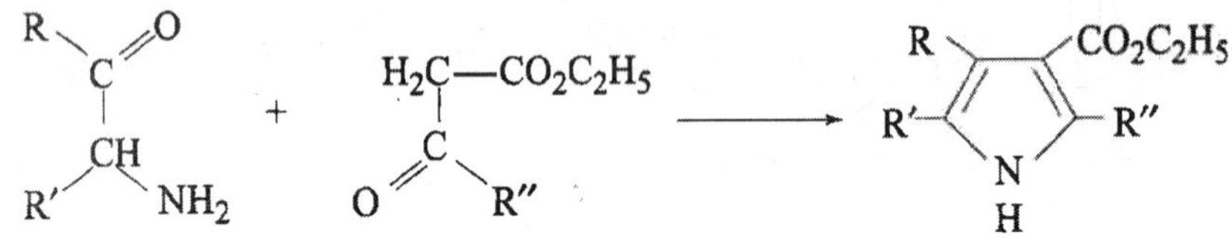

α-amino ketone — Carbonyl compound containing an active methylene group — Substituted pyrrole

The Knorr synthesis is illustrated by the preparation of 3,5-diethoxycarbonyl-2,4-dimethyl pyrrole.

71.1 2,5-Dimethyl pyrrole

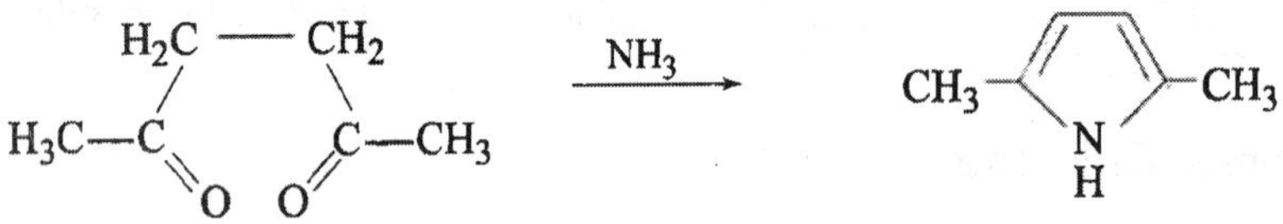

Hexane-2,5-dione — 2,5-dimethyl pyrrole

Step 1 ***Hexane-2,5-dione (acetonylacetone).***

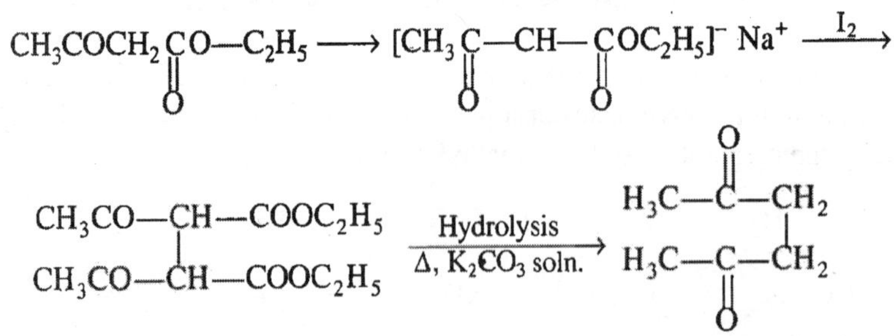

(hexane-2,5-dione)

Ethylacetoacetate	: 31.75 ml
Sodium	: 5.75 g
Ether (dry)	: 480 ml
Iodine	: 31.75 g

Pulverised sodium (5.75 g, 0.25 mole) is taken in a three necked R.B. flask (500 ml capacity) fitted with a sealed stirrer, reflux condenser and a dropping funnel. Sodium is covered with dry ether (120 ml) and the flask cooled (ice-water). A solution of ethylacetoacetate (31.75 ml, 0.25 mol) in dry ether (180 ml) is added slowly (dropping funnel) to the stirred suspension. The stirring is continued till all the sodium has reacted. A solution of iodine (31.75 g. 0.125 mol) in anhydrous ether (180 ml) is added (dropping funnel) till the iodine colour persists. The separated sodium iodide is filtered, washed with ether and the combined filtrate evaporated. The diethyl 2,3-diacetosuccinate obtained as a sticky mass is refluxed for one hour with potassium carbonate solution (20%, 125 ml). The clear yellow solution is cooled, potassium carbonate (50 g) added and extracted with ether (4 × 25 ml). The combined ether extract is dried (anhydrous sodium sulphate), distilled and the fraction between 185–192°C is collected. The yield is 3.7 g (26%).

Step 2 2,5-Dimethylpyrrole

Hexane-2,5-dione : 25 g
Ammonium carbonate : 50 g

A mixture of hexane-2,5-dione (25 g, 0.22 mol) and ammonium carbonate (50 g) is heated in a R.B. flask (250 ml capacity, fitted with an air condenser) at 100°C (oil bath) until efferversence ceases (1–1.5 hours). The mixture is then refluxed (use a reflux condenser) (oil bath, 115°C) for 30 minutes. The reflux condenser is washed with hot water (5 ml). The reaction mixture is cooled and extracted with chloroform (2 × 25 ml). The chloroform extract is dried (anhydrous sodium sulphate) and distilled using a small fractionating column under nitrogen atmosphere. The fraction between 78–80°C/25 mm is collected. The yield is 18 g (86%). The product is stored in a sealed container under nitrogen atmosphere.

71.2 3,5-Diethoxycarbonyl-2,4-dimethyl Pyrrole

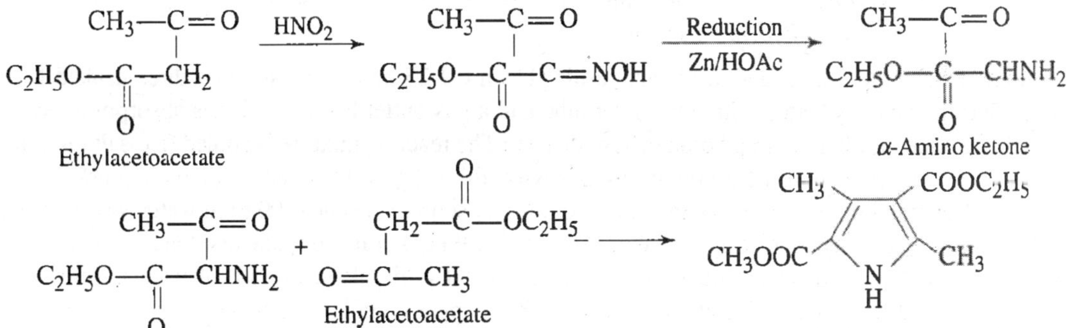

3,5-Diethoxy carbonyl-2,4-Dimethyl pyrrole

Ethylacetoacetate : 47.5 ml
Acetic acid (glacial) : 110 ml
Sodium nitrite : 13 g
Zinc dust : 25 g

Ethylacetoacetate (47.5 ml, 0.375 mol) and glacial acetic acid (110 ml) are taken in a three necked R.B. flask (500 ml capacity) fitted with a sealed stirrer and a dropping funnel. The mixture is cooled to 5°C (ice-salt mixture) and a cold solution of sodium nitrite (13 g, 0.187 mol) in water (20 ml) is added dropwise (dropping funnel) to the stirred mixture. The rate of addition is so adjusted (30 minutes) that the temperature remains between 5–7°C. The mixture is stirred for 30 minutes more and kept at room temperature for 4 hours. The reaction vessel is fitted with a reflux condenser and zinc dust (25 g, 0.375 mol) added in small portions (to the stirred solution) so that the mixture boils. If the reaction becomes too vigorous, it is cooled (ice-water). When all the zinc has been added (40–45 minutes), the mixture is stirred for one hour. More acetic acid is added in case the stirring becomes difficult. While still hot, the clear liquid is poured by decantation into water (1.2 litre) with stirring. The zinc residue is washed with acetic acid (2 × 20 ml) and the washings added to the water solution. The aqueous solution is left overnight and the separated product filtered, washed with water and dried (air). It is recrystallised from ethanol. The yield is 28.5 g (64%); m.p. is 136–137°C.

71.3 3,5-Dimethylpyrazole

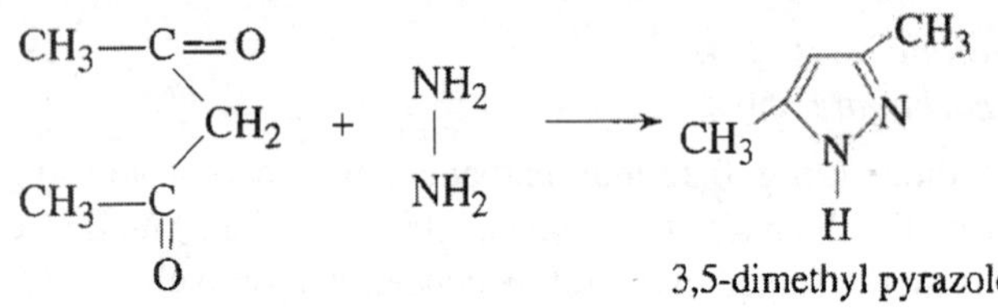

3,5-dimethyl pyrazole

Step 1 Acetyl acetone (Pentane-2,4-dione)

$$CH_3-\underset{\underset{O}{\|}}{C}-CH_3 + (CH_3CO)_2O \xrightarrow{BF_3} CH_3-\underset{\underset{O}{\|}}{C}-CH_2-\underset{\underset{O}{\|}}{C}-CH_3 + CH_3COOH$$

Acetone Acetic anhydride Acetyl acetone

Acetone : 36.5 ml
Acetic anhydride : 118 ml

Acetone (36.5 ml, 0.5 mol) and acetic anhydride (118 ml, 1.25 mol) are taken in a three neoked R.B. flask (500 ml capacity, fitted with a gas inlet tube and a gas outlet leading to a gas absorption device containing aqueous alkali to trap boron trifluoride gas.) The reaction mixture is cooled (ice-salt mixture) and a slow stream of boron trifluoride gas is passed till 125g is absorbed (5 hours, 2 bubbles per second). The reaction mixture is poured into a solution of sodium acetate (200 g) in water (400 ml) and the resulting mixture steam distilled till about 500 ml distillate is collected. The distillate is treated with a hot solution (85°C) of clear copper acetate solution (60 g dissolved in 750 ml water). Allow the mixture to stand for 3 hours (ice-chest). The separated copper complex is filtered, washed with water and air dried. It is treated with dilute sulphuric acid (20%, 200 ml) and ether (75 ml). The ether layer is removed (separatary funnel) and the aqueous layer extracted with ether (2 × 70 ml). The combined ether solution is dried (anhydrous sodium sulphate) and distilled using a short fractionating column. The yield is 40 g, (80%); b.p. is 134–136°C.

Note: It is convenient to use boron trifluoride-etherate (BF_3 $(C_2H_5)_2O$) which is commercially available and contains 48 per cent w/w of boron trifluoride. It is purified before use by adding 2 per cent by weight of dry ether and distilling from calcium hydrides under reduced pressure (46°C/10 mm).

Step 2 3,5-Dimethyl pyrazole

Acetyl acetone : 25 g
Hydrazine sulphate : 32.5 g

Hydrazine sulphate (32.5 g, 0.25 mol) is dissolved in sodium hydroxide solution (2.5 M, 200 ml) in a three necked flask (500 ml capacity) equipped with a mechanical stirrer, a thermometer and a dropping funnel. The flask is cooled to 15°C (ice bath) and acetyl acetone (25 g, 0.25 mol) added dropwise (30 minutes) (dropping funnel) with stirring. The reaction mixture is stirred for one hour at 15°C and treated with water (100 ml). The mixture is extracted with ether (2 × 50 ml). The combined ether extract is washed with saturated sodium chloride solution, dried (anhydrous potassium carbonate) and distilled. The resultant 3,5-dimethyl pyrazole is crystallised from petroleum ether. The yield is 18 g (75%); m.p. is 107–108°C.

71.4 3-Methyl-1-Phenylpyrazol-5-one

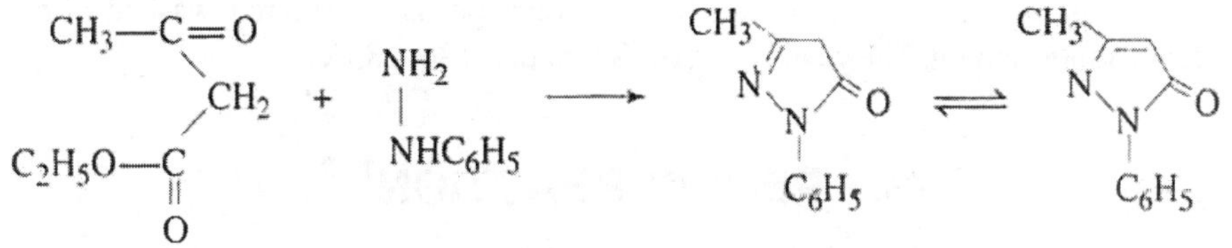

Ethylacetoacetate   Phenyl hydrezine   3-Methyl-1-phenylpyrozol-5-one

Ethyl acetoacetate : 4.9 ml
Phenyl hydrazine : 3.65 ml

A mixture of ethyl acetoacetate (4.9 ml, 0.0384 mol) and phenylhydrazine (3.65 ml. 0.037 mol) is heated (boiling water bath) in a conical flask for 2 hours. The reaction mixture is stirred with a glass rod. To the cooled reaction mixture is added ether (20 ml) and stirred. The separated product is filtered, washed with ether and recrystallised from dilute ethanol (1:1). The yield is 5.2 g (80%); m.p. is 127°C.

72. PECHMANN CONDENSATION[1]

The reaction of phenol with a β-ketoester in presence of concentrated sulphuric acid or polyphosphoric acid to give coumarin is known as Pechmann condensation. Thus,

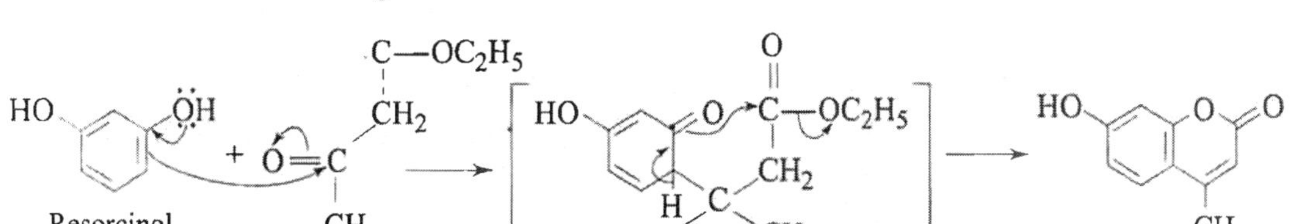

Resorcinol   7-Hydroxy-4-methyl coumarin

72.1 7-Hydroxy-4-methylcoumarin

Resorcinol : 11 g

1. H.V. PENCHMANN and C. DUISBERG, *Ber.*, 1883, **16**, 2119.

Ethylacetoacetate : 13 g
Polyphosphoric acid : 160 g

A mixture of polyphosphoric acid (160 g, obtained by dissolving 126 g of phosphorous pentoxide in 70 g of commercial orthophosphoric acid, d. 1.75), resorcinol (11 g, 0.1 mol) and ethyl acetoacetate (13 g, 0.1 mol) is stirred and heated at 75–80°C for 20 minutes. It is poured onto ice-water. The separated product is filtered, washed with cold water and crystallised from dilute alcohol. The yield is 17 g (97%); m.p. is 184–185°C.

It can also be prepared by using concentrated sulphuric acid in place of polyphosphoric acid as follows. A solution of resorcinol (11 g, 0.1 mol) in ethyl acetoacetate (13 g, 0.1 mol) is added to cold concentrated sulphuric acid (100 ml, temp. 10°C) with stirring. The reaction mixture is kept for 20 hours at room temperature and poured with vigorous stirring on to clushed ice. The separated product is filtered, washed with water and crystallised from dilute alcohol. The yield is 16 g (91%); m.p. is 184–185°C.

73. PERKIN REACTION[1, 2]

The condensation of an aromatic aldehyde with an acid anhydride in presence of sodium or potassium salt of the corresponding acid gives an α, β-unsaturated acid. This is known as Perkin reaction

$$\text{ArCHO} + (\text{CH}_3\text{CO})_2\text{O} \xrightarrow{\text{CH}_3\text{CO}\overset{-}{\text{O}}\overset{+}{\text{N}}\text{a}} \text{AR—CH=CHCOOH} + \text{CH}_3\text{COOH}$$

Mechanism:

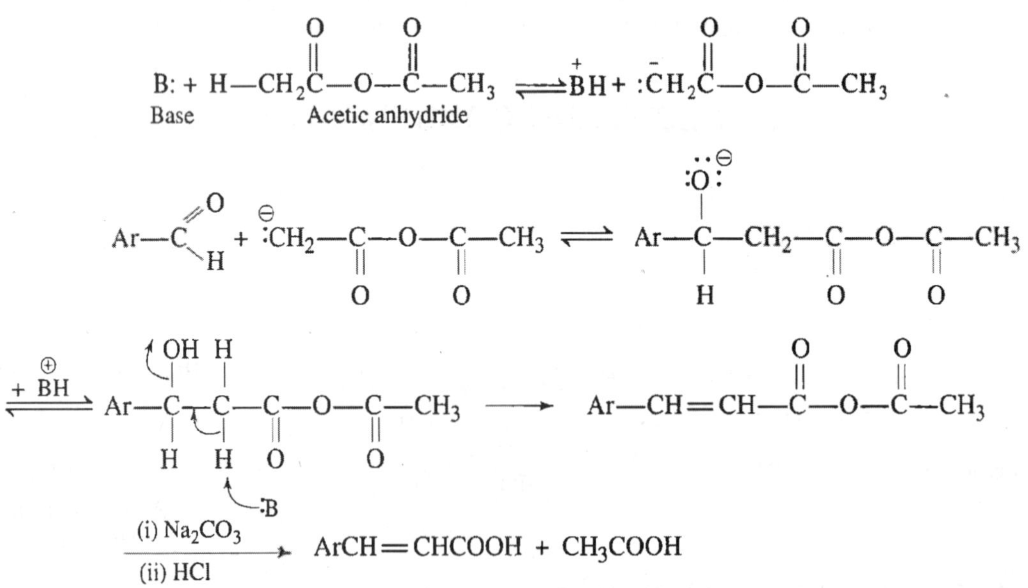

1. W.H. PERKINS, *J. Chem. Soc.*, 1868, **21**, 53, 181.
2. W.H. PERKINS, *J. Chem. Soc.*, 1877, **31**, 388.

73.1 Cinnamic Acid

Benzaldehyde : 10 ml
Acetic anhydride : 14 ml
Fused sodium acetate : 6 g

A mixture of pure benzaldehyde (10.5 g, 10 ml, 0.15 mol), acetic anhydride (15 g, 14 ml,0.15 mol) and freshly fused and finely powdered sodium acetate (6 g, 0.066 mol) is heated in a R.B. flask (100 ml capacity) (fitted with an air condenser) at 160°C (oil bath) for one hour. The temperature of the oil bath is raised to 170–180°C and heating continued for 3 hours more. The hot mixture is poured into water (50 ml). The solution is rendered alkaline by addition of sodium carbonate. The alkaline solution is extracted with ether (30 ml) to remove the unreacted benzaldehyde. The remaining clear alkaline solution is acidified with hydrochloric acid and cooled. The separated cinnamic acid is filtered, washed with cold water and crystallised from hot water. The yield is 9 g (62%); m.p. is 133°C. The cinnamic acid obtained is more stable trans isomer.

73.2 Coumarin

It is obtained by the Perkin reaction of salicylaldehyde with acetic anhydride.

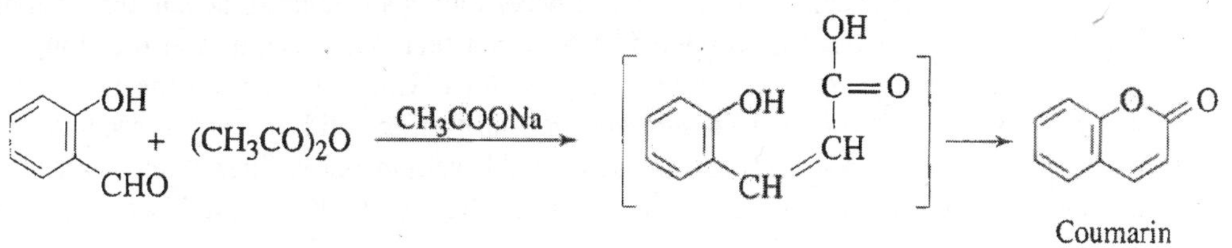

Salicyclic aldehyde : 14.8 g
Acetic anhydride : 20 ml
Sodium acetate (fused) : 15 g
Iodine : 0.5 g

A mixture of salicylic aldehyde (14.8 g, 0.12 mol), acetic anhyride (20 ml, 0,2 mol), freshly fused and powdered sodium acetate (15 g) and iodine (0.5 g) is heated for 2 hours in an oil bath at 120°C. The temperature of the oil bath is raised to 180–190°C and heating continued for 4 hours more. The reaction mixture is poured into cold water containing sodium bisulphite (10 g, to remove iodine). The separated product is filtered, washed with cold water and crystallised from dilute methanol. The yield is 6.8 g (39%); m.p. is 68–69°C.

74. PHOTOCHEMICAL REACTIONS

Compared to a thermal reaction, which is promoted by heat, a photochemical reaction derives its energy from the absorption of light radiation. The absorption of light raises an individual molecule to an excited electronic state without immediately effecting the surrounding molecules. This selective excitation of the individual molecule is a unique character of photochemical reactions. In a photochemical reactions, the chemistry of the excited molecules differs from the chemistry of those in the ground state in the thermochemical reactions, and so the products of photochemical reactions are quite different. Thus, it is possible to change the course of a reaction in a photochemical reaction.

74.1 Benzopinacol

It is obtained by photoreduction of benzophenone by irradiation in propan-2-ol. Sunlight or a medium pressure mercury arc lamp can be used. The mechanism involved in this transformation is given below. The propane-2-ol acts as a hydrogen donor and itself is oxidised to acetone.

$$(C_6H_5)_2C{=}O \xrightarrow{h\nu} (C_6H_5)_2C{=}O^* \text{ (singlet)} \longrightarrow (C_6H_5)_2C{=}O^* \text{ (triplet)}$$

$$\underset{\text{(triplet)}}{(C_6H_5)_2C{=}O^*} + \underset{\text{(isopropanol)}}{(CH_3)_2CHOH} \longrightarrow (C_6H_5)_2\dot{C}OH + (CH_3)_2\dot{C}OH$$

$$(CH_3)_2\dot{C}OH + (C_6H_5)_2C{=}O \longrightarrow (CH_3)_2C{=}O + (C_6H_5)_2\dot{C}OH$$

$$2(C_6H_5)_2\dot{C}OH \longrightarrow (C_6H_5)_2\underset{OH}{C}-\underset{OH}{C}(C_6H_5)_2$$

Benzophenone : 10 g
Isopropanol : 60 ml

Benzophenone (10 g, 0.05 mol) is dissolved by warming in isopropanol (60 ml) in a R.B. flask (100 ml capacity) and one drop of acetic acid is added. More isopropanol is added until the solution is about 1 cm below the flask joint. The flask is stoppered (rubber stopper) and the joint and the stopper are covered with aluminium foil. The flask is placed in direct sunlight. Colourless crystals begin to separate within 20 hours. It is allowed to remain in sunlight until no further solid appears to separate (8–10 days). The solution is cooled (ice-water) and the separated benzopinacol is filtered and washed with cold isopropanol (10 ml). The yield is 9.2 g (98%); m.p. is 180–182°C. It is crystallised from glacial acetic acid; m.p. is 185–186°C.

Note: Benzopinacol is also obtained by reductive dimerisation of benzophenone with zinc and acetic acid (see page 50).

74.2 Conversion of trans-Azobenzene into cis-Azobenzene

The azobenzene as ordinarily obtained, is the *trans*-isomer. It can be photochemically isomerised to the *cis* form. This conversion can be achieved by irradiation with a fluorescent lamp or even by sunlight.

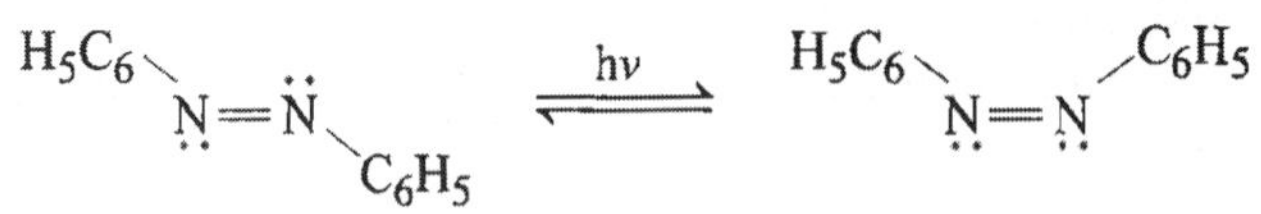

Trans azobenzene Cis-azobenzene

A olution of azobenzene (*trans* form) (1 g) is dissolved in petroleum ether (b.p. 40–60°C, 50 ml) in a beaker (250 ml capacity). It is irradiated for 30 minutes with ultraviolet light by supporting a Hanovia fluorescent lamp above the surface of the liquid in the beaker. In the meanwhile, a chromatographic column (approx. 20 cm × 1.8 cm) from activated alumina (grade I, 50 g) is prepared by making a slurry in petroleum ether and pouring into the column. A circular well fitting filter paper is placed at the top of the column. The irradiated solution is poured on the filter paper in the column with the help of a glass rod. Necessary care is taken to see that the column is not disturbed. The column chromatogram is taken to see that the column is not disturbed. The column chromatogram is run with petroleum ether (b.p. 40–

60°C, 100 ml). A sharp coloured band of the *cis* form (about 2 cm) appears at the top of the column while a diffuse coloured region of the *trans* form moves down the column. The upper band is protected from light by covering with a carbon paper. This will prevent the reconversion of *cis* into *trans* form. The column is extruded and the upper band is cut and shaken with petroleum ether (150 ml) containing methanol (1.5 ml). The solution is filtred and the filtrate washed with water (2 × 15 ml). The extract is dried (anhydrous sodium sulphate) and the solvent evaporated. The residual coloured product (m.p. 71.5°C) is pure *cis*-azobenzene.

The purity of *cis*-azobenzene is ascertained by recording the ultraviolet absorption spectra in ethanol solution as soon as possible after its isolation *cis*-azobenzene has λ_{max} : 281 nm (ε 5260), and *trans*-azobenzene has λ_{max}: 320 nm (ε 21300) in ethanol solution.

74.3 1,4-Naphthaquinone Photodimer

It is obtained by the photochemical reaction of 1,4-naphthaquinone in benzene solution by irradiation with uv light.

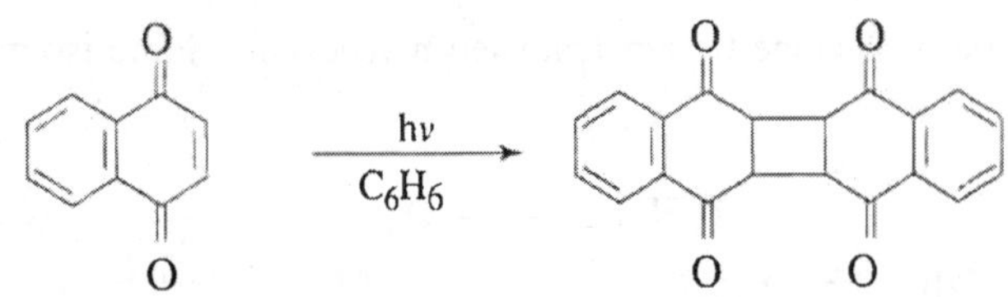

1,4-Naphthaquinone 1,4-Naphthaquinone photodimer

A solution of 1,4-naphthaquinone (3 g, 0.019 mole) in thiophen free benzene (50 ml) is taken in a R.B. flask (50 ml capacity). A slow stream of nitrogen is passed through the solution to remove any air or oxygen. The flask is stoppered and placed in sunlight. The rubber stopper is wrapped with aluminium foil. A solid product slowly separates from the solution after 4–5 days. The separated product is filtered, and the filtrate again exposed to sunlight after passing nitrogen gas (as earlier). The separated product after 4–5 days is again collected. The process is repeated 5–6 times (total of approx. 30 days exposure). Total yield is 0.44 g (13%); m.p. is 244–246°C.

Alternatively photodimerisation can be done by using a 100 hour medium pressure mercury arc lamp in 6 hours. The yield is about 12%.

75. PINACOL-PINACOLONE REARRANGEMENT

A 1,2-glycol, R R′C(OH) C(OH)R″R‴, e.g., pinacol on heating with sulphuric acid or phosphoric acid is transformed into methyl *tert.* butyl ketone or pinacolone.

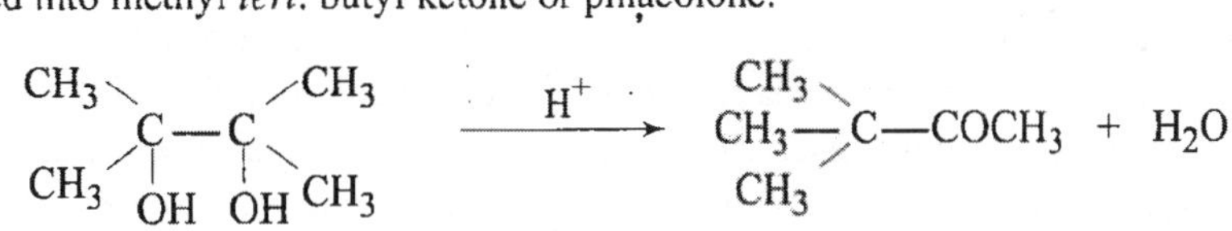

Pinacol Pinacolone

This rearrangement is known as pinacol-pinacolone rearrangement,

Mechanism:

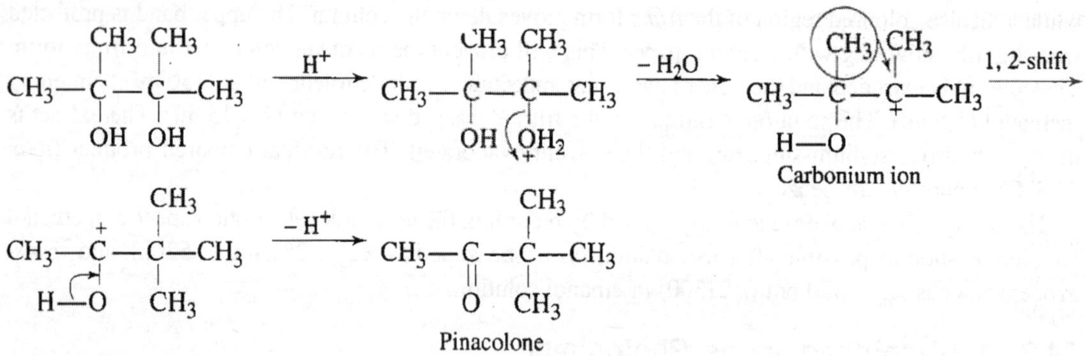

75.1 Pinacolone

Step 1 Pinacol

It is prepared by reduction of acetone by amalgamated magnesium. Some isopropyl alcohol is also formed.

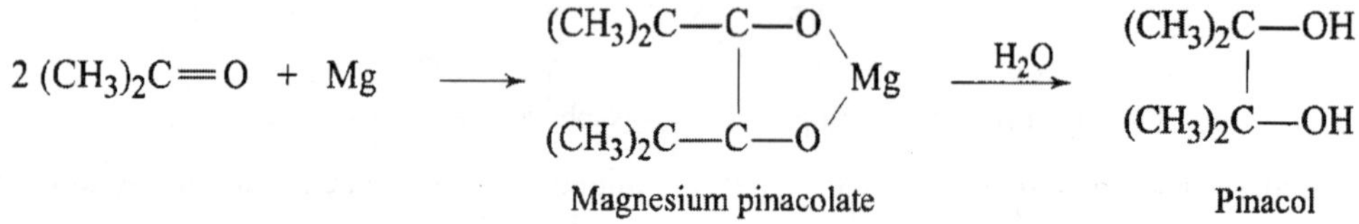

The amalgamated magnesium is obtained by dissolving mercuric chloride in acetone and then adding to a suspension of magnesium in an inert solvent.

Magnesium powder	: 7.5 g
Benzene (anhydrous)	: 100 ml
Mercuric chloride	: 8.4 g
Acetone	: 76 ml

A solution of mercuric chloride (8.4 g) in pure acetone (48 ml, 0.654 mole) is added dropwise through a dropping funnel to a suspension of dry megnesium powder (7.5 g) in anhydrous benzene (75 ml) contained in a three necked R.B. flask (500 ml capacity) fitted with a reflux condenser and a dropping funnel. In case the reaction does not commence, the reaction mixture is warmed (water bath). Once the reaction has started, no further heating is necessary. In case the reaction becomes too vigorous it is necessary to cool the flask by using a cold water bath. After the addition of mercuric chloride solution is complete, a mixture of dry acetone (28 ml) and dry benzene (25 ml) is added. When the reaction slows down, the mixture is heated in a warm water bath for one hour. Magnesium pinacolate swells up. Water (25 ml) is added through the condenser and the mixture shaken. The mixture is heated for 30 minutes on a boiling water bath. During this time the magnesium pinacolate is converted into pinacol (soluble in benzene) and magnesium hydroxide is precipitated. The mixture is cooled (50°C) and filtered. The solid residue is refluxed with benzene (45 ml) for 10 minutes and filtered. The combined benzene solution is concentrated to one half of the original volume and the residual benzene solution treated with water (25 ml) and cooled (10–15°) with stirring. The formed pinacol hydrate is filtered, washed with benzene (15 ml) and dried. The yield is 30 g (67.4%); m.p. is 45–46°C.

Step 2 Pinacolone

Pinacol hydrate : 20 g
Sulphuric acid (6N): 60 ml

A mixture of pinacol hydrate (20 g 0.147 mole) and sulphuric acid (6N, 60 ml) is refluxed gently for 10 minutes. The reaction mixture is cooled and water (20 ml) added. The mixture is distilled (using the distillation arrangement) until the upper layer of the distillate no longer increases in volume (10–15 minutes are required). The pinacolone layer is separated from the distillate, dried (anhydrous magnesium sulphate) and distilled. Pinacolone is collected at 104–106°C. The yield is 10g (68%).

76. POLYMERISATION

The term polymerisation was originally used to indicate the process that took place when a single substance—the monomer—give products having the same empirical formula but different molecular weights. each being a multiple of that of the monomer. It was subsequently found that a number of monomer molecules on suitable treatment (polymerisation) did not always have exactly the same empirical formula as the parent monomer. This led to a change in the definition of the term polymer and polymerisation. According to Carothers (1931), polymeristion is best defined as intermolecular combinations that are capable of procceeding indefinitely. However, in practice the polymer ceases to grow for various reasons.

There are two types of polymerisation. These are addition polymerisation and condensation polymerisation—the polymers obtained are called addition polymers and condensation polymers respectively. Typical examples of addition polymers are polystyrene and that of condensation polymers are phenol-formaldehyde resin, thiokol rubbers and nylon 66.

The mechanism of polymerisation involve three steps.

(i) *Initiation step* Also called the activation step. The substance used for initiation is known as initiator.

$$\underset{\text{(monomer)}}{M} \xrightarrow{\text{activation}} \underset{\text{(active monomer)}}{M^*}$$

(ii) *Propagation step*. Active monomer unit reacts with other monomer units.

$$\underset{\text{(active monomer)}}{M^* + nM} \longrightarrow \underset{\text{(active polymer)}}{M^*\text{—}(M)_n}$$

(iii) *Termination step*. The polymer chain is terminated.

$$\underset{\text{(active polymer)}}{M^* - (M)_n} \longrightarrow \underset{\text{(polymer)}}{(M)_n}$$

76.1 Polystyrene

It is an addition polymer and is obtained by free radical polymerisation of styrene using benzoylperoxide as a radical initiator.

$(RCO)_2 O_2^{\cdot} \longrightarrow RCO_2^{\cdot} \longrightarrow 2R^{\cdot} + 2CO_2$ ... (Initialtion)

organic peroxide

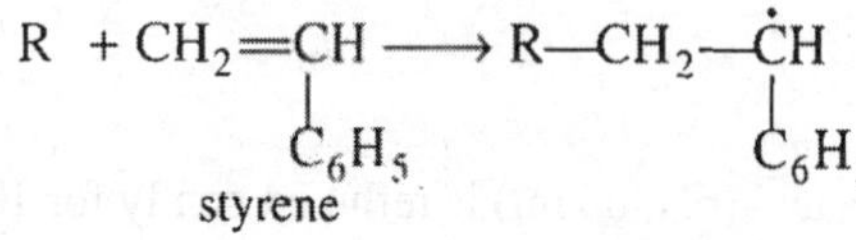

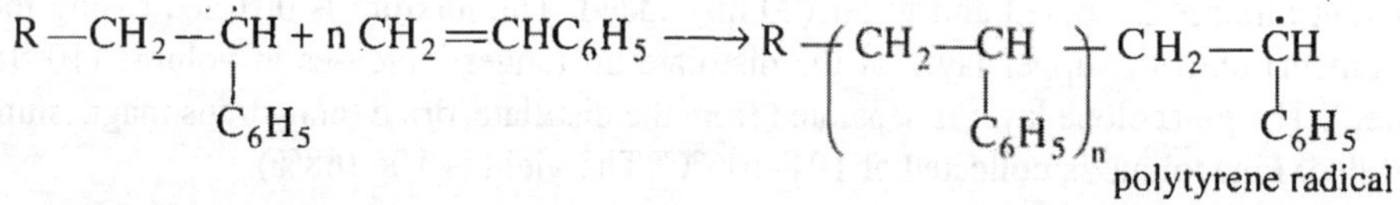

The large polymer radicals combine to stop the chain (termination).

Styrene : 20 ml
Benzoyl peroxide : 0.4 g

The commercial styrene is purified of antioxidants, usually *tert.* butyl phenol (which is present in it), as follows: the sytrene (20 ml, 0.175 mole) is shaken with water (30 ml) and aqueous sodium hydroxide (10%, 10 ml). The aqueous layer is discarded and the remaining styrene is washed with water (3 × 15 ml). It is dried over anhydrous calcium chloride, in a vacuum, dessicator.

The purified styrene, benzoyl peroxide (0.4 g) and toluene (30 ml) is heated in a boiling water (90–95°C) for one hour. The mixture is cooled at room temperature and poured into methyl alcohol (250 ml). The white precipitate of polystyrene is filtered, washed with a little alcohol and dried in air.

76.2 Nylon-66

It is a condensation polymer and is obtained by the condensation of adipoyl chloride with hexamethylenediamine in presence of sodium hydroxide.

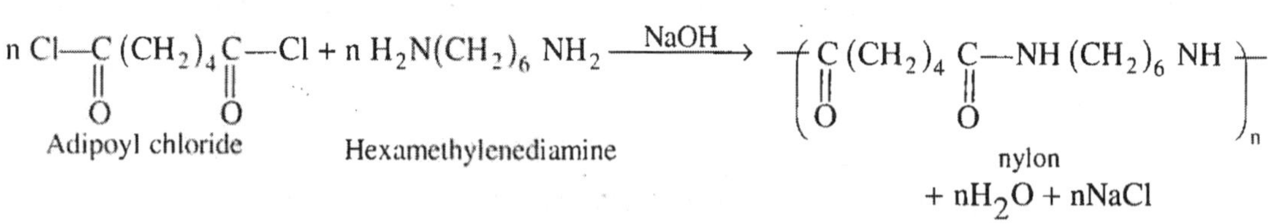

Adipoyl chloride : 1.5 ml
Hexamethylenediamine : 1.5 g
Sodium hydroxide : 1.2 g

A solution of hexamethylenediamine (1.5 g, 0.013 mole) in sodium hydroxide solution (1.2 g in 40 ml water) is added slowly to a solution of adipoyl chloride (1.5 ml, 0.01 mole) in hexane (75 ml) in a beaker. A polymeric film starts forming at the interface of the two liquids. The polymer film is removed with a forcep or a spatula as a continuous forming rope. The polymer is washed with water and then with aqueous acetone (50%). It is dried and placed on a watch glass and melted carefully on a hot plate so that charring does not taken place. The molten polymer is drawn with a glass rod into a fibre.

76.3 Phenol-formaldehyde resin (Bakelite)

Amongst the oldest of the synthetic polymer, and still extremely important, is that resulting from reaction between phenol and formaldehyde, the phenol-formaldehyde resin (Bakelite). When phenol is heated with formaldehyde in presence of alkali or acid there is obtained a high molecular weight substance in which many phenol rings are joined together by—CH_2– groups:

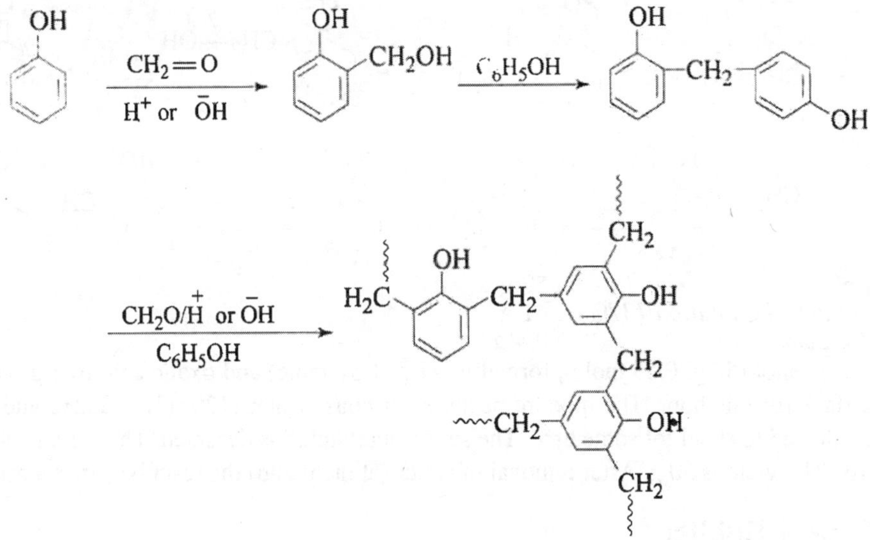

Phenol formaldehyde resin

Mechanism:

In acidic medium:

$$H_2C{=}O + H^+ \rightleftharpoons H_2C{=}\overset{+}{O}H$$

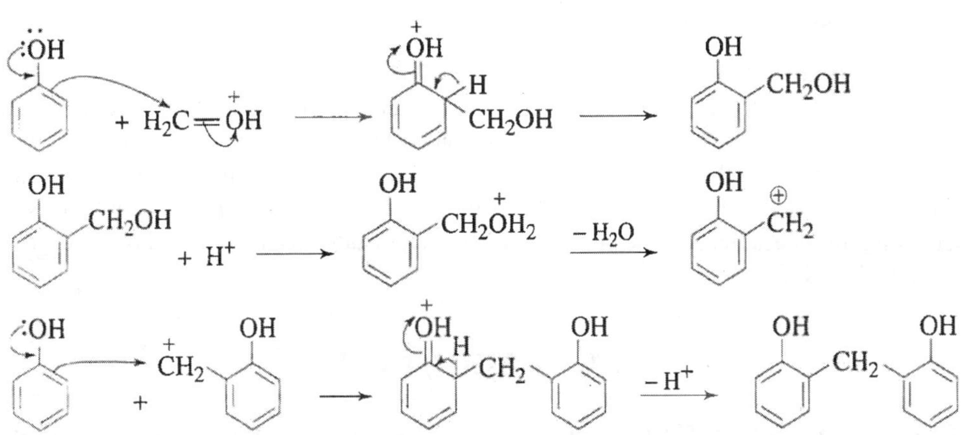

In basic medium:

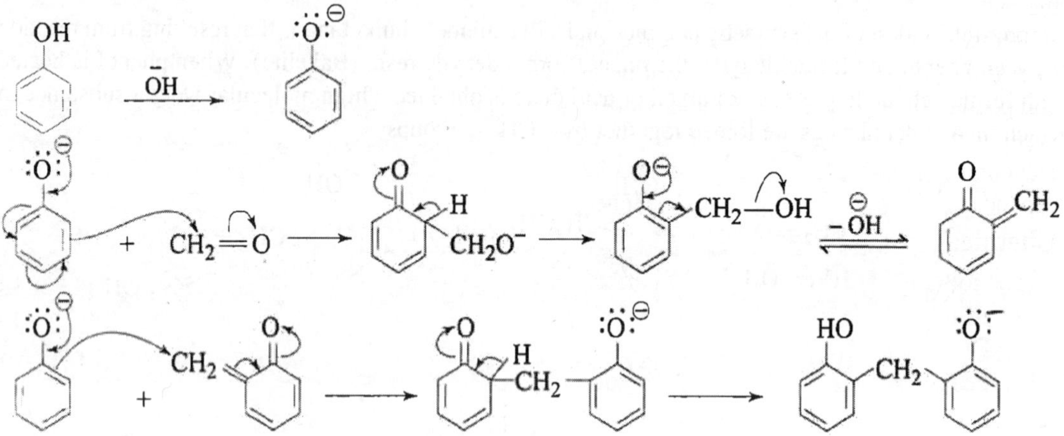

Phenol	: 52 g
Formaldehyde solution (37%)	: 41 g
Oxalic acid	: 0.4 g

A mixture of phenol (52 g, 0.55 mole), formalin (41 g, 1.37 mole) and oxalic acid (0.4 g) is refluxed in an R.B. flask for one hour. The mixture becomes viscous. Water (120 ml) is added and the flask cooled and allowed to stand for some time. The supernatant liquid is decanted. The resin is obtained as a pasty mass. The yield is 50 g. After removal of water (at the pump) the resin becomes a brittle solid.

76.4 Thiokol Rubber

It is obtained by the reaction of 1, 2-dichloroethane with sodium polysulphide.

$$nClCH_2CH_2Cl + nNa^{+}\,{}^{-}S(=S)\text{—}S(=S)^{-}\,Na^{+} \xrightarrow{\Delta} \left(CH_2CH_2\text{—}S(=S)\text{—}S(=S) \right)_n + nNaCl$$

1,2 dichloro ethane; Sodium polysulphide; Thiokol rubber

1, 2-dichloroethane (ethylene chloride)	: 15 ml
Sodium hydroxide	: 3 g
Powdered sulfur	: 6 g

Sulphur (powdered, 6 g) is added in small lots with constant stirring to a warm solution of sodium hydroxide (3 g) in water (60–70 ml). A deep red solution is obtained. If some sulfur remains undissolved the solution is filtered. To the clear solution is added, with stirring, 1, 2-dichloroethane (15 ml, 0.19 mole). The mixture is stirred for 15–20 minutes, while a rubber polymer separates out as a lump. The supernatant liquid is decanted and the product washed under a tap. It is dried within the folds of a filter paper. The yield is 2.2 g.

77. REDUCTION

Most of the organic synthesis involves reduction at some stage. Generally speaking, by reduction we mean the addition of hydrogen to an unsaturated group such as carbon-carbon double bond, a carbonyl group or an aromatic nucleus. It is also used to bring about fission of a bond between two atoms as in the reduction of a disulphide to a thiol or an alkyl halide to a hydrocarbon.

Reductions are generally effective either chemically or catalytically. Each method has its advantages. In many reductions either method can be used equally well. Complete reduction of an unsaturated compound can be achieved without any difficulty. The most important is the selective reduction of one group in a molecule in presence of other unsaturated groups. Both catalytic and chemical methods of reduction offer considerable scope in this direction. However, the method of choice in a particular case depends on the selectivity and the stereochemistry of the desired products.

Chemical Reductions

The following reagents are used for effecting reductions chemically

(i) Aluminium isopropoxide
(ii) Dimide
(iii) Ferrous sulphate-ammonium hydroxide
(iv) Iron-hydrochloric acid
(v) Hydrazine
(vi) Lithium aluminium hydride
(vii) Sodium-ethyl alcohol
(viii) Sodium amalgam
(ix) Sodium borohydride
(x) Sodium hydrogen sulphide
(xi) Sodium dithionite
(xii) Sodium sulphite
(xiii) Magnesium-alcohol
(xiv) Magnesium-amalgam
(xv) Photochemical reduction
(xvi) Tin-hydrochloric acid
(xvii) Anhydrous stannic chloride-hydrogen chloride
(xviii) Zinc-acetic acid
(xix) Zinc-sodium hydroxide
(xx) Zinc amalgam

77.1 Aluminium Isopropoxide

It is used for the reduction of aldehydes or ketones with aluminium isopropoxide to the correspondings alcohol.

$$\underset{R'}{\overset{R}{>}}C{=}O + CH_3{-}\underset{\displaystyle OH}{\underset{|}{CH}}{-}CH_3 \xrightarrow{Al\,[O\,CH(CH_3)_2]_3} \underset{R'}{\overset{R}{>}}CHOH + CH_3COCH_3$$

For details see Mererwein-Pondorf-Verley reduction (page 144).

77.2 Diimide

The reagent diimide is generated in situ by copper catalysed oxidation of hydrazine in presence of an oxidising agent like hydrogen peroxide. Thus,

$$H_2N{-}NH_2 \xrightarrow[Cu^{2+}]{H_2O_2} HN{=}NH$$

The reagent reduces double bonds between the same atoms. Thus, C≡C, C=C and N=N will be reduced, while C≡N or C=O are not reduced.

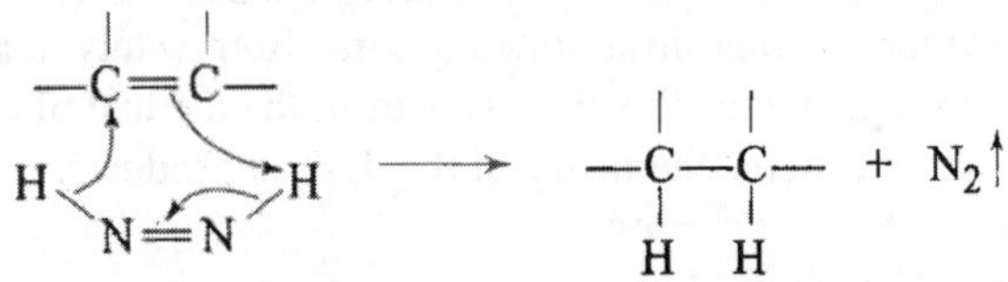

The reduction involves a stereoselective addition of hydrogen via a non-polar cyclic transition state. In the above reaction both *cis-* and *trans-* forms of diimide are formed but the former is involved in reduction.

77.2.1 Hydrocinnamic acid (β-phenylpropionic acid)

It is obtained by the reduction of cinnamic acid with diimide

$$\underset{\text{cinnamic acid}}{C_6H_5CH{=}CH\,COOH} \xrightarrow{HN{=}NH} \underset{\text{Hydrocinnamic acid}}{C_6H_5CH_2CH_2COOH} + N_2\uparrow$$

Cinnamic acid : 4 g
Hydrazine hydrate : 8 g
Cupric sulphate : 50 mg
Hydrogen peroxide (30%) : 10 g (9 ml)

To a clear solution of cinnamic acid (4 g, 0.027 mole) in hydrazine hydrate (8 g), water (50 ml) is added. The solution is shaken, cooled in an ice bath and few crystals of cupric sulphate (50 mg) added. Hydrogen peroxide (30%, 10g, 9 ml) is added slowly to the above cooled solution so that the temperature remains below 30°C. After the addition is complete (20–25 minutes) the solution is allowed to stand for 30 minutes in an ice bath followed by 10 minutes at room temperature. The solution is again cooled (ice bath) and hydrochloric acid (1:1) added with stirring. The separated hydrocinnamic acid (it solidifies on cooling) is filtered and recrystallised from moist ether. The yield is 1.6 g (39.5%); m.p. is 48–49°C.

Note: Hydrocinnamic acid is also obtained by catalytic reduction of cinnamic acid in presence of Adam's catalyst (see page 200).

77.3 Ferrous sulphate-ammonia

Aromatic nitro compounds can be reduced into aromatic amino compounds on reduction with ferrous sulphate-ammonia.

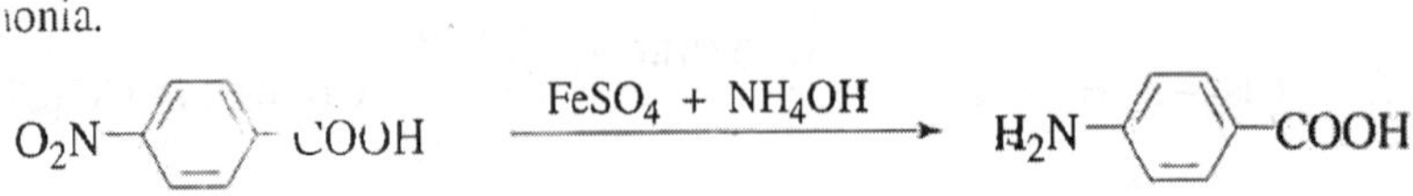

p-Nitrobenzoic acid   *p*-Aminobenzoic acid

77.3.1 p-Aminobenzoic acid

p-Nitrobenzoic acid : 20.5 g
Ferrous sulphate : 250 g
Ammonia (sp g. 0.88) : 80 ml

Ferrous sulphate crystals (250 g) are washed with cold water (minimum amount) to make it free from any basic ferric sulphate and dissolved in water (500 ml) in a beaker. The solution is heated to boiling and stirred vigorously by a mechanical stirrer. A solution of *p*-nitrobenzoic acid (20.5 g, 0.123 mole) in dilute ammonia [80 ml liquor ammonia (sp. g. 0.88) diluted with 42 ml water] is added dropwise (dropping funnel) to the stirred and hot solution. The boiling solution turns black and turbid. Liquor ammonia is then added to keep the solution slightly alkaline. When all the acid has been added (40–45 minutes), the stirring and boiling is continued for 15 minutes and the precipitate of ferric hydroxide filtered and washed with water. The combined filtrate is concentrated to about half the volume and cooled. The clear solution (filter if necessary) is acidified carefully with dilute acetic acid (approx. 11.2 ml of 4 N strength). The separated *p*-aminobenzoic acid is filtered and washed with a small amount of cold water. Some more of acid is obtained by concentrating the mother liquor. The yield is 10.5 g (62.5%); m.p. is 187–188°C.

77.4 Iron-hydrochloric Acid

Iron and hydrochloric acid like tin-hydrochloric acid (see page 196) is used for the reduction of a nitro compound to an amino compound. It is a cheaper reagent (compared to tin-hydrochloric acid) and is used on a commercial scale for the reduction of nitrobenzene to aniline.

$$C_6H_5NO_2 + 2Fe + 6HCl \longrightarrow C_6H_5NH_2 + 2FeCl_3 + 2H_2O$$

or

$$C_6H_5NO_2 + 3Fe + 6HCl \longrightarrow C_6H_5NH_2 + 3FeCl_2 + 2H_2O$$

The amount of hydrochloric acid employed is less than 5 per cent of the amount indicated by either of the above equations. The explanation advanced is that the following reaction is catalysed by acid or by hydroxonium ions.

$$C_6H_5NO_2 + 2Fe + 4H_2O \xrightarrow{H_3O^+} C_6H_5NH_2 + 2Fe(OH)_3$$

The reduction of nitrobenzone to aniline is carried out in the same way as described on page 196.

77.5 Hydrazine

Aldehydes and ketones are reduced to the corresponding hydrocarbons with hydrazine and potassium hydroxide.

$$\begin{matrix} R \\ R' \end{matrix}\!\!>\!C{=}O + NH_2NH_2 \longrightarrow \begin{matrix} R \\ R' \end{matrix}\!\!>\!C{=}NNH_2 \xrightarrow[\Delta]{KOH} \begin{matrix} R \\ R' \end{matrix}\!\!>\!CH_2$$

For details see Wolff-Kishner reduction (page 220).

77.6 Lithium Aluminium Hydride, $LiAlH_4$

Lithium aluminium hydride is a convenient and useful reagent for the selective reduction of the carbonyl group and other polar functional groups. The reductions are generally carried out in dry ether, tetrahydrofuran or dioxane. The complex hydride ion, $[AlH_4]^-$ acts as a carrier or donor of hydride ions and all four hydrogen atoms are ultimately available as H^-.

Following are given some of the common functional groups which are reduced by lithium aluminium hydride.

Table 1

Group	*Product*	*Theoretical amount of LAH required (mole)*
—CHO	$—CH_2OH$	0.25
>C=O	>CHOH	0.25
—C—C— (bridged by O)	$—CH_2CH_2OH$	0.25
—COOR′	$—CH_2OH + R'OH$	0.5
Lactone	diol	0.5
—COOH	$—CH_2OH$	0.75
—CO—O—CO— (cyclic anhydride)	$—CH_2OH$ / $—CH_2OH$	0.5
$—CH_2CONH_2$	$—CH_2CH_2NH_2$	1.0
—C≡N	$—CH_2NH_2$	1.0
—COCl	$—CH_2OH$	0.5
R—Cl	R—H	0.25

It is commonly used for the reduction of esters, acids, aldehydes and ketones into the corresponding alcohols.

Mechanism; Some plausiable mechanisms for some typical reductions are given below:

(i) *Aldehyde or ketone* ⟶ *alcohol*

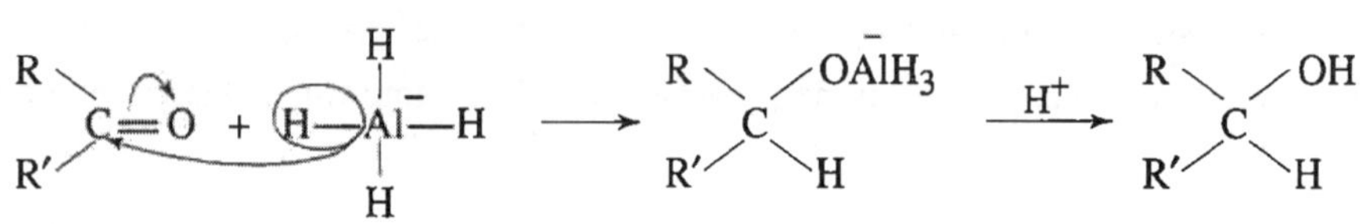

$$4RR'C{=}O + \bar{A}lH_4 \longrightarrow Al(OCHRR')_4 \xrightarrow{H^+} 4RR'CHOH + Al^{3+}$$

(ii) *Ester* ⟶ *alcohol*

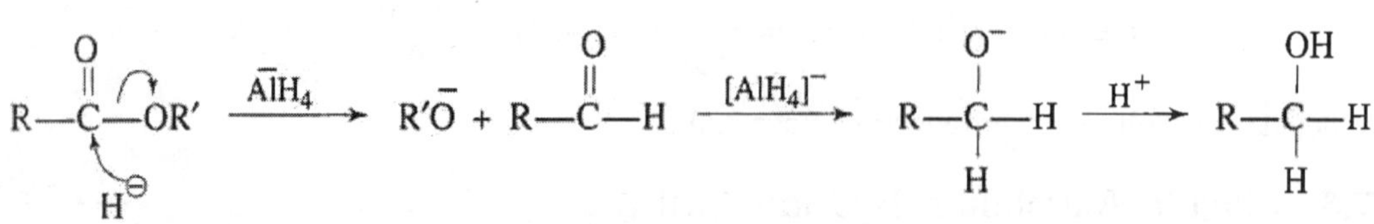

(iii) *Alkyl halide* ⟶ *hydrocarbon*

$$4R{-}Br \xrightarrow{\bar{A}lH_4} 4R{-}H + [AlBr_4]^-$$

(iv) *Nitrile* ⟶ *Amine*

$$2R{-}C{\equiv}N \xrightarrow{LiAlH_4} 2R{-}CH{=}NLi \xrightarrow{LiAlH_4} (RCH_2{-}N)_2\,\bar{A}l\overset{+}{Li} \xrightarrow[NaOH]{H_2O} RCH_2NH_2$$

Lithium aluminium hydride reacts violently with water and dilute mineral acid and must be used with extreme care.

$$LiAlH_4 + 4H_2O \longrightarrow LiOH + Al(OH)_3 + 4H_2$$

The reduction must be conducted in a dry apparatus using dry solvents.

77.6.1 Benzylamine

It is obtained by the reduction of benzonitrile with lithium aluminium hydride in dry ether.

$$C_6H_5C{\equiv}N \xrightarrow{LiAlH_4} C_6H_5CH_2NH_2$$

Benzonitrile : 7.5 ml
Lithium aluminium hydride : 2.75 g
Dry ether : 120 ml

Powdered lithium aluminium hydride (2.75 g) and dry ether (120 ml) are taken in a two necked R.B. flask (250 ml capacity) fitted with a reflux condenser and a dropping funnel. Benzonitrile (7.5 ml, 0.073 mole) in dry ether (20 ml) is added dropwise through a dropping funnel during 5–10 minutes. The reaction mixture is shaken frequently to ensure thorough mixing. If the reaction becomes vigorous, it is cooled for a few seconds in an ice bath. After the addition of nitrile is over, the reduction is allowed to proceed for 10 minutes. Ethyl acetate (1–2 ml) is added through the condenser to decompose any excess of lithium aluminium hydride. Sodium hydroxide solution (20%, 2.2 ml), followed by water (10 ml), is then added to the reaction mixture. The ether solution is separated and the residue extracted with more ether (2 × 20 ml). The combined ether extract is dried (potassium hydroxide pellets) and the solvent is evaporated. Benzyl amine is finally obtained by distillation of residual liquid. The yield is 5.2 g (66.6%); b.p. is 185–187°C.

77.7 Sodium-ethyl Alcohol

Esters on reduction with sodium-ethyl alcohol give primary alcohols

$$RCOOC_2H_5 + 4[H] \xrightarrow{Na/C_2H_5OH} RCH_2OH + C_2H_5OH$$

For details see Bouveault-Blanc reduction (see page 37).

77.8 Sodium Amalgam

It is useful for the reduction of oximes to amines.

$$RCHO \xrightarrow{NH_2OH} RCH{=}NOH \xrightarrow[{[H]}]{Na\text{-}Hg} RCH_2NH_2$$

77.8.1 Piperonylamine

It is obtained from piperonal by conversion into oxime followed by reduction with sodium amalgam.

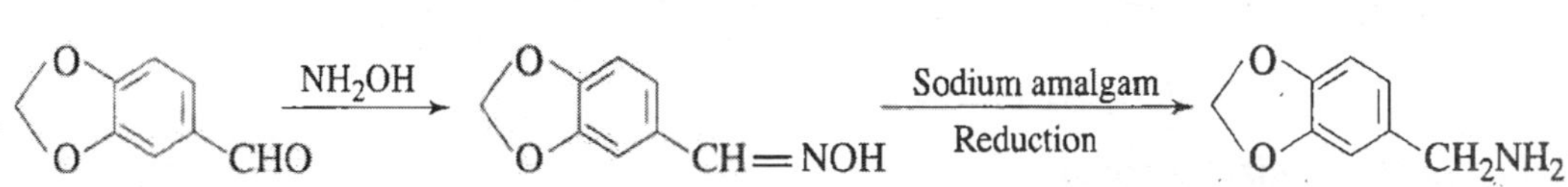

Piperonal

Step 1 Piperonal oxime

Piperonal : 10 g
Hydroxylamine hydrochloride : 5 g
Sodium hydroxide : 2.5 g

A solution of hyroxylamine hydrochloride (5 g, 0.072 mole) in water (10 ml) is added to a solution of piperonal (10 g, 0.067 mole) in alcohol (15 ml). Sodium hydroxide solution (2.5 g in 2 ml water) is then added followed by addition of water to make the solution clear. The solution is refluxed for 3 hours on a water bath and the mixture poured into cold water. The separated oxime is filtered, washed with cold water and crystallised from boiling water. The yield is 10 g (91%); m.p. is 109–110°C.

Step 2 Piperonylamine

Piperonal oxime : 8 g
Sodium amalgam (5%) : 100 g (obtained as given below)

Sodium amalgam (5%, 100 g) is added to a solution of piperonal oxime (8 g, 0.048 mole) in alcohol (50 ml) at 60–70°C. The reaction mixture is kept acidic by the gradual addition of dilute acetic acid (50%). Alcohol is removed by distillation, the residue diluted with water and filtered.The clear filtrate is rendered alkaline with solid potassium hydroxide. The oil is extracted with ether (2 × 30 ml), the ether extract dried (anhydrous potassium carbonate) and the solvent evaporated. The piperonyl amine is obtained by distilling the residual oily product. The yield is 6 g (82.2%); b.p. is 136–138°C/10 mm.

Note: The sodium amalgam required is obtained as follows; clean sodium (5.5 g) is placed in a three necked R.B. flask (250 ml capacity) fitted with a dropping funnel, an inlet for nitrogen and an outlet tube. The air is displaced with nitrogen. Mercury (5 g) is added and the reaction flask warmed over a free flame or until the reaction commences. The flame is removed and the reaction is continued by the slow addition of mercury. When approximately half the mercury has been added, the amalgam tends to solidify. It is kept molten by heating with occasional shaking. After the addition of all the mercury (total 187.5 g), the hot amalgam is poured in a mortar and powdered. It is kept in a tightly stoppered bottle under nitrogen atmosphere. The sodium amalgam thus obtained contains 1.2 per cent sodium.

77.9 Sodium Borohydride

Sodium or potassium borohydride has a considerable degree of selectivity. Thus, aldehydes and ketones are reduced to alcohol while the halogen, cyano, nitro, amido and alkoxycarbonyl groups remain uneffected. The reagent is used in aqueous or aqueous alcoholic solution.

Mechanism: The reaction involves a hydride transfer to the carbonyl carbon from the borohydride anion, which can reduce four moles of the carbonyl compound. The decomposition of the resulting complex liberates the desired alcohol.

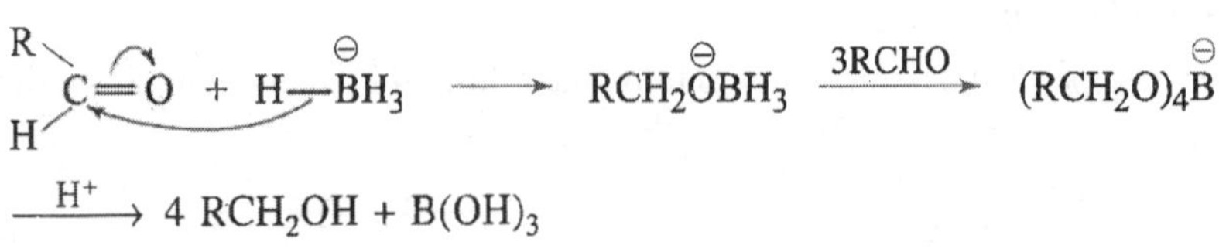

77.9.1 m-Nitrobenzyl alcohol

It is obtained by the reduction of *m*-nitro benzaldehyde with sodium borohydride.

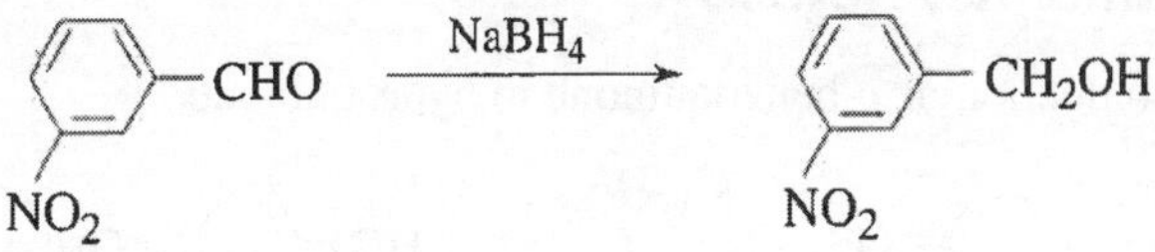

m-Nitrobenzaldehyde : 3.0 g
Sodium borohydride : 0.45 g

A solution of *m*-nitrobenzaldehyde (3 g, 0.033 mol) in methanol (75 ml) is placed in a three necked R.B. flask (500 ml capacity) fitted with a mechanical stirrer, a dropping funnel and a thermometer. The solution is stirred and a solution of sodium borohydride (0.45 g, 0.012 mol) in sodium hydroxide solution (2 M, 0.66 ml) diluted with 6 ml water, is added dropwise with stirring. The temperature of the mixture is maintained at 18–25°. When the reaction is complete (20–25 minutes), the methanol is removed by distillation (steam bath) and the residue diluted with water (40 ml). It is extracted with ether (2 × 30 ml). The ether extract is washed with water, dried (anhydrous magnesium sulphate) and the solvent evaporated. The residual pale yellow oil is distilled under reduced pressure. The yield is 4.5 g (85%); b.p. is 183–185°C/17 mm.; m.p. is 30°C.

77.10 Sodium Hydrogen Sulphide

It is used for the selective reduction of one nitro group in polynitro compound. Thus, *m*-dinitrobenzene on reaction with sodium hydrogen sulphide, only one of the nitro group is reduced and *m*-nitroaniline is obtained.

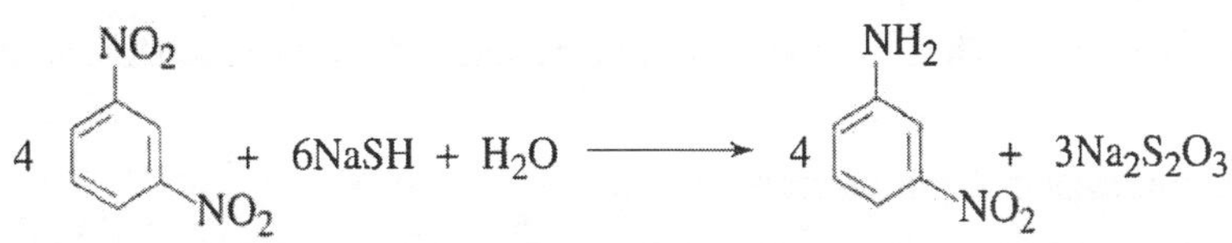

77.10.1 m-Nitroaniline

Dinitrobenzene : 5.6 g
Sodium hydrogen sulphide : 3.2 g (obtained as given below)

Sodium hydrogen sulphide required for reduction is obtained as follows: Sodium bicarbonate (5 g) is added to a solution of crystallised sodium sulphide ($Na_2S \cdot 9H_2O$, 15 g in 40 ml water) with constant stirring. After the bicarbonate has dissolved, methanol (40 ml) is added with stirring. The precipitated sodium carbonate is filtered and washed with methanol (2 × 5 ml). The filtrate and washings contain about 3.2 g sodium hydrogen sulphide in solution and is used for reduction.

$$Na_2S + NaHCO_3 \longrightarrow NaSH + Na_2CO_3$$

To the above solution of sodium hydrogen sulphide is added a solution of dinitrobenzene (5.6 g, 0.033 mole) in hot methanol (40 ml) with shaking. The mixture is gently refluxed for 20–25 minutes. Some more sodium carbonate may separate. Most of the methanol is distilled (steam bath) and the residual mixture poured into cold water (150 ml). The separated yellow precipitate of *m*-nitroniline is filtered, washed with water and crystallised from dilute methanol. The yield is 3.1 g; (67.4%); m.p. is 113–114°C.

77.11 Sodium Dithionite, $Na_2S_2O_4$

It is used mostly for the reduction of *p*-benzoquinone to hydroquinone.

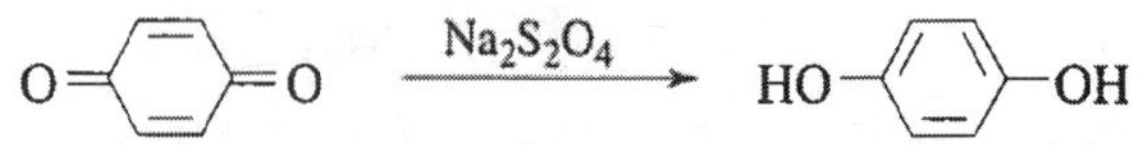

p-Benzaquionone : 2g
Sodium dithionite : 3g

The *p*-benzoquinone (2 g, 0.019 mole; see page 169) is dissolved or suspended in ether or benzene (20 ml) and shaken vigorously with a solution of sodium dithionite (3 g in 30 ml sodium hydroxide) until the colour of the quinone has disappeared. The alkaline solution of hydroquinone is separated, cooled and acidified with concentrated hydrochloric acid. The product is filtered. If soluble, it is extracted with ether and the extract worked up.

Sodium dithionite is also used for the reduction of azo-dyes. Thus,

$$R{-}N{=}N{-}R' + 2Na_2S_2O_4 + 4H_2O \longrightarrow RNH_2 + R'NH_2 + 4NaHSO_3$$

See also reduction with stannous chloride-hydrogen chloride (page 197).

77.12 Sodium Sulphite

It is used for the reduction of diazonium salts to yield the corresponding hydrazines. For details see page 62.

77.13 Magnesium-alcohol

It is a mild reducing agent. Thus nitrobenzene can be reduced to hydrazobenzene or azobenzene depending on the reaction conditions.

77.13.1 Hydrazobenzene (sym.diphenylhydrazine)

It is obtained by the reduction of nitrobenzene with magnesium turnings and methyl alcohol.

$$2C_6H_5NO_2 + 5Mg + 10CH_3OH \longrightarrow C_6H_5NHNHC_6H_5 + 5Mg(OCH_3)_2 + 4H_2O$$

Nitrobenzene : 6 g (5.05 ml)
Methyl alcohol : 120 ml
Magnesium turnings : 9 g

A solution of nitrobenzene (5.05 ml, 0.049 mole) in methanol (120 ml) is taken in a R.B. flask (250 ml capacity) fitted with a reflux condenser. Magnesium turnings (2 g) are introduced into the flask. In case the reaction does not commence within 2–3 minutes, the flask is warmed (water bath). In case the reaction becomes vigorous, it is moderated by cooling for a few seconds in a cold water bath. When the reaction has subsided, more magnesium turnings (2–3 g) are added. The process is repeated till all the magnesium turnings (total 9 g) have been introduced. The reaction mixture is finally heated on a water bath until the reaction mixture is pale yellow or colourless. In case the colour is deep yellow (due to little azobenzene), more magnesium turnings are added. The reaction mixture is filtered or centrifuged. The clear solution is poured into a separatory funnel (250 ml capacity) containing ammonium chloride solution (20%, 120 ml). The mixture is shaken for 5–10 minutes. The magnesium methoxide dissolves and hydroazobenzene is precipitated. It is filtered and crystallised from alcohol (90%) containing a little dissolved sulphur dioxide. The yield is 2.6 g (28.8%). It is stored in a vacuum desiccator under nitrogen.

Alternatively, hydrazobenzene can be obtained by reduction of azobenzene (2 g) in methanol (80 ml) with magnesium turnings (1g) and proceeding in the same way as given above.

77.13.2 Azobenzene

It is obtained by the reduction of nitrobenzene in methyl alcohol with magnesium powder.

$$2C_6H_5NO_2 + 4Mg + 8\ CH_3OH \longrightarrow C_6H_5N{=}NC_6H_5 + 4\ Mg(OCH_3)_2 + 4H_2O$$

Excess of magnesium is avoided, since this will give hydrazobenzene (see above).

Nitrobenzene : 6.2 g (5.2 ml)
Methanol : 110 ml
Magnesium turnings: 6 g

Magnesium turnings (about 3 g) are added to a solution of nitrobenzene (6.2 g, 5.2 ml, 0.05 mole) in methanol (110 ml) contained in a R.B. flask (250 ml capacity) fitted with a reflux condenser. A small crystal of iodine is then added. If the reaction does not commence in 2–3 minutes, the mixture is warmed (water bath) to start the reaction. If the reaction becomes too vigorous, it is moderated by cooling in a cold water bath for a few seconds. When most of the magnesium has reacted, the remaining magnesium turnings (total 6 g) are added and the reaction allowed to proceed (as above). Finally it is heated for 30 minutes on a water bath. It is cooled, poured into water (200 ml) and acidified with glacial acetic acid until the mixture is acidic to congo red. The mixture is cooled and the separated abzobenzene filtered. It is crystallised from ethanol (90%). The yield is 3.1 g (34.1%); m.p. is 67–68°C.

77.14 Magnesium Amalgam

It is used for the reduction of acetone to a bimolecular reduction product, tetramethylene glycol or pinacol.

$$2(CH_3)_2C{=}O + Mg\,(Hg) \longrightarrow \begin{array}{l}(CH_3)_2C{-}O\\ \quad\ \ |\\ (CH_3)_2C{-}O\end{array}\!\!>Mg \xrightarrow{H_2O} \begin{array}{l}(CH_3)_2\,C{-}OH\\ \quad\ \ \ |\\ (CH_3)_2\,C{-}OH\end{array}$$

For details see the preparation of pinacol (page 182).

77.15 Photochemical Reduction

Reduction of benzophenone to benzopinacol (see page 50).

77.16 Tin-hydrochloric Acid

This reagent is used for the reduction of nitro group to amino group.

$$C_6H_5NO_2 + 6H \xrightarrow{Sn\ +\ HCl} C_6H_5NH_2 + 2H_2O$$

The steps involved in the reduction are shown below.

$$Sn + 2HCl \longrightarrow SnCl_2 + 2H$$

Stannous chloride itself is an excellent reducing agent.

$$SnCl_2 + 2HCl \longrightarrow SnCl_4 + 2H$$

The aniline obtained forms a complex, which on decomposition with alkali gives the free base.

$$SnCl_4 + 2HCl \longrightarrow \underset{\text{Chlorostannic acid}}{H_2[SnCl_6]} \xrightarrow{2\,C_6H_5NH_2} [(C_6H_5\,NH_2)_2]\,SnCl_6$$

$$\xrightarrow{8\,NaOH} 2C_6H_5NH_2 + Na_2\,SnO_3 + 6NaCl + 5H_2O$$

On a commercial scale, the cheaper iron and hydrochloric acid is employed as a reducing agent. However if a small quantity is to be reduced and the time is important and cost is a secondary consideration, tin and hydrochloric acid is employed.

77.16.1 Aniline

Nitrobenzene : 15 g (12.6 ml)
Granulated tin (mossy) : 24 g
Conc. hydrochloric acid : 60 ml

Concentrated hydrochloric acid (about 9 ml) is added to a mixture of nitrobenzene (15 g, 12.6 ml, 0.123 mole) and mossy granulated tin (24 g) placed in a R.B. flask (250 ml capacity) fitted with a reflux condenser. In case the reaction becomes vigorous, it is moderated by cooling for a few minutes in a cold water bath. When the reaction subsides, more hydrochloric acid (9–10 ml) is added. The mixture is shaken and cooled if the reaction becomes too vigorous. The process is repeated till all the hydrochloric acid (total 60 ml) has been added (20–25 minutes). The mixture is finally heated on a boiling water bath until the odour of nitrobenzene is not perceptable (30 minutes) and a few drops of the reaction mixture on dilution with water gives a clear solution. The reaction mixture is cooled (a complex of aniline chlorostannate may separate) and sodium hydroxide solution (45 g in 75 ml water) is added carefully with stirring. The hydroxide of tin dissolves and an oily product (aniline) separates. The mixture is steam distilled till no more oily drops pass over. The steam distillate is saturated with sodium chloride and extracted with ether (2 × 50 ml). The clear ether extract is dried (anhydrous potassium carbonate), the solvent is evaporated and the residual product is distilled. The fraction between 180–184°C is collected. The yield is 10.5 g (93%).

Note: Since Sn≡44, this must be kept in mind while calculating the quantities required for the reaction.

77.16.2 Anthrone

It is obtained by the reduction of anthraquinone with tin-hydrochloric acid. The probable path way is,

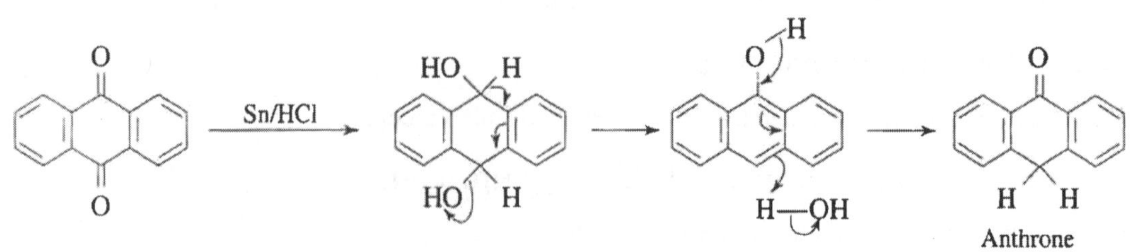

Anthraquinone : 10.4 g
Granulated tin : 10 g
Acetic acid (glacial) : 40 ml
Conc. hydrochloric acid : 12.5 ml

Concentrated hydrochloric acid (12.5 ml) is added slowly (1.5 to 2 hours) from a dropping funnel to a refluxing mixture of anthraquinone (10.4 g, 0.042 mol) and glacial acetic acid (40 ml) contained in a

R.B. flask (250 ml capacity) fitted with a reflux condenser. All the anthraquinone dissolves. If not, more tin and hydrochloric acid are added. The reaction mixture is filtered (sintered glass), water (5 ml) added to the filtrate and cooled. The separated anthrone is filtered, washed with water, dried and crystallised from 'benzene-petroleum ether mixture. The yield is 3 g (61%); m.p. is 154–155°C.

77.17 Stannous Chloride-dry hydrogen Chloride

It is used for reduction of nitriles to aldehydes.

$$R{-}C{\equiv}N + HCl \xrightarrow{SnCl_2} [R{-}C{\equiv}\overset{+}{N}H]\,Cl^- \xrightarrow{[H]} [RCH{=}\overset{+}{N}H_2]Cl^- \xrightarrow{H_2O} RCHO$$

For details see stephen's reaction (page 212).

Stannous chloride and hydrochloric acid is also used for the reduction of azo dyes. Thus,

$$R{-}N{=}N{-}R' + 2SnCl_2 + 4HCl \longrightarrow RNH_2 + R'NH_2 + 2SnCl_4$$

77.17.1 p-Aminodimethylanilinc

It is obtained by the reduction of methyl orange with stannous chloride and hydrochloric acid,

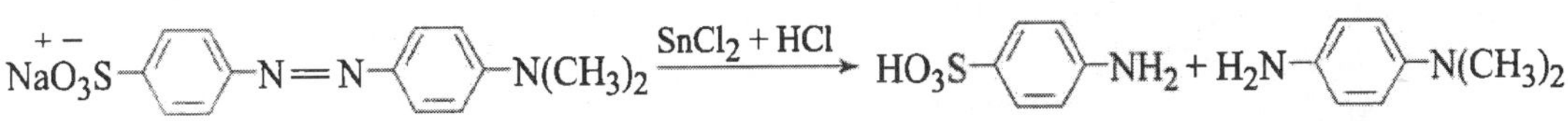

Methyl orange — Sulphanilic acid — *p*-amino dimethylaniline

Methyl orange : 5 g
Stannous chloride : 7.5 g
Conc. hydrochloric acid: 25 ml

A solution of stannous chloride (7.5 g) in concentrated hydrochloric acid (25 ml) is added slowly to a hot solution of methyl orange (5 g, 0.015 mole), in water (40 ml) until decolourisation take place. The solution is boiled if necessary. The resulting solution is cooled, aqueous sodium hydroxide (10%) added until the precipitate of tin hydroxide redissolves and the solution is extracted with ether (3 × 30 ml). The combined ether extract is dried (anhydrous potassium carbonate) and the solvent is evaporated to give *p*-aminodimethyl aniline. The m.p. is 41°C; yield is 2 g (95%).

Note: The reduction of —N=N— can also be effected with sodium dithionite (see page 194).

77.18 Zinc-acetic Acid

It is used for the reductive coupling of ketones (also see formation of pinacol from acetone with magnesium amalgam).

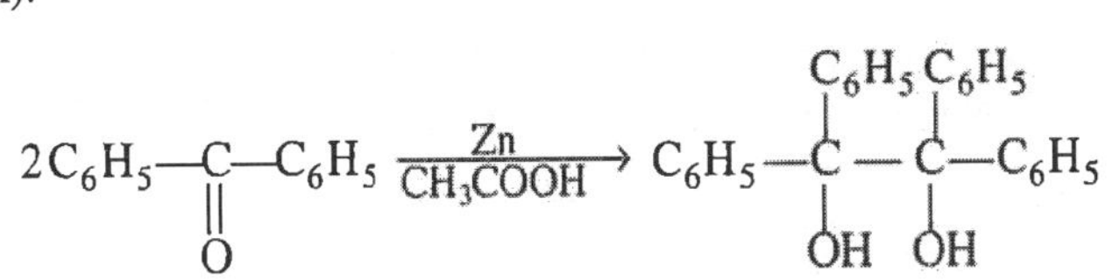

Benzophenone — Benzopinacol

77.18.1 Benzopinacol (see page 50)

77.19 Zinc-sodium Hydroxide

It is a mild reducing agent. Thus, reduction of nitrobenzene in alcoholic sodium hydroxide with zinc powder leads to the formation of azobenzene or hydrazobenzene according to the proportion of the zinc powder employed.

Aromatic ketones can also be reduced with zinc powder and alcoholic sodium hydroxide to give a secondary alcohol.

$$R-\underset{\underset{O}{\|}}{C}-R' + 2H \longrightarrow R-\underset{\underset{OH}{|}}{CH}-R'$$

77.19.1 Azobenzene

It is obtained by the reduction of nitrobenzene with zinc powder and sodium hydroxide.

$$2C_6H_5NO_2 + 4Zn + 8NaOH \longrightarrow C_6H_5-N{=}N-C_6H_5 + 4\ Na_2\ [ZnO_2] + 4H_2O$$

Nitrobenzene	: 25 g (20.75 ml)
Zinc powder (90%)	: 29.5 g
Sodium hydroxide	: 37.5 g
Methanol	: 250 ml

A mixture of nitrobenzene (25 g, 20.75 ml, 0.20 mol), methanol (250 ml) and aqueous sodium hydroxide (37.5 g in 75 ml water) is stirred and refluxed (water bath) in a three necked R.B. flask fitted with a sealed mechanical stirrer and a reflux condenser. Zinc powder (29.5 g, 0.45 mol) is added and the mixture refluxed with stirring for 10 hours. If the odour of nitrobenzene persists, the mixture is refluxed for 2 hours more. The hot mixture is filtered, and the precipitate of sodium zincate washed with methanol (2 × 25 ml). The strongly alkaline solution (which is turbid) is rendered neutral by cautious addition of concentrated hydrochloric acid and filtered. The clear filtrate is distilled to remove methanol and cooled. The separated azo benzene is filtered and is purified by warming at 70°C with dilute hydrochloric acid (2%, 50 ml). The azobenze-melts. The mixture is stirred for 5–10 minutes and cooled (ice + water). The separated azobenzene is filtered and crystallised from alcohol (90%). The yield is 15.5 g (86%); m.p. is 67–68°C.

Note: The commercial zinc powder is usually 90 per cent and is purified as follows:
It is treated with 10 per cent hydrochloric acid for 1–2 minutes (to remove surface oxide), filtered, washed with water and finally with acetone.

77.19.2 Hydrazobenzene (sym. diphenylhydrazine)

It is obtained by the reduction of nitrobenzene with zinc powder and sodium hydroxide (compared to azobenzene the amount of zinc powder and alkali are more).

$$2C_6H_5NO_2 + 5Zn + 10NaOH \longrightarrow C_6H_5-NH-NH-C_6H_5 + 5\ Na_2[ZnO] + 4H_2O$$

Nitrobenzene	: 25 g (20.75 ml)
Zinc powder (90%)	: 39 g
Sodium hydroxide	: 42 g
Methanol	: 250 ml

A mixture of nitrobenzene (25 g, 20.75 ml, 0.20 mol), aqueous sodium hydroxide solution (42 g in 92 ml water) and methanol (250 ml) is refluxed while stirring with zinc powder (90%, 39 g, 0.53 mol) in a three necked R.B. flask (1 litre capacity). The mixture is refluxed for 10 hours. The solution becomes reddish due to the formation of azobenzene and then the solution becomes near colourless due to further reduction. In case the colour is not completely discharged, the mixture is refluxed for 3 hours more with the addition of zinc powder (5.5 g, 0.052 mol, 90%). The hot solution is filtered, and the sodium zincate washed with hot methanol (2 × 25 ml). The filtrate is cooled (freezing mixture) and the separated hydrazobenzene is filtered and washed with dilute methanol (50%) containing a little sulphurous acid and dried in a vacuum desiccator. The yield is 7.5 g (40%); m.p. is 125–126°C.

77.19.3 Benzhydrol (diphenylcarbinol)

It is obtained by the reduction of benzophenone with zinc powder and alcoholic sodium hydroxide.

$$\underset{\text{Benzophenone}}{C_6H_5COC_6H_5} + 2H \longrightarrow \underset{\text{Benzhydrol}}{C_6H_5-\underset{\displaystyle OH}{\underset{|}{C}H}-C_6H_5}$$

Benzophenone : 6.25 g
Zinc powder : 6.2 g
Rectified spirit : 60 ml
Sodium hydroxide : 6.25 g

A mixture of benzophenone (6.25 g, 0.034 mole), rectified spirit (60 ml), sodium hydroxide pellets (6.25 g) and zinc powder (90%, 6.2 g) is mixed by shaking in a R.B. flask (250 ml capacity) fitted with a reflux condenser. The mixture is warmed for 90 minutes on a water bath. The reaction mixture is filtered and the residue washed with hot alcohol (2 × 10 ml), (the residual zinc dust is inflammable and should be destroyed by treatment with mineral acid.) The filtrate is poured into ice cold water (250 ml) containing concentrated hydrochloric acid (12 ml). The separated benzhydol is filtered and crystallised from ethyl alcohol. The yield is 3.9 g (61.7%); m.p. is 68°C.

77.20 Zinc-amalgam

It is comparatively a strong reducing agent and is used to convert

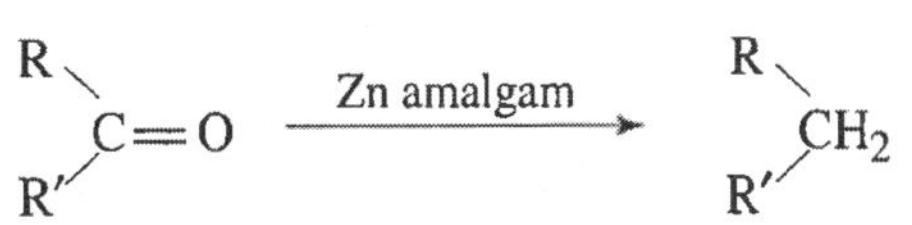

For details see clemmenson reduction

77.21 Catalytic Hydrogenation

Catalytic hydrogenation is one of the most convenient method for reduction of organic compounds. It results simply in the addition of hydrogen to one or more unsaturated groups in the molecule or it may be accompanied by fission of a bond between atoms. The latter process is known as hydrogenolysis. Most of the common unsaturated groups, e.g.

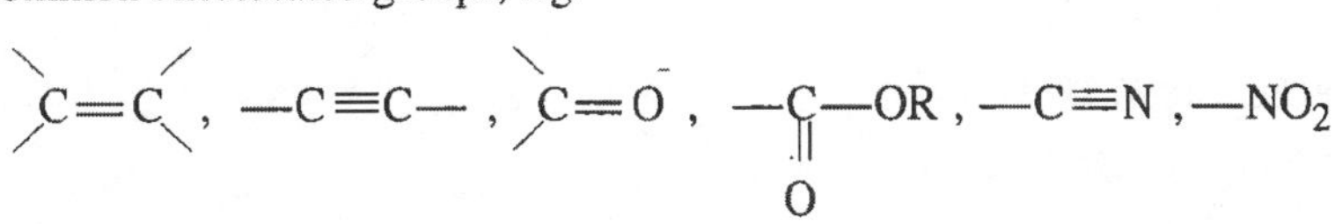

and the aromatic and heteroaromatic nuclei can be reduced catalytically under appropriate conditions. The most commonly used catalysts in the laboratory are the platinum metals (platinum, palladium), nickel and copper chromite.

The following are some preparations which can be easily carried out in the laboratory.

77.21.1 Hydrocinnamic acid

It is obtained by catalytic reduction of cinnamic acid with Adam's catalyst (PtO).

$$C_6H_5—CH{=}CH—COOH + 2H_2 \xrightarrow{PtO} C_6H_5CH_2CH_2COOH$$

Cinnanic acid : 7.4 g
Adam's catalyst : 0.05 g

Cinnamic acid (7.4 g, 0.05 mol) is dissolved in ethanol (50 ml) and hydrogenated in presence of Adam's catalyst (0.05 g) until the hydrogen uptake ceases. The hydrogenation is carried out in the hydrogenation apparatus. After the hydrogenation is complete, the mixture is filtered and the filtrate evaporated. The resulting product is coystallised from petroleum ether. The yield is 7.15 g (95%); m.p. is 47–48°C.

Alternatively, cinnamic acid (7.4 g, 0.05 mol) is dissolved in an inert high b.p. solvent like tetralin (20 ml) and refluxed for 1–2 hours with Pd-C (0.1 g of 30% Pd on activated carbon). The reaction mixture is cooled, ether (25 ml) added and filtered. The filtrate is distilled and the residue crystallised from petroleum ether. The yield is 7.15 g (95%); m.p. is 47–48°C.

77.21.2 Catalytic reduction of acid chlorides

Acid chlorides are selectively hydrogenated in presence of palladium on barium sulphate catalyst.

$$ArCOCl + H_2 \xrightarrow{Pd - BaSO_4} ArCHO + HCl$$

For details see Rosenmund reduction (Page 203).

77.21.3 Catalytic reduction of alkenes

Alkenes can be reduced to alkanes by catalytic reduction in presence of Adams platinic oxide catalyst.

77.21.4 Catalytic reduction of alkynes

Partial reduction of —C≡C— to —CH=CH— is easily effected by Lindlar's catalyst (Pd-$BaSO_4$) in presence of quinoline.

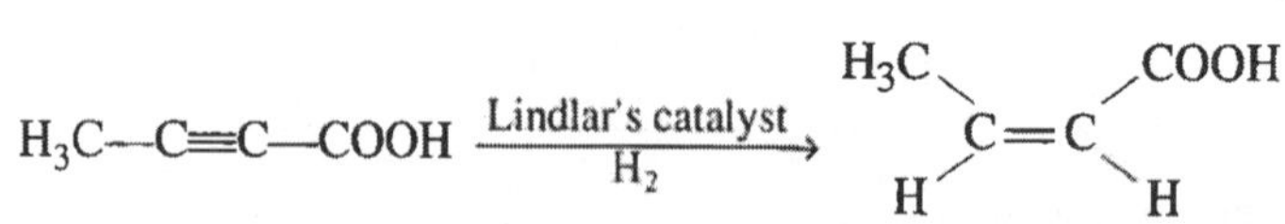

In this case *cis* alkenes are obtained.

78. REFORMATSKY REACTION[1,2]

The reaction of *α*-haloesters (usually *α*-bromoester) with carbonyl compounds such as aldehydes and ketones in presence of zinc in an inert solvent (e.g., ether, benzene) to produce *β*-hydroxy esters is called Reformatsky reaction.

1. S. Reformatskii, *Ber.*, 1887, **20**, 1201.
2. J. Russ. *Phys. Chem. Soc.*, 1890, **22**, 44.

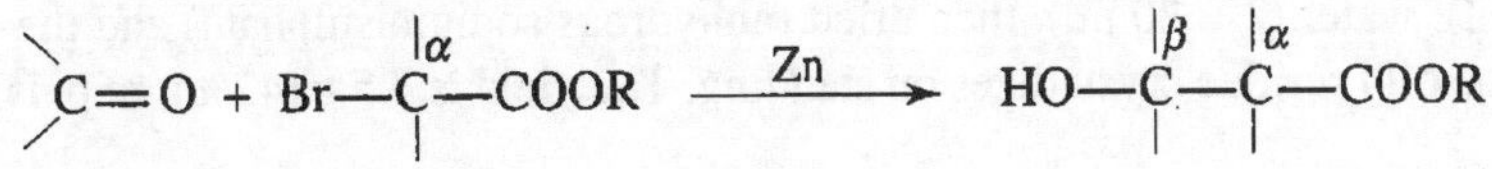

Mechanism: The reaction involves the formation of an intermediate organo zinc compound which then adds to the carbonyl group

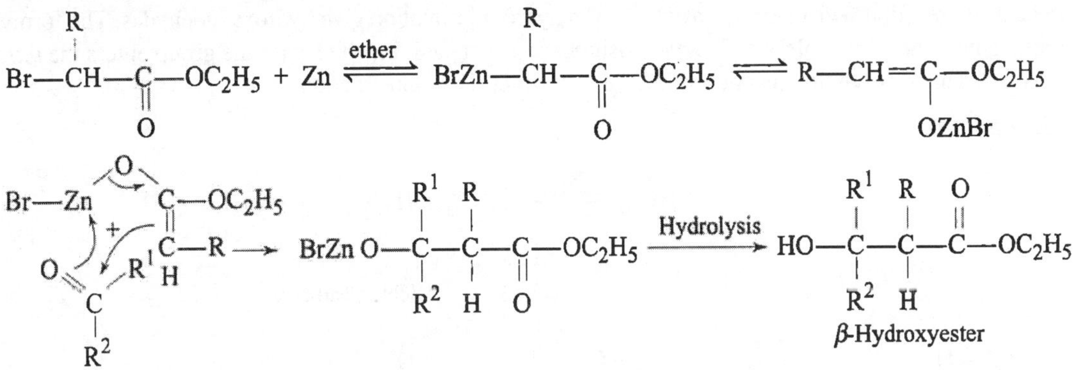

β-Hydroxyester

The organozinc reagent is formed in the first step and is similar to the Grignard reagent. However, it is lower in reactivity.

78.1 Ethyl 3,3-diphenyl-3-hydroxypropionate

It is obtained by the reaction of ethyl bromoacetate with benzophenone in presence of zinc in dry benzene.

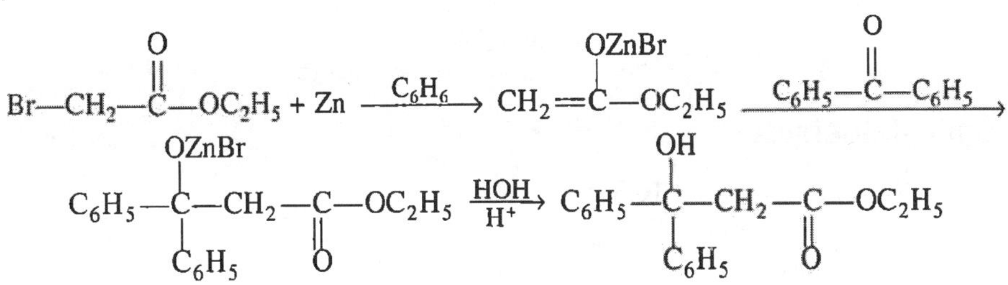

Ethyl 3,3- dipheny l 3-hydroxy propionate

Ethyl bromoacetate : 10 ml
Benzophenone : 12.1 g
Dry benzene : 70 ml
Zinc turnings : 5.7 g

Zinc turnings (5.7 g) are placed in a three necked R.B. flask fitted with a sealed stirrer, a reflux condenser and a dropping funnel. A solution of ethyl bromoacetate (10 ml, 0.09 mole) and benzophenone (12.1 g, 0.066 mole) in dry benzene (70 ml) is added slowly (10–15 ml initially) through a dropping funnel. The reaction mixture is warmed until the reaction starts. Subsequently the mixture is stirred and the remaining solution is added from the dropping funnel at such a rate that only gentle refluxing occurs and the reactions does not become violent. When the reaction subsides, the reaction mixture is refluxed for one hour (water bath), cooled (ice bath) and acidified with ice-cold dilute sulphuric acid (10%, 35 ml) with vigorous stirring. The upper benzene layer is separated (separatory funnel), washed with sodium carbonate solution

(10%, 2 × 20 ml), water (2 × 20 ml), then dried (anhydrous sodium sulphate) and the solvent distilled (water bath). The oily residue crystallizes on standing. The yield is 7.5 g (42%); m.p. is 87–88°C.

79. REIMER-TIEMANN REACTION[1]

Phenols on reaction with chloroform and sodium hydroxide solution give hydroxy aldehydes. The formyl group enters the *ortho* unless both *ortho* positions are occupied, in which case the group enters the *para* position. This formylation reaction is known as Reimer-Tiemann reaction.

Mechanism:

$$CHCl_3 \xrightarrow{\bar{O}H} :\bar{C}Cl_3 + H_2O$$

$$:\bar{C}Cl_3 \xrightarrow{-Cl^-} :CCl_2$$

Dichlorocarbene

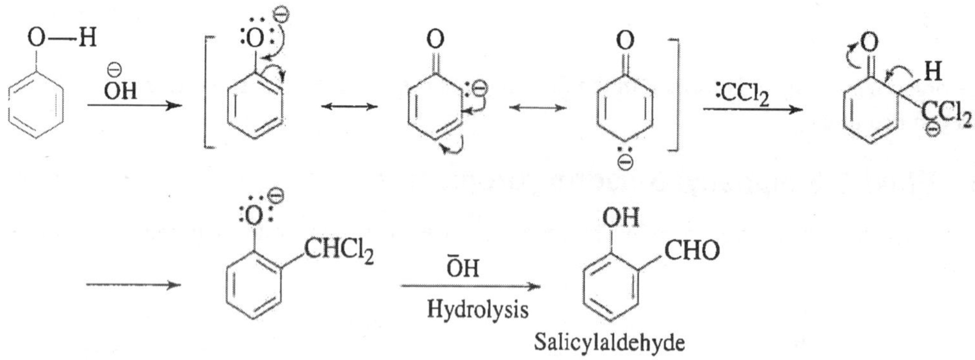

79.1 Salicylaldehyde

Phenol	: 16.3 g
Sodium hydroxide	: 53.3 g
Chloroform	: 40 g (27.3 ml)

To a solution of pure phenol (16.3 g, 0.173 mole) in sodium hydroxide solution (53.3 g in 50 ml water) at 60–65°C (warming on a water bath is necessary) in a 500 ml R.B. flask is dropwise added chloroform (27.3 ml, 0.34 mole) in four portions during 10–15 minutes by means of a dropping funnel. The mixture is shaken during addition and the temperature maintained between 65–70°C. The mixture is finally heated on a boiling water bath for 30 minutes. Excess chloroform is removed by steam distillation. The remaining aqueous alkaline solution is acidified cautiously with dilute sulphuric acid and again steam distilled until no more oily drops are collected. The distillate containing salicylaldehyde is extracted with ether (2 × 25 ml). The ether is removed by distillation. The residue containing unreacted phenol and salicylaldehyde is shaken vigorously for 30 minutes with a saturated solution of sodium bisulphite and allowed to stand for one hour. The bisulphite adduct is filtered, washed with a little alcohol and

1. R. Reimer and F. Tiemann, *Ber.*, 1876, **9**, 824, 1268, 1285.

finally with ether. It is decomposed by warming on a water bath with dilute sulphuric acid. The cooled mixture is extracted with ether, the extract dried (anhydrous sodium sulphate) and the ether evaporated. The residue is distilled. Salicylaldehyde is collected at 195–197°C. The yield is 5.5 g (26%).

The *p*-hydroxybenzaldehyde obtained as a by product is isolated as follows. The residue from the first steam distillation is filtered hot. The clear solution is cooled, extracted with ether and the solvent removed. The yellow solid obtained is crystallised from aqueous sulphurous acid. The yield is 1.5 g (7.1%) m.p. is 116–117°C.

79.2 β-Hydroxynaphthaldehyde

It is obtained by the reaction of *β*-naphthol with chloroform and sodium hydroxide solution.

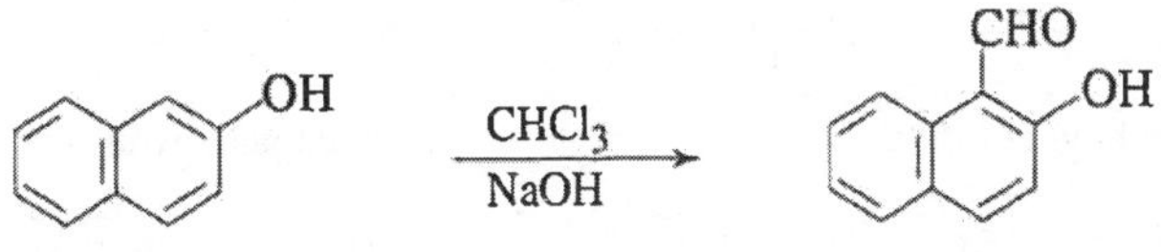

β-Naphthol *β*-Hydroxynaphthaldehyde

β-Naphthol	: 12 g
Sodium hydroxide	: 24 g
Chloroform	: 8.4 ml
Ethyl alcohol	: 45 ml

Chloroform (8.4 ml, 0.105 mole) is added dropwise (dropping funnel) to a solution of *β*-naphthol (12 g, 0.083 mole) in aqueous sodium hydroxide (24 g dissolved in a mixture of 45 ml water and 45 ml ethyl alcohol) in a R.B. flask (250 ml capacity). The addition is completed in 15–20 minutes and the mixture shaken for 45–50 minutes more (exothermic reaction). The alcohol is distilled (water bath) and the residual solution acidified with dilute hydrochloric acid. It is extracted with ether, the extract dried (anhydrous sodium sulphate) and the ether is evaporated. The residue is crystallised from alcohol, the yield is 10 g (70%); m.p. is 80–81°C.

80. RILEY OXIDATION[1] (SELENIUM DIOXIDE OXIDATION)

It is used for the oxidation of active methylene groups to carbonyl groups. Thus,

$$CH_3CH_2CHO + SeO_2 \longrightarrow CH_3COCHO + H_2O + Se$$

For the mechanism see page 167.

81. ROSENMUND REDUCTION[2,3]

Acid chlorides can be selectively hydrogenated in presence of a catalyst (palladium on barium sulphate). The reaction involves hydrogenolysis of the carbon-halogen bond and is known as the Rosenmund reduction.

1. H.L. Riley, J.F. Morley and N.A.C. Friend, *J. Chem. Soc.*, 1932, 1875.
2. R.W. Rosenmund, *Ber.*, 1918, **15**, 585.
3. R.W. Rosenmund and F. Zetzschi, *Ber.*, 1921, **54**, 425.

$$\text{Ar—}\overset{\displaystyle O}{\overset{\|}{C}}\text{—Cl} + H_2 \xrightarrow{\text{Pd–BaSO}_4} \text{Ar—}\overset{\displaystyle O}{\overset{\|}{C}}\text{—H} + \text{HCl}$$

81.1 2-Naphthaldehyde

It is obtained by reduction of 2-naphthoyl chloride with hydrogen in presence of Pd–$BaSO_4$ catalyst.

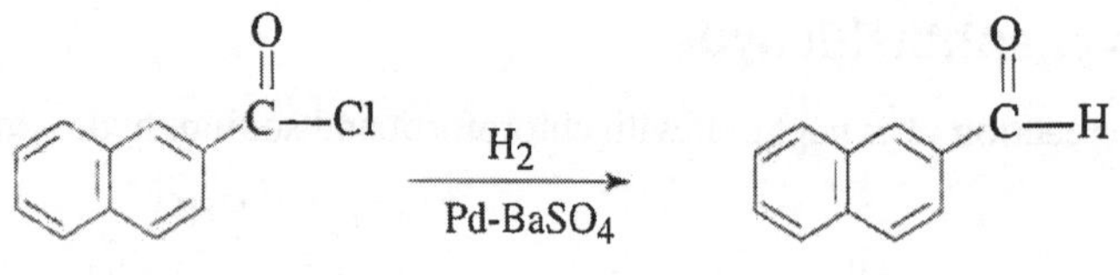

2-Naphthoyl chloride 2-Naphthaldehyde

2-Naphthoyl chloride	: 14.25 g
Dry xylene	: 50 ml
Palladium-barium sulphate catalyst	: 1.5 g

A mixture of 2-naphthoyl chloride (prepared as given in the *note* below, 14.25 g, 0.075 mol), sodium dried xylene (50 ml), palladium-barium sulphate catalyst (1.5 g) and poison solution (1.5 ml, prepared as given in the note below) is taken in a three necked R.B. flask (250 ml capacity) fitted with a sealed stirrer, a gas inlet tube extending to a point just above the bottom of the stirrer and a reflux condenser. The top of the condenser is connected by a rubber tube and stopper to a glass tube extending to the bottom of a conical flask (250 ml capacity) containing distilled water (200 ml) and a few drops of phenolphthalein indicator. Sodium hydroxide solution (1M) is prepared and kept in a burette for addition to the conical flask.

Hydrogen gas (from a hydrogen cylinder) is passed into tin reaction flask to displace the air. The flask is heated to 140–150°C (oil bath) and the mixture stirred vigorously. The passage of hydrogen gas is continued at a slow rate (1–2 bubbles per second). The course of the reaction is followed by the rate of hydrogen chloride evolution. The first 25 ml of 1 M alkali should be neutralised in 15–20 minutes and the reaction should be completed in 1.5–2 hours. The reaction is stopped when about 92 per cent of the theoretical amount of hydrogen chloride (corresponding to 71.25 ml of 1M NaOH solution) is recovered. The reaction mixture is cooled, stirred with decolourising carbon (1 g) and filtered. Xylene is removed by distillation under reduced pressure using an oil bath. 2–Naphthaldehyde distils at 147–149°C/11 mm. The product solidifies on cooling. The m.p. is 59–60°C; yield is 9.5 g (81%).

Note: (1) 2-Naphthoyl chloride is prepared from 2-naphthoic acid (14.28 g, 0.082 mol) and phosphorus pentachloride (17.25 g, 0.082 mol) following the procedure described for 3, 5-dinitrobenzoyl chloride. The product is obtained by distillation and the fraction between b.p. 160–162°C/11 mm is collected. It solidifies on cooling. The yield is 15 g (95%); m.p. is 51–52°C.

(2) For use in Rosenmund reduction, the catalyst Pd-$BaSO_4$ is moderated by the addition of an appropriate quantity of a quinoline-sulphur poison prepared by refluxing sulphur (1 g) with quinoline (6 ml) for 5 hours and diluting to 70 ml with xylene (sodium dried).

82. SANDMEYER REACTION (GATTERMANN REACTION, KÖRNER-CONTARDI REACTION)[1,-3]

The replacement of diazonium groups in aromatic compounds by halo or cyano group in presence of cuprous salt [copper powder (Gattermann) or cupric salt (Körner-Conterdi)] is known as Sandmeyer reaction.

$$C_6H_5N_2^+ X^- + KCN \xrightarrow{Cu} C_6H_5CN + N_2 + KCl$$

82.1 *p*-Chlorotoluene

See page 60.

82.2 *p*-Bromotoluene

See page 61.

82.3 *p*-Tolunitrile

It is obtained by diazotisation of *p*-toluidine followed by treatment of the diazotised solution with cuprous cyanide solution.

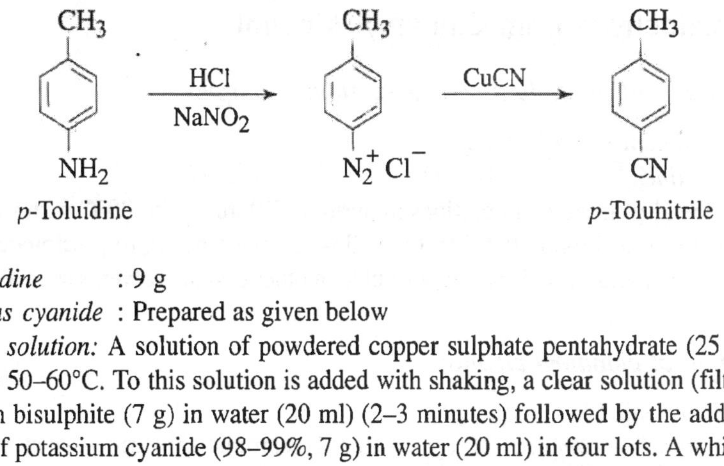

p-Toluidine : 9 g
Cuprous cyanide : Prepared as given below

Cuprous cyanide solution: A solution of powdered copper sulphate pentahydrate (25 g) in water (80 ml) is warmed to 50–60°C. To this solution is added with shaking, a clear solution (filter if necessary) (60°C) of sodium bisulphite (7 g) in water (20 ml) (2–3 minutes) followed by the addition of a warm solution (60°C) of potassium cyanide (98–99%, 7 g) in water (20 ml) in four lots. A white precipitate of cuprous cyanide is formed. The mixture is allowed to stand for 10–15 minutes and filtered. The product is washed with hot water (4 × 10 ml). It is dissolved in a solution of sodium cyanide (10 g) in water (25 ml).

$$2CuSO_4 + 2NaCN + NaHSO_3 + H_2O \longrightarrow 2CuCN + 3NaHSO_4$$

---

1. T. SANDMEYER, *Ber.*, 1884, **17**, 1633, 2650.
2. L. GATTERMANN, *Ber.*, 1890, **23**, 1218.
3. G. KORNER, A. CONTARDI, *Atti. Acad. Nazl. Lincei*, 1914, **23II**, 462 (CA, 1915, **9**, 1478).

p-Tolunitrile *p*-Toluidine (9 g, 0.084 mole) is diazotised (for details see preparation of *p*-chlorotoluene, page 60), using a solution of sodium nitrite (6 g) in water (112 ml). The cold diazonium salt solution is added in small lots to a warm solution (60°C) of cuprous cyanide solution (prepared as given above). The mixture is shaken vigorously after each addition and the temperature of the reaction mixture is kept at 60–70°C. It is then heated on a boiling water bath for 15–20 minutes. The mixture is then steam distilled until no more oily liquid passes over. The distillate is cooled and the solidified *p*-tolunitrile is filtered and dried. It is purified by distillation. The b.p. is 215–219°C, m.p. is 29°C, yield is 6.1 g (62.2%).

83. SARETT OXIDATION[1,2]

Sarett oxidation is the oxidation of *primary* and *secondary* alcohols to aldehydes and ketones by means of CrO_3-pyridine complex.

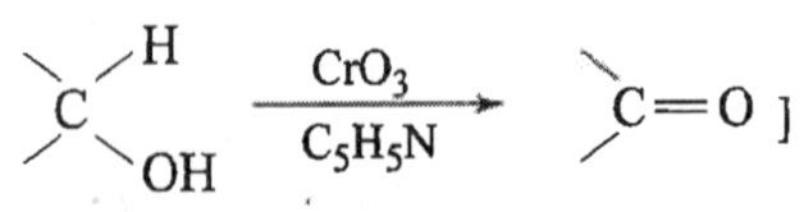

This reagent does not attack olefins, ketals or the thioether group. It oxidises lactols to lactones and is particulary useful[2] for oxidising allylic and benzylic alcohols to the corresponding aldehydes in 50–80 per cent yield.

83.1 Cinnamaldehyde from Cinnamyl Alcohol

Step 1 Preparation of chromium trioxide-pyridine complex

Chromium trioxide : 1 g
Pyridine : 10 ml

Chormic anhydride (1 g) is added in portions to pyridine (10 ml) at 15–20°C with stirring. After about one third addition of the chromic trioxide, the yellow complex began to precipitate. This complex is moderately soluble in pyridine and sparingly soluble in other organic solvents such as dioxane, benzene and acetone.

Step 2 Oxidation of cinnamyl alcohol

Cinnamyl alcohol : 1 g
CrO_3-pyridine complex : (Prepared from 1g chromic anhydride–see step 1)
Pyridine : 10 ml

A solution of cinnamyl alcohol (1 g, 0.007 mole) in pyridine (10 ml) is mixed with CrO_3-pyridine complex prepared from chromic anhydide (1 g) and pyridine (10 ml). The reaction flask is stoppered and the contents are mixed thoroughly. The reaction mixture is allowed to stand overnight. It is then poured into water and extracted with ether. The ether extract is dried over magnesium sulphate and concentrated and finally distilled. The cinnamaldehyde is collected at 248°C; the yield is 0.68 g (70%).

1. G.I. Poss, G.E. Arth, A.E. Beyler and L.H. Sarett, *J. Am. Chem. Soc.*, 1953, **75,** 422.
2. J.R. Holum, *J. Org.*, 1961, **26,** 4814.

84. SCHOTTEN-BAUMANN REACTION

In the Schotten-Baumann procedure for the benzoylation of amines, a mixture of amine, cold aqueous base (sodium hydroxide) and benzoyl chloride is shaken together. The aqueous alkali acts as a scavenger for the HX produced. Thus,

$$R{-}NH_2 + C_6H_5COCl \xrightarrow{NaOH} RNHCOC_6H_5 + NaCl + H_2O$$

Mechanism:

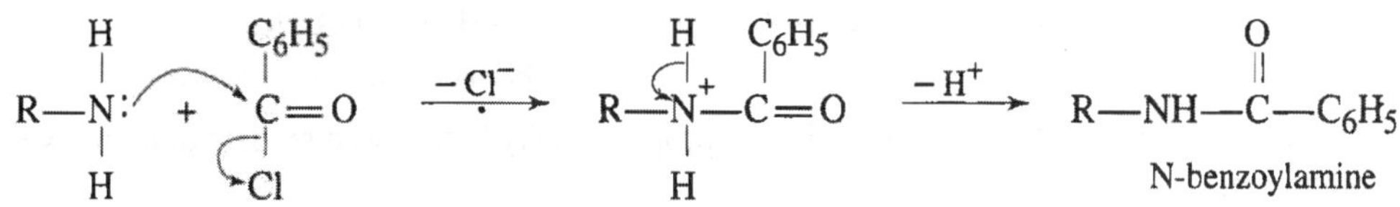

84.1 Hippuric Acid (Benzoyl glycine)

It is obtained by the reaction of glycine with benzoyl chloride in presence of sodium hydroxide solution.

$$\underset{\text{Glycine}}{NH_2CH_2COOH} + \underset{\text{Benzoyl Chloride}}{C_6H_5COCl} + NaOH \longrightarrow \underset{\text{Hippuric acid}}{C_6H_5CONHCH_2COOH} + NaCl + H_2O$$

Glycine	: 5 g
Sodium hydroxide	: 10%, 50 ml
Benzoyl chloride	: 10.8 g (9 ml)
Carbon tetrachloride	: 40 ml

Benzoyl chloride (9 ml, 0.078 mole) is added in two lots to a solution of glycine (5 g, 0.067 mole) in sodium hydroxide solution (10%, 50 ml) in a conical flask. After each addition the flask is stoppered and shaken vigorously. The mixture is cooled (by adding crushed ice) and acidified carefully to congo red with concentrated hydrochloric acid. The separated hippuric acid contaminated with little benzoic acid is filtered, washed with cold water and air dried. It is heated with carbon tetrachloride (2 × 20 ml) for 5 minutes each time and filtered. It is finally crystallised from boiling water with the addition of a little decolourising carbon if necessary. The yield is 9 g (75.6%); m.p. is 187–188°C.

85. SKRAUP SYNTHESIS[1]

It consists in the synthesis of quinolines from aromatic amines, glycerol, an oxidizing agent and sulphuric acid.

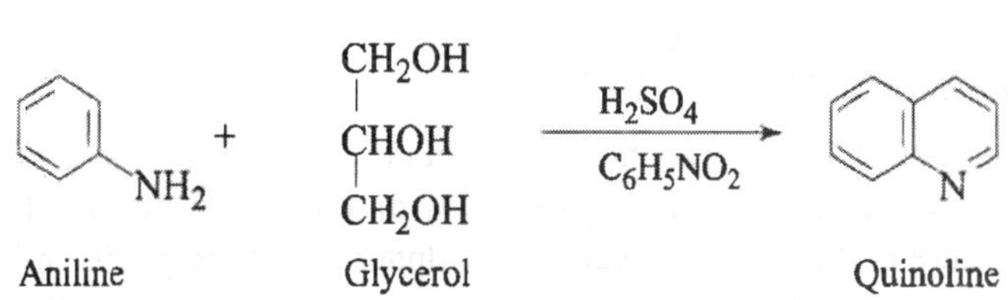

1. Z.H. Skraup, *Ber.*, 188, **13**, 2086.

Mechanism The quaternary hexamine salt may undergo hydrolysis to the *primary* amine, formaldehyde and ammonia. The subsequent step may involve a hydride ion transfer from banzylamine to the protonated aldimine, derived from formaldehyde and ammonia. Hydrolysis of the resulting aldimine gives the required aldehyde.

$$\underset{\text{Quaternary hexamine salt}}{[ArCH_2\overset{+}{N}(CH_2)_6N_3]\,\overset{-}{X}} \xrightarrow{H_2O} \underset{\text{Primary amine}}{Ar\,CH_2NH_2} + 4HCHO + 3NH_3$$

$$HCHO + NH_3 \longrightarrow H_2C\begin{matrix}NH_2\\ OH\end{matrix} \xrightarrow[+H^+]{-H_2O} \underset{\text{Protonated aldimine derivative}}{H_2C{=}\overset{+}{N}H_2}$$

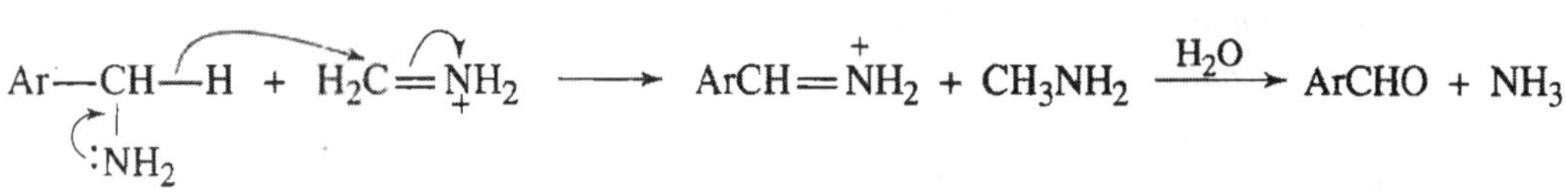

Primary amine

Note: See also Delepine reaction; the same quaternary salt is an intermediate in the Delepine reaction

86.1 Naphthaldehyde

It is obtained by the reaction of 1-(chloromethyl) naphthalene with hexamine in dilute acetic acid.

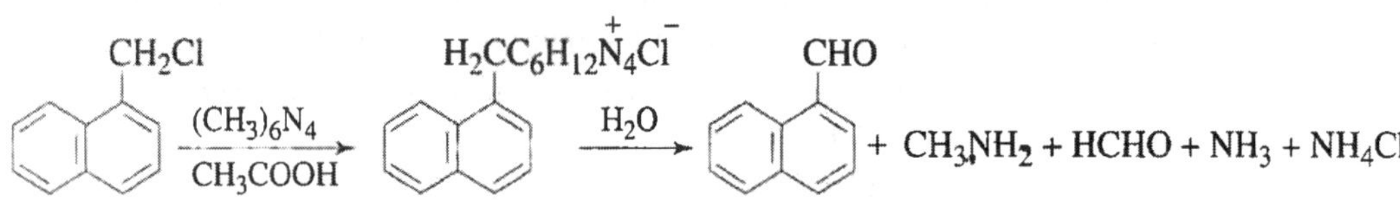

1-(Chloromethyl) naphthalene	: 26.5 g
Hexamethylenetetramine (Hexamine)	: 42 g
Acetic acid (aqueous) (50%)	: 125 ml

A mixture of 1-(chloromethyl) naphthalene (26.5 g, 0.15 mol) hexamethylenetetramine (hexamine) (42 g, 0.3 mol) and aqueous acetic acid (50%, 125 ml) is refluxed for 2 hours in a round bottom flask (250 ml capacity). The mixture becomes homogeneous and oil begins to separate. Concentrated hydrochloric acid (50 ml) is added and the mixture refluxed for 15 minutes more. The cooled mixture is extracted with ether (2 × 30 ml). The ether extract is washed with water (2 × 10 ml) and then with sodium carbonate solution (10%, 20 ml) and finally with water. The etheral solution is dried (anhydrous magnesium sulphate) and the ether evaporated on a water bath. The residual product is distilled under reduced pressure. 1-Naphthaldehyde is collected at 160–162°C/18 mm. The yield is 19 g (81%).

87. SCHIEMANN REACTION[1] (BALZ-SCHIEMANN REACTION)

Thermal decomposition of aromatic diazonium fluoroborates to yield an aromatic fluoride, boron trifluoride and nitrogen is known as Schiemann reaction. Thus,

$$ArNH_2 \xrightarrow{\text{Diazotisation}} Ar\overset{+}{N}_2Cl^- \xrightarrow{NaBF_4} ArN_2^+BF_4^- \xrightarrow{\Delta} ArF + BF_3 + N_2$$

Mechanism

$$Ar\overset{+}{N}_2\overset{-}{B}F_4 \longrightarrow \overset{+}{Ar} + N_2 + [F{-}BF_3]^-$$

$$\overset{+}{Ar} + [F{-}BF_3]^- \longrightarrow Ar{-}F + BF_3$$

87.1 4,4′-Difluorobiphenyl

It is obtained from benzidine by diazo reaction followed by treatment with sodium borofluoride to give the corresponding diazoniumfluoroborate, which on heating decomposes to give 4,4′-difluorobiphenyl.

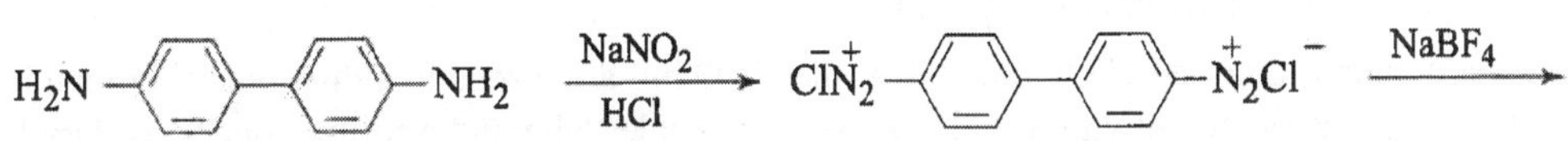

Benzidine   Diazo compound

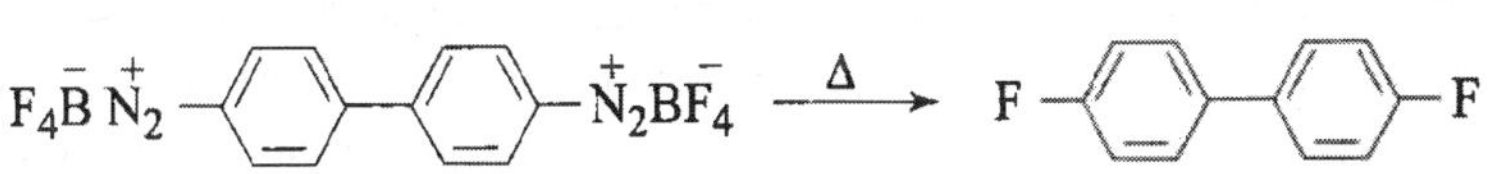

Diazoniumfluoroborate   4,4′-Difluorobiphenyl

Benzidine	: 13.8 g
Sodium nitrite	: 10.5 g
Sodium borofluoride	: 25.5 g

Benzidine (13.8 g, 0.075 mole) is dissolved in a mixture of concentrated hydrochloric acid (45 ml) and water (45 ml). The solution is cooled (ice bath) and bisdiazotised with a solution of sodium nitrite (10.5 g) in water (18 ml) (for detail see the method of diazotisation in the preparation of iodobenzene on page 59). The diazo solution is filtered and to the clear filtrate is added a solution of sodium borofluoride (25.5 g) in water (45 ml). The mixture is stirred for 5–10 minutes and the precipitated bis-diazonium flouroborate is filtered, washed with water and dried at 90–100°C for 20–25 minutes. The dry salt is heated at 150°C (oil bath) in a round bottom flask (250 ml capacity) fitted with an air condenser. The temperature of the oil bath is raised slowly. When the decomposition of the salt is complete, the mixture is steam distilled when the 4,4′-difluorobiphenyl is obtained in the steam distillate. It is filtered and crystallised from ethanol. The yield is 5.7 g (40%); m.p. is 92–93°C.

1. C. Balz and G. Schiemann, Ber., 1927, **60** 1186.

88. STEPHEN REACTION[1,2]

The nitriles on reduction with stannous chloride in ether saturated with hydrogen chloride give the intermediate aldimine salts, which on hydrolysis give the corresponding aldehydes. Thus,

$$R{-}C{\equiv}N \xrightarrow[SnCl_2\,/\,HCl]{Redox} \left[R{-}C{\equiv}\overset{+}{N}H\right]Cl^- \xrightarrow{[H]} R{-}CH{=}\overset{+}{N}H_2\ Cl^- \xrightarrow{H_2O} RCHO$$

The above sequence of reaction is known as Stephen reaction. Best results are obtained in the aromatic series.

88.1 Hexanal

It is obtained by the reduction of hexane nitrile with stannous chloride and hydrogen chloride in anhydrous ether followed by hydrolysis of the formed intermediate.

$$CH_3(CH_2)_3CH_2CN \xrightarrow[SnCl_2]{HCl} \left[CH_3(CH_2)_3CH_2CH{=}\overset{+}{N}H_2\right]Cl^- \xrightarrow{H_2O} CH_3(CH_2)_3CH_2CHO$$

Hexane nitrile : 9.75 g
Stannous chloride : 28.5 g
Ether (anhydrous) : 100 ml

Anhydrous stannous chloride (28.5 g) and dry ether (100 ml) are placed in a three-necked round bottom flask (250 ml capacity) fitted with a sealed stirrer, a gas inlet tube and a reflux condenser. Dry hydrogen chloride gas is passed through the stirred mixture until saturated. Two layers are separated. The lower viscous layer consists of stannous chloride dissolved in ethereal hydrogen chloride. The mixture is stirred and hexane nitrile (9.75 g, 0.1 mol) is added through a dropping funnel. The stirring is continuous for 15–20 minutes. The separated aldimine hydrochloride is filtered and refluxed with water (25 ml) till hydrolysis is complete. The cooled mixture is extracted with ether, the extract dried (anhydrous magnesium sulphate) and the solvent removed by distillation. The residual product is distilled and the fraction between 127–129°C is collected. The yield is 9.5 g (95%).

Note: Hexane nitrile is obtained as follows:

Step 1: Hexanoyl chloride

A mixture of hexanoic acid (29 g, 0.25 mol) and thionyl chloride (36 g, 21.5 ml, 0.3 mol) (added dropwise during 45 minutes) is heated in a round bottom flask on a water bath for 30 minutes. The haxanoyl chloride is isolated by distillation. The yield is 28 g (83%) b.p. is 150–155°C.

Step 2: Hexanamide

Hexanoyl chloride (28 g, 0.21 mol) is added to a stirred and cooled ammonia solution (d. 0.880, 62.5 ml) using a dropping funnel. The separated amide is filtered and crystallised from hot water. The yield is 15 g (63%); m.p. is 98–99°C.

Step 3: Hexane nitrile

Hexanamide (14.5 g, 0.125 mol) is refluxed (water bath) for one hour with thionyl chloride (22.5 g, 13.75 ml, 0.19 mol) in a round bottom flask. The nitrile is obtained by distillation. The yield is 10.5 g (86%); b.p. is 161–163°C.

1. H. Stephen, *J. Chem. Soc.*, 1925, **127**, 1875.
2. T. Stephen and H. Stephen, *J. Chem. Soc.*, 1956, 4695.

89. STRECKER AMINO ACID SYNTHESIS[1]

The original Strecker procedure consists in the reaction of an aldehyde with ammonia to form the corresponding addition products (aldehyde ammonia), followed by reaction with hydrogen cyanide to form α-amino nitrile. This intermediate can also be obtained by the treatment of the aldehyde with hydrogen cyanide followed by the reaction of the formed cyanohydrin with ammonia. The latter is known as Erlenmeyer modification. However, a more convenient route is to treat the aldehyde in one step with ammonium chloride and sodium cyanide (this mixture is equivalent to ammonium cyanide). This procedure is referred to as the Zelinsky-Stadnikoff modification. The final step is the hydrolysis of the intermediate α-amino nitrile under acidic or basic conditions to give the corresponding α-amino acid.

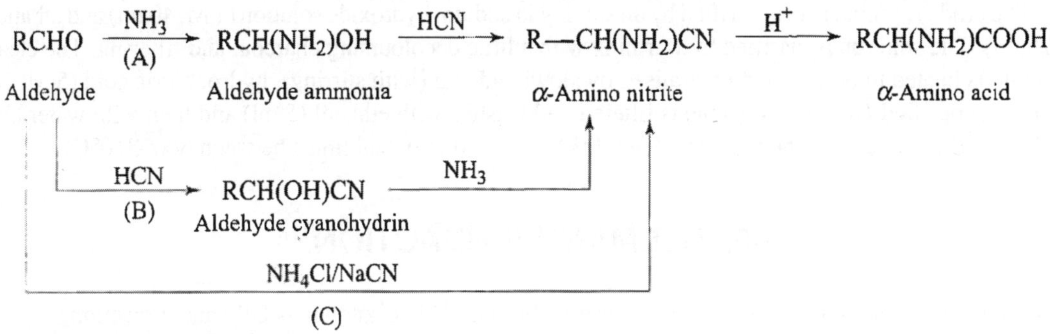

A–Original Strecker procedure
B–Erlenmeyer modification
C–Zelinsky-Stadnikoff modification

Mechanism

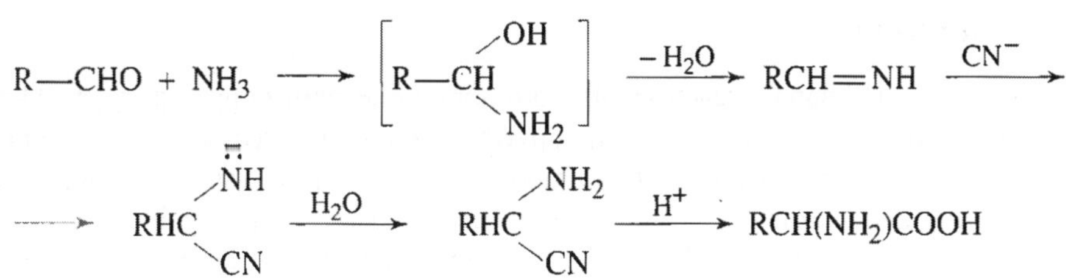

89.1 α-Phenylglycine (DL-2-Aminophenylacetic acid)

It is obtained by the reaction of benzaldehyde with a solution of sodium cyanide and ammonium chloride to give the corresponding amino nitrile followed by final hydrolysis with acid.

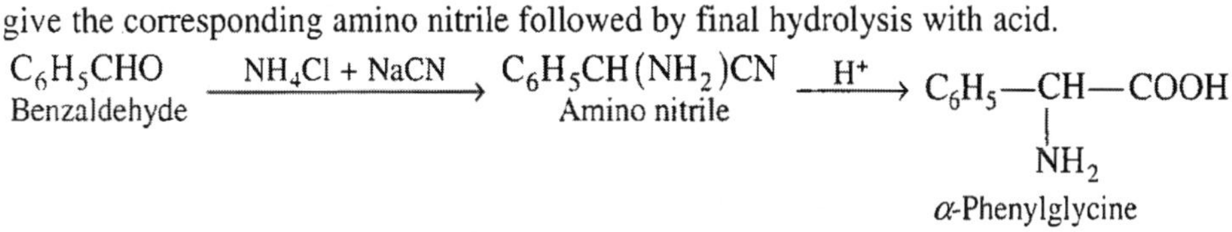

Benzaldehyde	: 10.5 g
Ammonium chloride	: 5.5 g
Sodium cyanide	: 5 g

1. A. Strecker, *Ann.*, 1850, **75**, 27; 1854, **91**, 349.

Hydrochloric acid : 5M, 60 ml

A solution of benzaldehyde (10.5 g, 0.1 mol) in methanol (20 ml) is added with vigorous stirring to a solution of sodium cyanide (5 g, 0.1 mol) in water (20 ml) to which ammonium chloride (5.5 g, 0.1 mol) has been added. The mixture is vigorously shaken and left for 2 hours at room temperature. Water (50 ml) is then added and the oily aminonitrile is extracted with toluene (2 × 25 ml). The combined toluene extract is washed with water and extracted with hydrochloric acid (5 M, 2 × 30 ml). The clear acidic extract is refluxed for 18–20 hours (*caution:* hydrogen cyanide is evolved), filtered through a small cotton plug to remove tarry matter if present, and then basified with concentrated ammonia solution (d. 0.88, 20 ml). The precipitated α-amino acid is filtered, washed with cold water (50 ml) and then with hot alcohol (7 ml).

The crude α-amino acid is purified by dissolving in sodium hydroxide solution (1 M, 40 ml) and ethanol (25 ml). The solution is clarified by warming with a little decoloursing charcoal and filtering. The clear filtrate is heated to boiling and neturalised by slowly adding (with stirring), hydrochloric acid (5 M, 16 ml). The purified *DL*-phenylglycine is filtered and washed with ethanol (5 ml) and then with water (10 ml). It is dried at 50°C. The yield is 4.5 g (30%). It has no m.p. (sublimes between 300–310°C).

90. ULLMANN'S REACTION[1,2]

Aryl halides on heating with copper powder give biaryls. This is known as Ullmann's reaction.

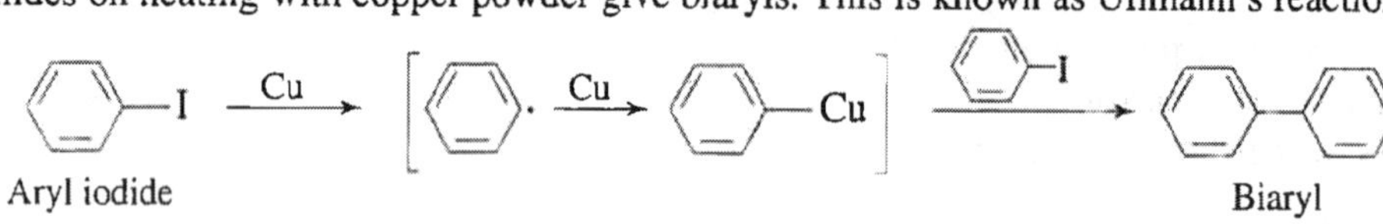

90.1 2,2′-Dipyridyl

It is obtained by the Ullmann's reaction of 2-bromopyridine with copper powder. The required 2-bromopyridine is obtained by treating 2-aminopyridine in concentrated aqueous hydrobromic acid with bromine and then with sodium nitrite. Diazotisation of the first formed 2-aminopyridinium perbromide complex gives pyridinium diazonium dication, which readily undergoes nucleophilic displacement of the diazonium group with bromide ion. The 2-bromopyridine is obtained in the form of a complex with bromine. The complex on treatment with alkali gives 2-bromopyridine.

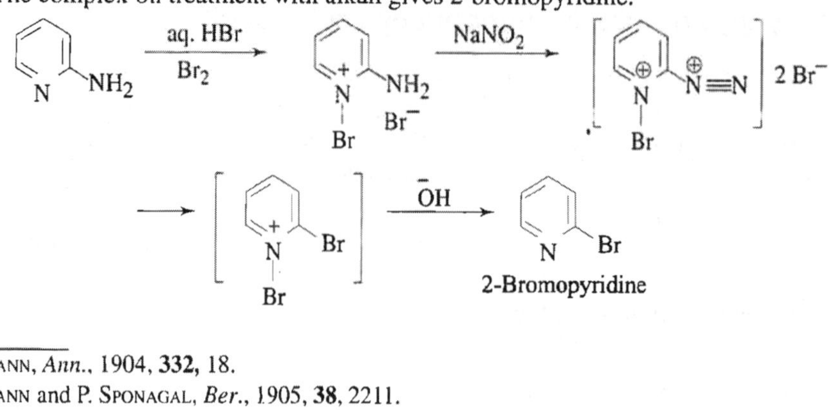

1. F. ULLMANN, *Ann.*, 1904, **332,** 18.
2. F. ULLMANN and P. SPONAGAL, *Ber.*, 1905, **38**, 2211.

Step 1: 2-Bromopyridine

2-Aminopyridine : 15 g
Hydrobromic acid 48% : 79 ml
Bromine : 24 ml
Sodium nitrite : 28 g
Sodium hydroxide : 60 g

2-Aminopyridine (15 g, 0.16 mole) is added dropwise (separatory funnel) during 10 minutes to stirred and cooled (10–15°C, ice bath) aqueous hydrobromic acid (48%, 79 ml) contained in a three necked flask (1 litre capacity) fitted with a mechanical stirrer. The reaction mixture is cooled to 0°C (ice-salt mixture) and bromine (24 ml, 0.47 mole) is added dropwise (separatory funnel) with stirring. During addition of half the bromine (30 minutes) the reaction mixture thickens due to the formation of yellow-orange perbromide; the second half of bromine is introduced more rapidly (10–15 minutes). A solution of sodium nitrite (28 g) in water (40 ml) is now added dropwise (2 hours) while keeping the reaction mixture at 0°C. The stirring is continued for 30 minutes and a solution of sodium hydroxide (60 g) in water (60 ml) is added at such a rate that the temperature does not rise above 20–25°C. The reaction mixture is extracted with ether (4 × 50 ml). The combined ether extract is dried (potassium hydroxide pellets) and the ether is evaporated. 2-Bromopyridine is obtained by distillation under reduced pressure. The yield is 22 g (87.6%), b.p. is 74–75°C/13 mm; b.p. 193–195°C.

Step 2: 2,2′-Dipyridyl

2-Bromopyridine : 21 g
p-Cymene : 40 ml
Copper powder : 16 g

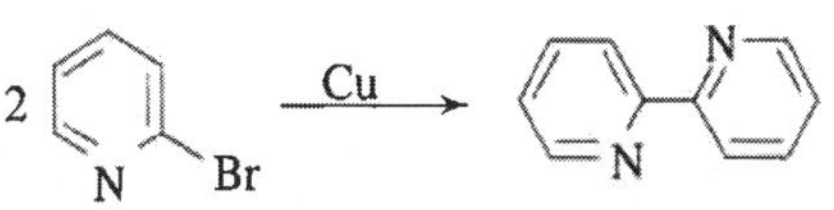

A mixture of *p*-cymene (40 ml, pure, b.p. 176–177°C) and copper powder (10 g) is refluxed in a three necked flask (500 ml capacity) fitted with a mechanical stirrer and a reflux condenser. 2-Bromopyridine (21 g, 0.133 mole) is added dropwise during one hour, to the refluxing and stirred mixture. Additional copper powder is added in three portions of 2 g each during the period. Heating and stirring is continued for 2.5 hours more. The reaction mixture is cooled and acidified with dilute hydrochloric acid. *p*-Cymene is removed from the reaction mixture by steam distillation. The residual solution is rendered strongly alkaline with concentrated sodium hydroxide solution. The steam distillation is done again till the distillate gives only a pale red colouration with $FeSO_4$ solution. The steam distillate is saturated with sodium chloride and extracted with ether (4 × 50 ml). The combined ether extract is dried (anhydrous potassium carbonate), the ether is evaporated and the residue distilled under reduced pressure. 2,2′-Dipyridyl is obtained at 147°C/10 mm, yield is 6.2 g (30%); m.p. is 69–70°C.

90.2 2,2′-Dinitrobiphenyl

o-Chloronitrobenzene : 25 g
Dimethyl formamide : 120 ml
Activated copper bronze : 50 g

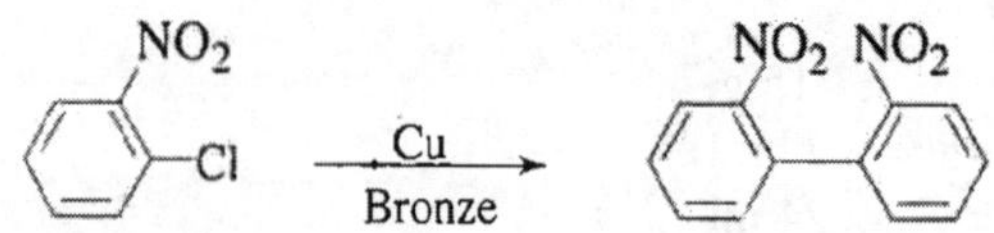

o-Chloronitrobenzene (25 g, 0.159 mole) and dry dimethylformamide (120 ml) are refluxed in a three necked round bottom flask (250 ml capacity) fitted with a reflux condenser and stirrer (mercury sealed). The mixture is refluxed and activated copper bronze (25 g) is added in one lot. After 4 hours of refluxing, more copper bronze (25 g) is added and refluxing continued for additional 4 hours. The reaction mixture is cooled and poured in water (2 litre). The separated product along with copper bronze is filtered (suction) and extracted with boiling alcohol (3 × 200 ml). The combined extract is concentrated and cooled. The crude product is recrystallised from alcohol (using decolourising charcoal). The yield is 14 g (36%), m.p. is 124–125°C.

Note: Activated copper bronze is obtained by treating 50 g copper bronze with iodine solution in acetone (2%, 500 ml) for 5–10 minutes. The product is filtered (suction) and washed by stirring with concentrated hydrochloric acid-acetone solution (1:1, 250 ml). This treatment dissolves the formed copper iodide. The remaining copper bronze is filtered, washed with acetone and dried (vacuum). The activated copper bronze should be used immediately.

91. VILSMEIER-HAACK REACTION[1]

Certain reactive aromatic or heterocyclic compounds can be formylated by dimethylformamide in presence of phosphorous oxychloride.

Mechanism

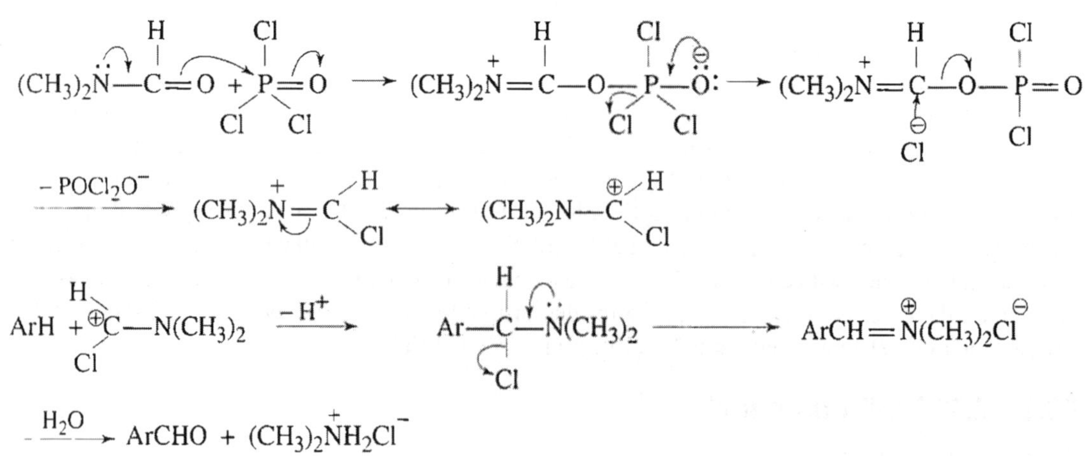

1. A. VILSMEIER and A. HAACK, *Ber.*, 1927, **60**, 119.

91.1 2,4-Dimethoxybenzaldehyde

It is obtained by formylation of resorcinol dimethyl ether with dimethylformamide in presence of phosphorus oxychloride

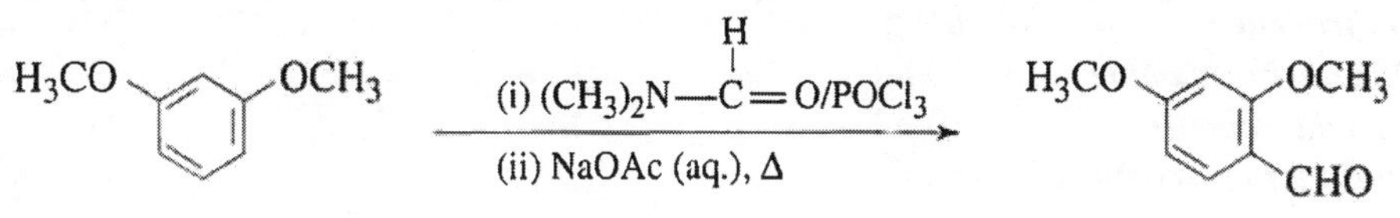

Resorcinol dimethylether : 14 g
(1,3-Dimethoxybenzene)
N,N-Dimethylformamide : 10 ml
Phosphorus oxychloride : 10 ml

Resorcinol dimethyl ether (14 g, 0.10 mol, see page 146) is added under anhydrous conditions to a mixture of *N, N*-dimethyl formamide (10 ml, 0.13 mol) and phosphorus oxychloride (10 ml, 0.10 mol) in small amounts while cooling the reaction mixture in ice-bath. The resulting reddish-orange viscous liquid is heated at 100°C (water bath) for 6 hours under anhydrous conditions. Excess phosphorus oxychloride is removed under reduced pressure and to the cooled reaction mixture is added a saturated solution of sodium acetate (30 g). The mixture is refluxed for 30 minutes and poured into ice-water. The separated product is filtered, washed with water and crystallised from dilute methanol. The yield is 10 g (62%); m.p. is 68–70°C.

91.2 2,4,5-Trimethoxybenzaldehyde (Asarlaldehyde)

It is obtained by the formylation of 1,2,4-trimethoxybenzene with dimethylformamide in presence of phosphorus oxychloride.

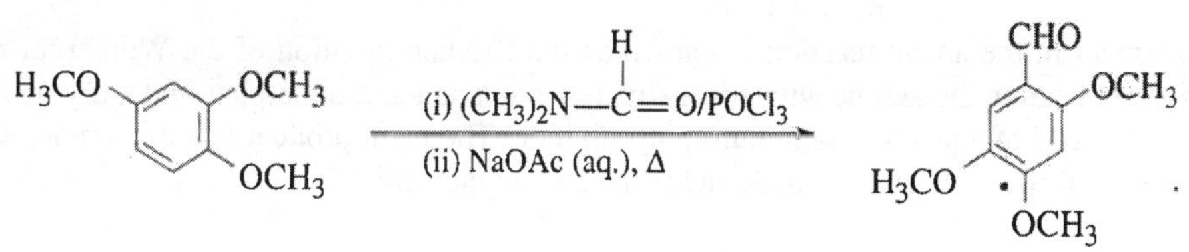

1,2,4-Trimethoxybenzene : 15 g
Dimethylformamide : 10.5 g
Phosphorus oxychloride : 15 ml

A mixture of 1,2,4-trimethoxybenzene (15 g, 0.09 mol, See page 146), dimethylformamide (10.5 g, 0.14 mol) and freshly distilled phosphorus oxychloride (15 ml, 0.16 mol) is heated on a boiling water bath for 6 hours under anhydrous conditions. A sturated solution of sodium acetate (5 ml) is then added and the mixture refluxed for 30 minutes. The solution is cooled and the separated product filtered and washed with cold water. It is crystallised from hot water. The yield is 10.5 g (60%); m.p. is 113–114°C.

91.3 9-Formylanthracene

It is obtained by the reaction of anthracene with dimethylformamide in presence of phosphorus oxychloride using *o*-dichlorobenzene as solvent.

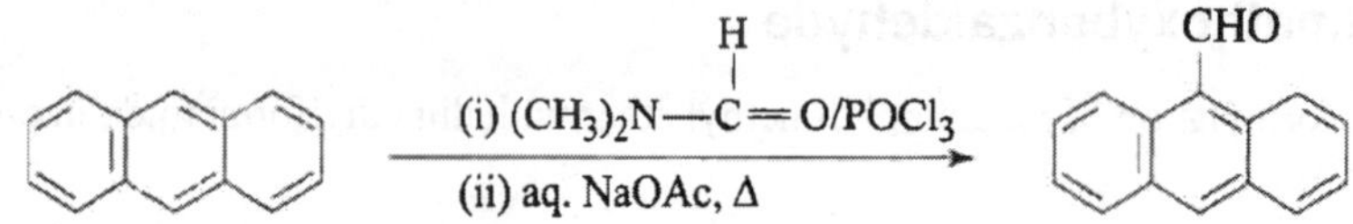

Anthracene	: 8.9 g
Dimethylformamide	: 10 ml
o-Dichlorobenzene	: 10 ml
Phosphorus oxychloride	: 8 ml

A mixture of anthracene (8.9 g, 0.05 mol), dimethyl-formamide (8.9 g, 10 ml, 0.13 mol) and *o*-dichlorobenzene (10 ml) is taken in a three necked round bottom flask (250 ml capacity) fitted with a sealed stirrer, a reflux condenser and a dropping funnel. Phosphorus oxychloride (13.5 g, 8 ml, 0.087 mol) is added slowly (using a dropping funnel) to the stirred mixture. It is then heated for 2 hours in a water bath. The reaction mixture is cooled (ice-salt bath) and neutralised to congo red by adding aqueous sodium acetate solution (approx. 50 g in 80 ml water). The mixture is diluted with more water (1 litre) and allowed to stand at 0°C for 2 hours. The separated 9-formylanthracene is filtered and washed with water. It is crystallised from dilute acetic acid. The yield is 6 g (58%); m.p. is 103–104°C.

92. WILLGERODT-KINDLER REACTION[1,2]

The conversion of a carbonyl compound by ammonium polysulphide solution into an amide with the same number of carbon atoms is known as Willgerodt reaction. The product isolated is the amide of ω-arylcarboxylic acid together with a smaller amount of the corresponding ammonium salt of the carboxylic acid. Thus,

$$ArCOCH_3 \xrightarrow[H_2O,\ 200-225^\circ]{(NH_4)_2S_4} ArCH_2CONH_2 + ArCH_2CO_2^- \overset{+}{N}H_4$$

An improvement in the above reaction is known as the Kindler variation of the Willgerodt reaction. This consists in heating the ketone with approximately equal amount of sulphur and a dry amine (e.g., morpholine) instead of aqueous ammonium polysulphide. The main product is a thioamide, which on subsequent hydrolysis with acid or alkali yields the carboxylic acid. Thus

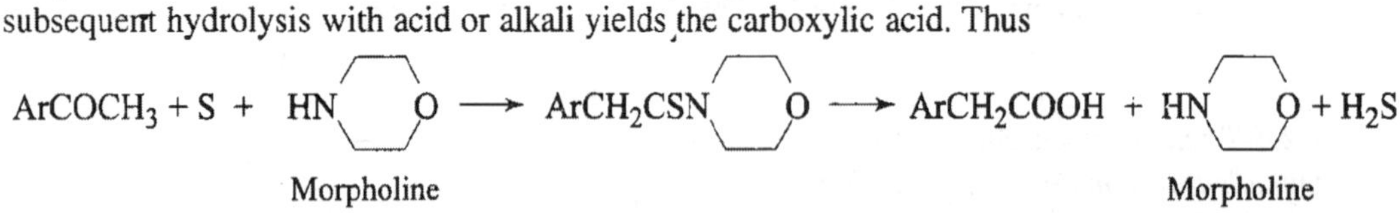

92.1 *p*-Methoxyphenylacetic acid

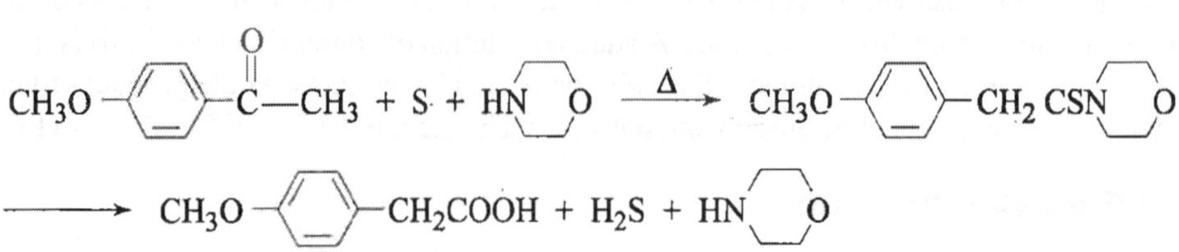

1. C. WILLGERODT, *Ber.*, 1887, **20**, 2467.
2. K. KINNDLER, *Ann.* 1923, **431**, 193.

p-Methoxyacetophenone	: 21 g
Sulphur	: 6.75 g
Morpholine	: 18 ml

A mixture of *p*-methoxyacetophenone (21 g, 0.14 mol), sulphur (6.75 g, 0.21 mol) and morpholine (18 ml, 0.21 mol) is refluxed (fume cupboard) in a round bottom flask for 5 hours. The reaction mixture is poured slowly into water. The separated thioacetomorpholide is filtered, washed with water and crystallised from dilute methanol. The yield is 28 g (80%); m.p. is 71–72°C.

The thioacetomorpholide (crude product can be used) (25 g) is refluxed for 10 hours with alcoholic sodium hydroxide solution (10%, 200 ml). The reaction mixture is diluted with water (50 ml) and acidified with concentrated hydrochloric acid. It is cooled and extracted with ether (3 × 30 ml). The combined ether extract is dried over anhydrous magnesium sulphate and the solvent evaporated. The residual product is crystallised from water or dilute alcohol. The yield is 13 g (63%); m.p. is 85–86°C.

93. WITTIG REACTION[1,2]

The aldehydes or ketones react with a phosphorus ylide (alkylidene phosphorane) to yield alkenes and triphenylphosphine oxide. The ylide (known as Witting reagent) is generally prepared by the treatment of a suitable quaternary phosphonium salt (obtained by the reaction of triphenyl phosphine and alkyl halide) with a suitable base. The Wittig reagent condenses with the carbonyl compound to give a betaine intermediate, which in turn yields alkene with loss of triphenyl phosphine oxide.

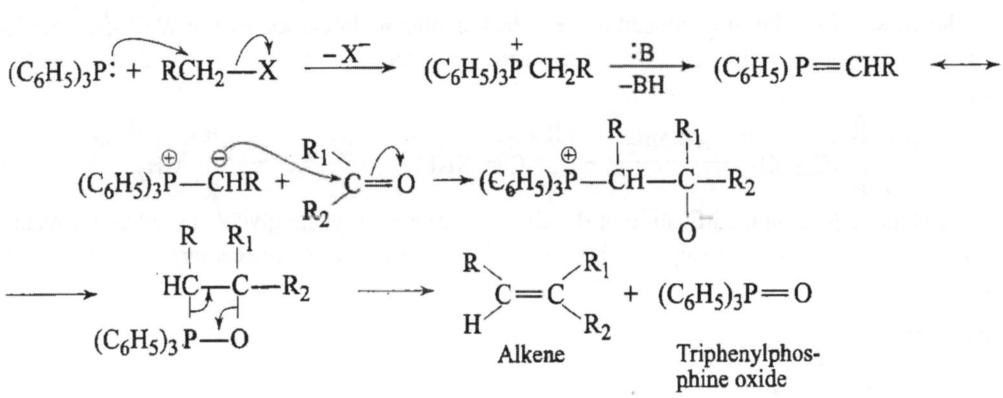

93.1 Trans-Stilbene

It is obtained by the reaction of triphenyl phosphonium salt with benzaldehyde.

$$C_6H_5CH_2Cl + (C_6H_5)_3P \xrightarrow{\overset{\ominus}{O}C_2H_5} (C_6H_5)_3P{=}CH{-}C_6H_5 \xrightarrow[\text{Warm}]{C_6H_5CHO}$$

1. C. Wittig and U. Schollkopf, *Ber.*, 1954, **87**, 1318.
2. G. Wittig and N. Haag, *Ber.*, 1955, **88**, 1654.

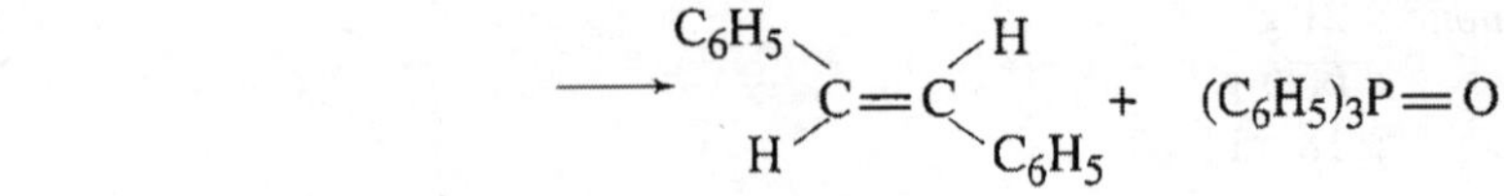

Triphenyl phosphine	: 14.3 g
Benzylchloride	: 4.4 ml
Xylene	: 100 ml
Sodium metal	: 0.88 g
Absolute ethanol	: 250 ml
Benzaldehyde	: 2.64 ml

A mixture of benzylchloride (4.4 ml, 0.038 mole), triphenylphosphine (14.3 g, 0.055 mole) and xylene (70 ml) is refluxed in a round bottom flask. The separated phosphonium salt (m.p. 310–311°C) is filtered, washed with xylene (30 ml) and dried in a vacuum dessicator.

Sodium metal (0.8 g) cut into small pieces is added to absolute ethanol (100 ml) in a round bottom flask (anhydrous conditions). The ethoxide solution thus obtained is added to a solution of phosphonium salt (obtained above) in absolute alcohol (150 ml) and benzaldehyde (2.64 ml, 0.026 mole) contained in another round bottom flask. The mixture is shaken during the addition of sodium ethoxide solution. The mixture is allowed to stand for 30 minutes. The separated product is filtered, washed with a little alcohol and crystallised from ethanol. The yield is 3.8 g (81.2%); m.p. is 124–125°C.

94. WOLFF-KISHNER REDUCTION[1-3]

The aldehydes and ketones are reduced to the corresponding hydrocarbon by the Wolff-Kishner reduction. It involves heating of the corresponding hydrazone with potassium hydroxide or with sodium ethoxide. Thus,

In the Hunang-Minlon modification of the above reaction diethylene glycol is used as a solvent. The reaction is carried out in one step (the isolation of the hydrazone is unnecessary) and the yields are excellent.

Mechanism

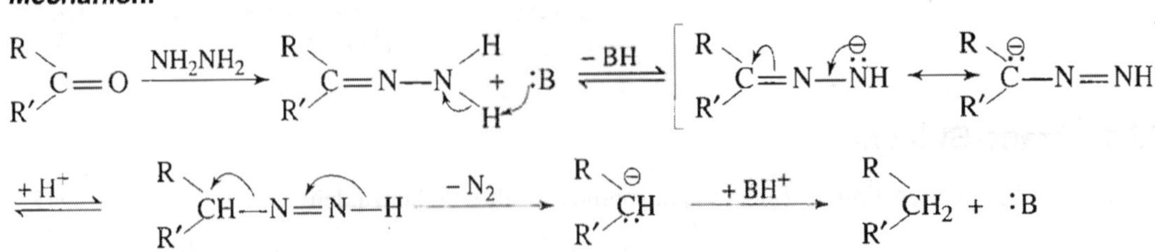

1. L. Wolff, *Ann.*, 1912, **394**, 86.
2. N. Kishner, *J. Russ. Phys. Chem. Soc.*, 1911, **43**, 582 (C.A., 1912, **6**, 347).
3. Huang Minlon, *J. Am. Chem. Soc.*, 1946, **68**, 2487.

94.1 Ethyl Benzene

It is obtained by the Huang-Minlon modification of the Wolff-Kishner reduction of acetophenone with hydrazine using diethylene glycol as solvent and potassium hydroxide as base.

$$C_6H_5COCH_3 \xrightarrow[KOH]{NH_2NH_2,\ \text{diethylene glycol}} C_6H_5CH_2CH_3$$

Acetophenone : 24 g
Diethylene glycol : 200 ml
Hydrazine hydrate (90%) : 20 ml
Potassium hydroxide : 27 g

A mixture of acetophenone (24 g, 0.2 mol), diethylene glycol (200 ml) and hydrazine hydrate (90%, 20 ml) is taken in a two necked round bottom flask (500 ml capacity) fitted with a reflux condenser and a thermometer, the bulb of which is dipping in the reaction mixture. The mixture is warmed (water bath) until most of the potassium hydroxide has dissolved and then refluxed for one hour (using a free flame or a heating mantle). The reaction mixture is distilled until the temperature of the liquid rises to 170–175°C. The upper hydrocarbon layer from the distillate is separated and the aqueous layer extracted with ether (2 × 25 ml). The combined upper layer and ether extract is dried (magnesium sulphate) and ether removed on a water bath. The residue is distilled. The yield is 13.4 g (62.5%); b.p. is 135–136°C.

Note: (1) For reduction of higher ketones refluxing for longer time (3–5 hours) is necessary. (2) See also Clemenson's reduction (page 45).

95. MISCELLANEOUS PREPARATIONS

95.1 *p*-Nitroaniline

It cannot be obtained by direct nitration of aniline. It is necessary to protect the amino group by acetylation. Nitration of acetanilide leads to *p*-nitroacetanilide (major product), which on hydrolysis gives *p*-nitroaniline.

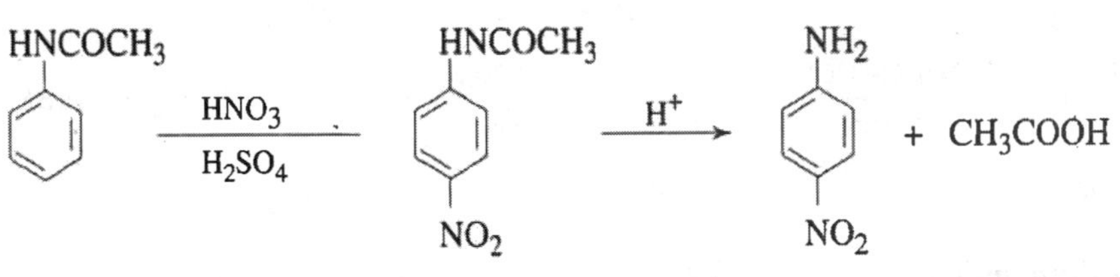

p-Nitroacetanilide : 5 g
Sulphuric acid (70%) : 30 ml

A mixture of *p*-nitroacetanilide (5 g, 0.028 mole see page 155), and sulphuric acid (70%, 30 ml) (obtained by cautiously adding 20 ml of concentrated sulphuric acid to 30 ml water) is refluxed for 25–30 minutes or till a test solution remains clear on dilution with water. The solution is poured into cold water (200 ml) and the solution rendered alkaline by adding excess sodium hydroxide solution (10%).

The separated *p*-nitroaniline is filtered, washed with water and crystallised from dilute alcohol (1:1). The yield is 3.5 g (92%); m.p. is 147–148°C.

95.2 *p*-Bromoaniline

It cannot be obtained by bromination of aniline, since the major product of bromination is 2, 4, 6-tribromo derivative. Therefore the amino group is protected and then bromination gives the *p*-bromoacetanilide as the major product, which on hydrolysis gives the required *p*-bromaniline.

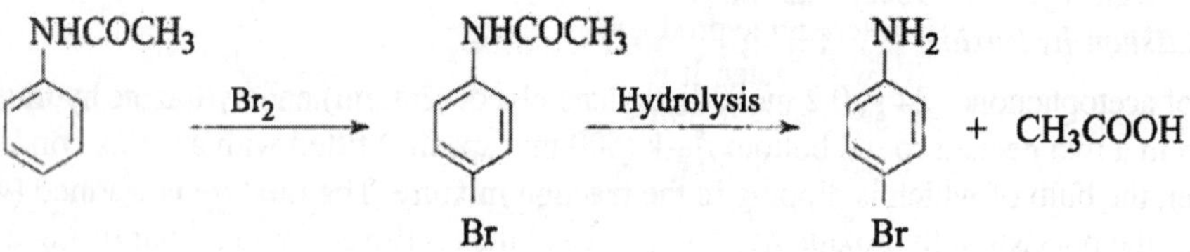

p-Bromoacetanilide : 5 g
Ethanol : 10 ml
Potassium hydroxide : 3 g

A solution of *p*-bromoacetanilide (5 g, 0.023 mole, see page 110) in ethanol (10 ml) and aqueous potassium hydroxide (3 g in 4 ml water) is refluxed for 30 minutes in a round bottom flask (100 ml capacity). To the reaction mixture is added water (40 ml) and the mixture distilled (wire guage) until about 35 ml of distillate (alcohol and water) is collected. The residual solution is poured into water (50 ml). *p*-Bromoaniline separates as an oil, which solidifies on cooling. It is filtered, washed with cold water and crystallised from dilute alcohol using decolorising charcoal. The yield is 3.9 g (97.5%); m.p. is 65–66°C.

95.3 *p*-Dinitrobenzene

It is obtained from *p*-nitroaniline by diazotisation followed by treatment with sodium borofluoride to give the corresponding diazonium fluoborate. It is treated with sodium nitrite solution in presence of copper powder. It is an interesting variation of Sandmeyer's reaction.

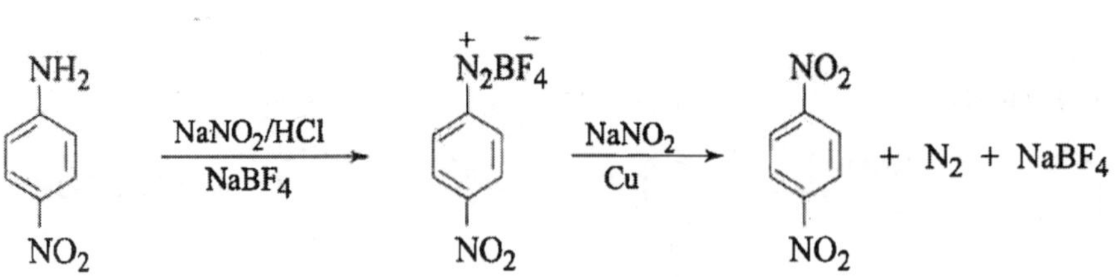

p-Nitroaniline : 6.3 g
Conc. Hydrocloric acid : 12 ml
Sodium nitrite : 41.4 g
Sodium borofluoride : 7.5 g
Copper powder : 7.5 g

p-Nitroaniline (6.3 g, 0.046 mole) is dissolved in a warm solution of concentrated hydrochloric acid (12 ml) and water (12 ml) in a beaker (400 ml capacity). The solution is cooled to 0–5°C (ice-salt bath), stirred mechanically and a cold solution of sodium nitrite (3.4 g) in water (8 ml) is added dropwise with stirring to an end point with potassium iodide-starch paper. During the initial stage amine hydrochloride separated in a finely divided form. The temperature is kept below 5°C. A solution of sodium borofluoride (7.5 g) in water (15 ml) is added. The precipitated diazonium fluoborate is filtered (sintered funnel), washed with cold sodium borofluoride solution (5%, 5 ml), followed by rectified spirit (2 × 5 ml) and finally with ether (4 × 5 ml).

The fluoborate (obtained above) is suspended in water (40 ml). The suspension is added slowly to a well stirred mixture of sodium nitrite solution (38 g in 90 ml water) and copper powder (7.5 g). Some ether (1–2 ml) is added occasionally to break the froth. When all the diazonium compound has been added, the reaction mixture is filtered, the product washed well with water, then with sodium hydroxide solution (5%, 5 ml) and finally with water. It is dried (100°C) and extracted with boiling benzene (3 × 15 ml) (extraction in a soxhlet apparatus is preferred). The benzene solution is distilled and the residue crystallized from glacial acetic acid (25–30 ml). The yield is 5.7 g (74%); m.p. is 171–172°C.

95.4 *p*-Nitrophenylhydrazine

It is obtained from *p*-nitroaniline by diazotisation, followed by reduction with sodium sulphite and subsequent acidification. The formed *p*-nitrophenylhydrazine hydrochloride on treatment with sodium acetate liberated the free base.

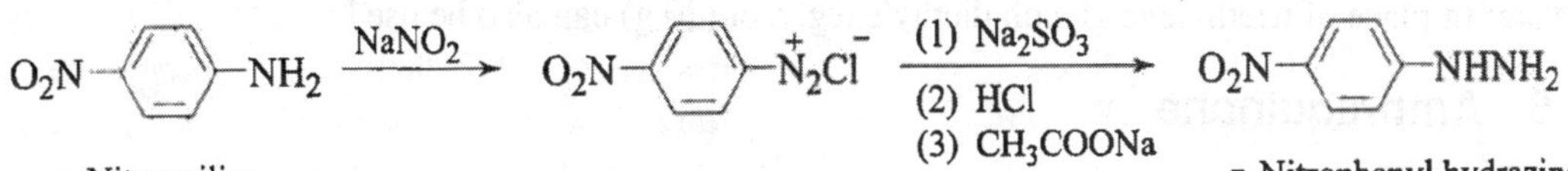

The mechanism is similar to the one described for the preparation of phenyl hydrazine (page 62).

p-Nitroaniline	: 8 g
Conc. hydrochloric acid	: 17.2 ml
Sodium nitrite	: 5 g
Sodium sulphite	: 33 g
Sodium hydroxide	: 3.2 g

p-Nitroaniline (8 g, 0.058 mole) is dissolved in a hot dilute hydrochloric acid (1:1, 34.4 ml). The solution is cooled (0–5°C) and diazotised with a solution of sodium nitrite (5 g) in water (10 ml) (as in the case of *p*-dinitrobenzene). The cold diazotised solution is filtered and the clear solution added through a separating funnel to an ice cold solution of sodium sulphite (90%, $Na_2SO_3 \cdot 7H_2O$, 33 g) in water (75 ml) containing sodium hydroxide (3.2 g). The diazo solution is added slowly (5–10 minutes). The reaction mixture is allowed to stand for 5–10 minutes, acidified with concentrated hydrochloric acid (57 ml) and heated on a water bath at 25°C for 3 minutes. The solution is left overnight and the separated product is filtered. It is heated with concentrated hydrochloric acid (15 ml) on a water bath for 5 minutes and cooled. The separated *p*-nitrophenylhydrazine hydrochloride is separated, dissolved in water and treated with a concentrated solution of sodium hydroxide. The formed *p*-nitrophenylhydrazine is filtered, washed with water and crystallised from alcohol. The yield is 5.3 g (59.7%); m.p. is 158°C (decomp.).

95.5 2,4-Dinitrophenylhydrazine

It is obtained by the reaction of 2,4-dinitrochlorobenzene with hydrazine solution.

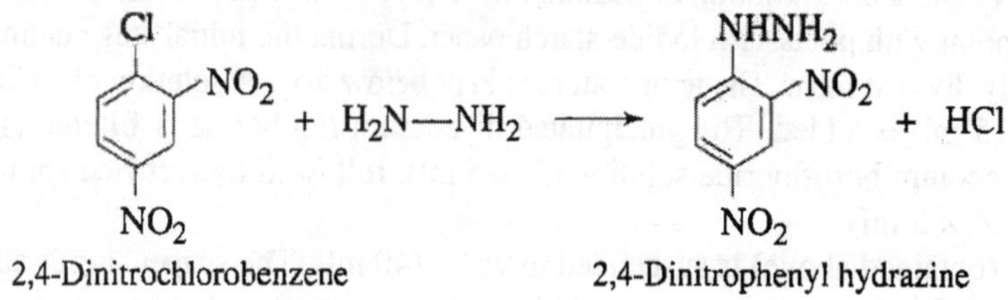

2,4-Dinitrochlorobenzene: 15 g
Triethylene glycol : 40 ml
Hydrazine solution (64%) : 12 ml

A solution of 2,4-dinitrochlorobenzene (15 g, 0.074 mole) in triethylene glycol (40 ml) (gentle warming is necessary) is taken in an evaporating basin (100 ml capacity) and cooled to 15°C (ice bath). Hydrazine solution (65%, 12 ml in 12 ml water) is added dropwise (10–15 minutes) with stirring. The temperature is maintained between 15–20°C. After the exothermic reaction is over, the resultant mixture is heated on a boiling water bath with methanol (15 ml) for 15–20 minutes. The reaction mixture is cooled and the separated 2,4-dinitrophenylhydrazine filtered and crystallised from *n*-butanol. The yield is 11.4 g (77.8%); m.p. is 192-193°C.

Note: In place of triethylene glycol, diethyleneglycol (30 g) can also be used.

95.6 Anthraquinone

It is obtained by the cyclisation of *o*-benzoylbenzoic acid with polyphosphoric acid.

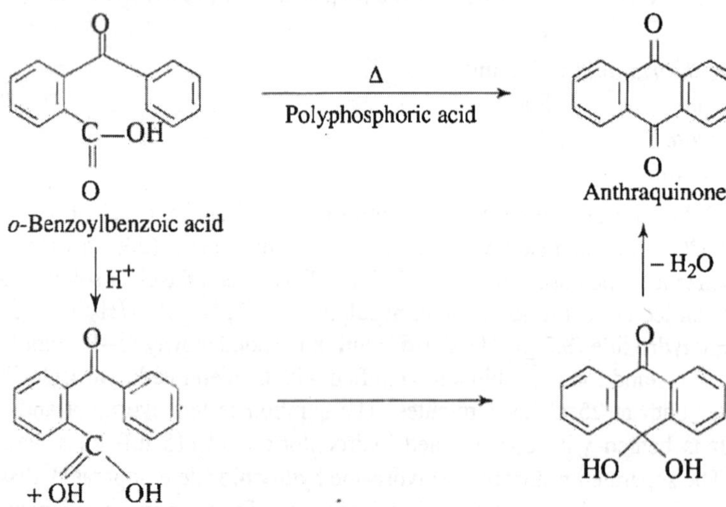

o-Benzoylbenzoic acid : 3 g
Polyphosphoric acid : 37 ml

A mixture of polyphosphoric acid (37 ml, obtained by dissolving 27 g phosphorus pentoxide in 15 g commercial orthophosphoric acid, d. 1.75) and *o*-benzoylbenzoic acid (3 g, 0.013 mole) is heated in a conical flask in a water bath for 2 hours. The cooled reaction mixture is poured with stirring into water

(50 ml). The separated product is filtered, boiled with concentrated ammonia solution (12 ml) for 5 minutes (to remove any unchanged acid) and filtered. It is crystallised from glacial acetic acid using decolourizing carbon (0.5 g). The yield is 2.7 g (98%); m.p. is 285–286°C.

Note: Anthraquinone can also be obtained by the oxidation of anthracene with chromium trioxide (see page 160)

95.7 2,4-Dichlorophenoxyacetic Acid

It is obtained by the condensation of 2,4-dichlorophenol with monochloroacetic acid in sodium hydroxide solution.

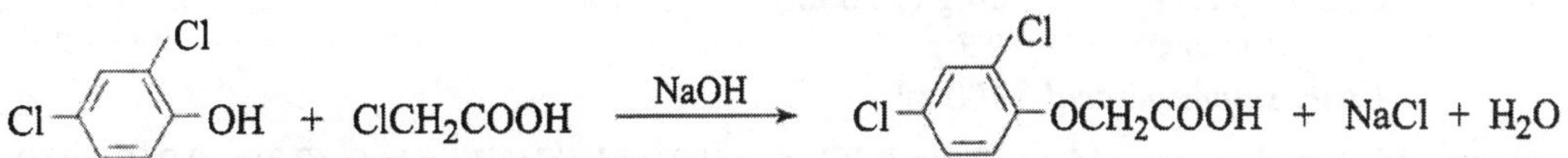

2,4-Dichlorophenol: 10 g
Chloroacetic acid : 5.9 g
Sodium hydroxide : 5.6 g

Sodium hydroxide solution (5.6 g in 30 ml water) is added slowly with stirring to a mixture of 2,4-dichlorophenol (10 g, 0.06 mole) and chloroacetic acid (5.9 g, 0.062 moles) contained in a beaker (400 ml capacity). An exothermic reaction commences. The mixture is heated on a sand bath until most of the liquid has evaporated. To the residue is added water (160 ml). The solution is cooled and acidified with dilute hydrochloric acid. The separated oil is extracted with ether (2 × 25 ml), the combined ether extract washed with water, dried (anhydrous magnesium sulphate) and ether evaporated. The residue is crystallised from benzene (35–40 ml). The yield is 7.2 g (53%); m.p. is 137–138°C.

95.8 *s*-Benzylisothiouronium Chloride

It is prepared by the reaction of benzyl chloride with thiourea in alcoholic solution.

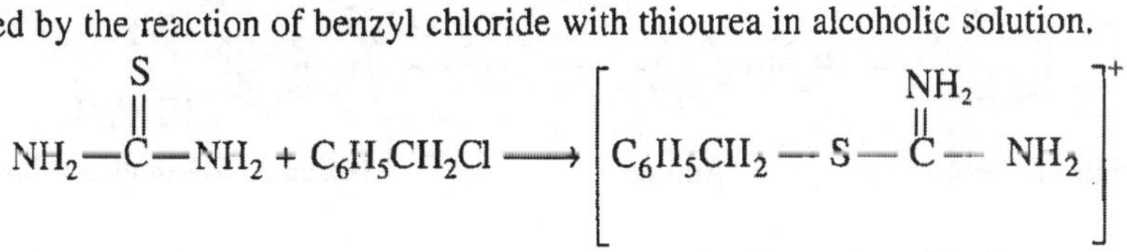

Thiourea   Benzyl chloride   S-Benzylisothiouronium chloride

Benzyl chloride : 15.20 g (13 ml)
Thiourea : 8.46 g
Alcohol : 20 ml

A mixture of redistilled benzyl chloride (15.2 g, 13 ml, 0.12 mole), thiourea (8.46 g, 0.11 mole) and alcohol (20 ml) is warmed in a R.B. flask (50 ml capacity) in a water bath. An exothermic reaction commences. The resulting solution is refluxed for 30 minutes and then cooled (ice bath). The separated S-benzylisothiouronium chloride is filtered and recrystallised from dilute hydrochloric acid (0.2 N, 45–50 ml). The yield is 18.5 g (82.2%); m.p. is 149–150°C.

Note: The product obtained above can be converted into the other form (m.p. 175°C) by dissolving in alcohol and seeding with a crystal of the form, m.p. 175°C. Both dimorphic forms give identical derivatives with carboxylic acids.

95.9 1,1-Bis(*p*-chlorophenyl)-2,2,2-trichloroethane (D.D.T.)

It is prepared by the condensation of chlorobenzene and chloral hydrate in presence of concentrated sulphuric acid.

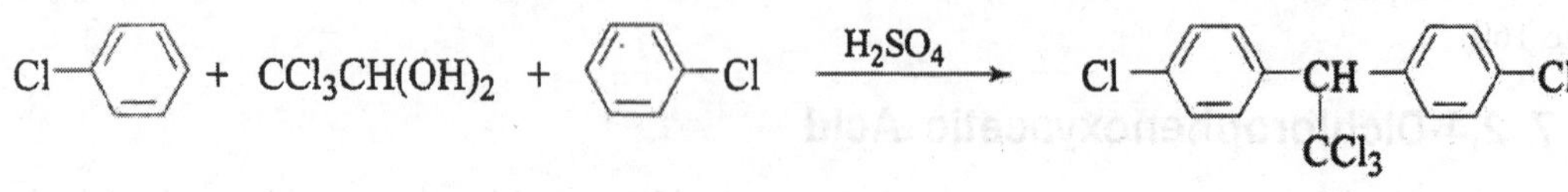

(D.D.T)

Chlorobenzene	: 14 g (12.6 ml)
Chloral hydrate	: 9.3 g
Conc. sulphuric acid	: 100 ml

A mixture of chlorobenzene (14 g, 12.6 ml, 0.124 mole) and chloral hydrate (9.3 g, 0.056 mole) is warmed in a R.B. flask in a water bath with occasional shaking till all crystals have dissolved. The reaction mixture is cooled (ice bath) and concentrated sulphuric acid (100 ml) added slowly. The flask is tightly stoppered and shaken for 1–1.5 hours and then allowed to stand for 1.5 minutes. The mixture is poured slowly with constant stirring to water (400 ml) and cooled. The separated D.D.T. is filtered (sintered funnel), washed with water, then with dilute sodium carbonate solution (4%) with stirring and finally with water till the washings are neutral to litmus. The product is thoroughly titurated with ethyl alcohol (100 ml) in a glass mortar and filtered. The yield is 7.8 g (39%); m.p. is 107°C. It can be recrystallised from *n*-propyl alcohol (5 ml per gram); m.p. is 108°C.

95.10 *N*-Phenyl-2,4-dinitroaniline

It is obtained by the condensation of 2,4-dinitrochlorobenzene with aniline in rectified spirit.

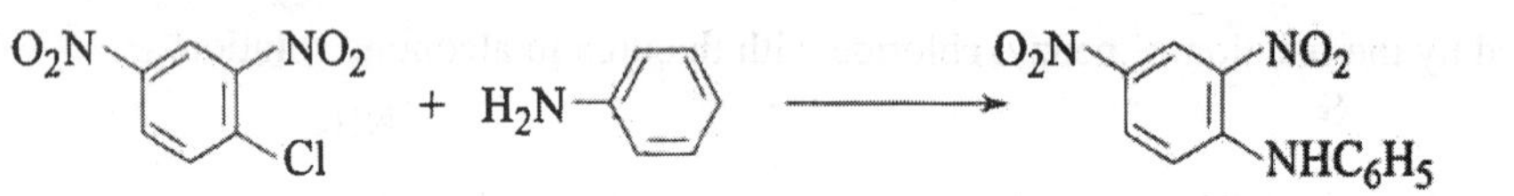

2,4-Dinitrochlorobenzene Aniline N-Phenyl-2,4-dinitroaniline

2,4-Dinitrochlorobenzene	: 5 g
Aniline	: 5 ml
Rectified spirit	: 50 ml

2,4-Dinitrochlorobenzene (5 g, 0.025 mole) is added in small lots (5–10 minutes) to a stirred solution (room temperature) of pure aniline (5 ml, 0.055 mole) in rectified spirit (50 ml). The reaction mixture is warmed to 30°C, stirred for 15 minutes and then cooled. The separated product is filtered, washed with water and crystallised from alcohol. The yield is 5.5 g (86%); m.p. is 154–156°C.

95.11 Hydantoin

It is obtained by the cyclisation of hydantoic acid with concentrated hydrochloric acid.

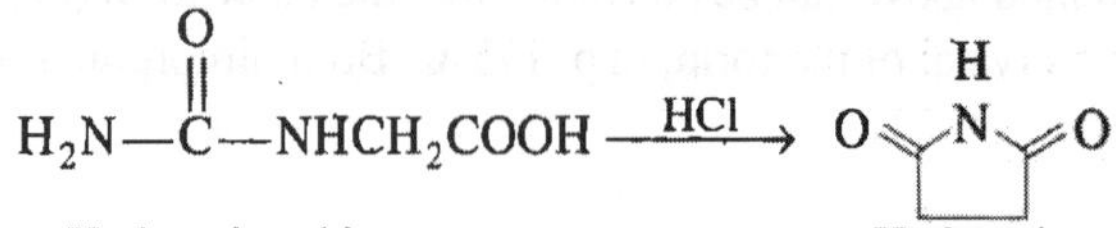

Hydantoic acid Hydantoin

Step 1: Hydantoic acid

It is obtained by the reaction of glycine with urea in presence of alkali.

$$\underset{\text{Glycine}}{NH_2CH_2COOH} + \underset{\text{Urea}}{H_2N-\overset{\overset{\displaystyle O}{\|}}{C}-NH_2} \xrightarrow{NaOH} \underset{\text{Hydantoic acid}}{H_2N-\overset{\overset{\displaystyle O}{\|}}{C}-NHCH_2COOH}$$

Glycine : 3.8 g
Urea : 6 g
Sodium hydroxide : 2 g

Aqueous sodium hydroxide solution (2 g in 3 ml water) is added to a mixture of glycine (3.8 g, 0.05 mole) and urea (6 g, 0.1 mole) in a R.B. flask. The mixture is shaken well and heated at 110–115°C (oil bath) for one hour. It is cooled to 60°C and then acidified with concentrated hydrochloric acid to congo red. It is cooled and the separated product is filtered and recrystallised from water. The yield is 4 g (67%); m.p. is 169–170°C.

Step 2: Hydantoin

Hydantoic acid : 3 g
Conc. hydrochloric acid : 5 ml

A mixture of hydantoic acid (3 g, 0.025 mole), concentrated hydrochloric acid (5 ml) and water (2 ml) is heated at 110–115°C (oil bath) for 15 minutes. The reaction mixture is cooled (ice bath). The separated product is filtered and crystallised from water. The yield is 2 g (78.7%); m.p. is 220–221°C.

95.12 ω-Nitrostyrene

It is obtained by the condensation of benzaldehyde with nitromethane in presence of alkali.

$$\underset{\text{Benzaldehyde}}{C_6H_5CHO} + CH_3NO_2 \xrightarrow{NaOH} \underset{\omega\text{-Nitrostyrene}}{C_6H_5CH{=}CHNO_2}$$

Benzaldehyde : 11.7 g
Nitromethane : 7.5 g
Methanol : 40 ml
Sodium hydroxide : 4.8 g

Benzaldehyde (11.7 g, 11.2 ml, 0.11 mole) and nitromethane (7.5 g, 6.65 ml, 0.123 mole) are dissolved in methanol (40 ml) in a round bottom flask. The solution is cooled in a freezing mixture of ice and salt. The mixture is stirred and a solution of sodium hydroxide in water (4.8 g in 12 ml) is added dropwise at such a rate that the temperature remains between 10–15°C. If the mixture becomes too viscous, more methanol is added. The mixture is stirred for 15–20 minutes more and cold water (70 ml) added. The solution is added slowly with stirring into dilute hydrochloric acid (25 ml of concentrated acid in 38 ml water). The separated nitrostyrene is filtered, washed with water and crystallised from alcohol. The yield is 12.5 g (76.2%); m.p. is 57–58°C.

95.13 Acetamide

It is obtained by heating ammonium acetate,

$$\underset{\text{Ammonium acetate}}{CH_3COONH_4} \xrightarrow{\Delta} \underset{\text{Acetamide}}{CH_3CONH_2} + H_2O$$

Ammonium acetate : 25 g
Acetic acid (glacial) : 20 ml

A mixture of dry ammonium acetate (25 g, 0.32 mole) and glacial acetic acid (20 ml) is refluxed for one hour in a R.B. flask (100 ml capacity) over a wire gauge. The mixture is distilled very slowly using a small fractionating column (5–6″), and taking care that the temperature does not rise above 104–105°C, till about 25 ml of distillate is collected (2–2.5 hours). Finally the liquid is distilled at a higher temperature and the fraction between 195–225°C is collected. This fraction solidifies on cooling. It is crystallised from benzene or ether. The yield is 14 g; (73.3%) m.p. is 81–82°C; b.p. is 222°C.

95.14 Benzamide

It is obtained by the reaction of benzoic acid with phosphorus pentachloride followed by treatment of the formed benzoyl chloride with ammonium hydroxide.

$$\underset{\text{Benzoic acid}}{C_6H_5COOH} + PCl_5 \xrightarrow{\Delta} \underset{\text{Benzoyl chloride}}{C_6H_5COCl} + POCl_3 + HCl$$

$$C_6H_5COCl + 2NH_4OH \xrightarrow{\Delta} C_6H_5CONH_2 + NH_4Cl + 2H_2O$$

Benzoic acid : 10.5 g
Phosphorus pentachloride : 19 g
Liquor ammonia : 38 ml

A mixture of benzoic acid (10.5 g, 0.086 mole) and phosphorus pentachloride (19 g) is mixed with a glass rod in a distilling flask (100 ml capacity). When the reaction slackens, the formed phosphorus oxychloride is distilled at 105–110°C. The residual benzoyl chloride is added dropwise (using a dropping funnel) to cooled liquor ammonia (38 ml). The solution is kept cold during the addition. The separated benzamide is filtered, washed with cold water and recrystallised from hot water. The yield is 7.5 g (72%); m.p. is 129–130°C.

95.15 Oxamide

It is obtained by the reaction of diethyl oxalate with ammonium hydroxide.

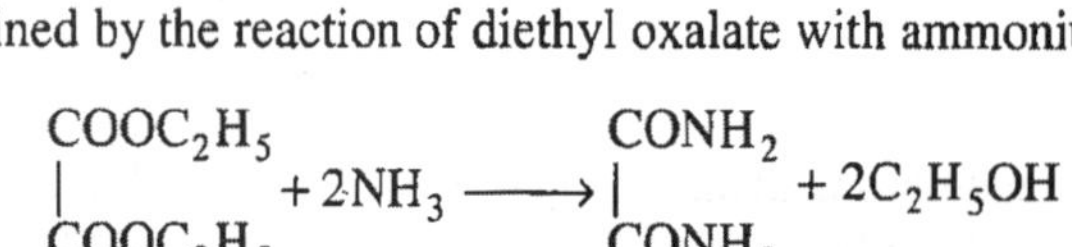

Diethyl oxalate   Oxamide
oxalate

Mechanism

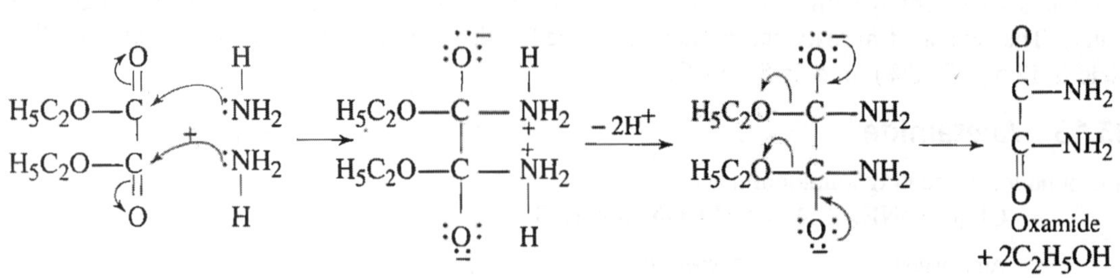

Diethyl oxalate : 7.5 ml
Ammonia : 24 ml

A mixture of diethyl oxalate (7.5 ml, 0.055 mole), water (18 ml) and ammonia solution (24 ml) is taken in a conical flask, which is stoppered and vigorously shaken for 30 minutes. The pressure inside the flask is released occasionally. The separated oxamide is filtered and washed with water. It is sufficiently pure. The yield is 4.5 g (92.6%); m.p. is 418°C (sealed tube, sublimes in open tube).

95.16 Phthalolylglycine

It is obtained by reaction of phthalic anhydride with glycine.

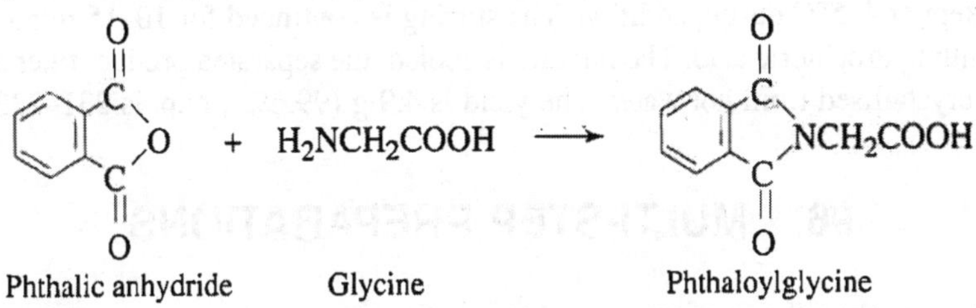

Phthalic anhydride : 9 g
Glycine : 4.5 g

A mixture of phthalic anhydride (9 g, 0.06 mole) and glycine (4.5 g, 0.06 mole) is fused in a boiling tube at 160–190°C (oil bath) for 20–30 minutes. The product obtained is cooled and crystallised from water. The yield is 10 g (81%); m.p. is 192–193°C.

95.17 Phthaloylglycylglycine

It is a dipeptide and is obtained by the reaction of phthaloylglycyl chloride with glycine.

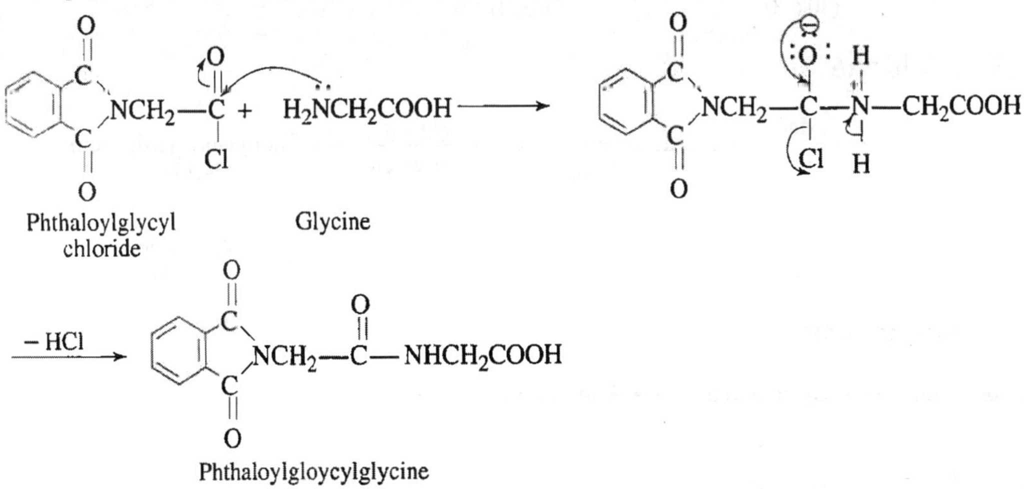

Step 1: Phthaloylglycine chloride

A mixture of phthaloyl glycine (8 g, 0.039 mole) and thionyl chloride (16 ml) is refluxed gently for 30 minutes in a R.B. flask fitted with a reflux condenser having a drying tube on the top. Excess thionyl chloride is removed by distillation under reduced pressure. The residual pathaloylglycine chloride is crystallised from petroleum ether. The yield is 4.5 g (51.6%); m.p. is 85–86°C.

Step 2: Phthaloylglycylglycine

Phthaloylglycyl chloride	: 4.2 g
Glycine	: 1.55 g
Dioxane	: 25 ml
Magnesium oxide	: 1.1 g

A solution of phthaloylglycyl chloride (4.2 g, 0.019 mole) in dioxane (25 ml) is added to a cooled (3–5°C) and stirred suspension of glycine (1.55 g) and magnesium oxide (1.1 g) in water (50 ml). The temperature is kept at 4–5°C during addition. The stirring is continued for 10–15 minutes at 25°C and then acidified with hydrochloric acid. The mixture is cooled, the separated product filtered, washed with cold water and crystallised from hot water. The yield is 4.9 g (99.6%); m.p. is 231–232°C.

96. MULTI-STEP PREPARATIONS

Given below are a number of sets of multi-step preparations.

96.1 Acetanilide

Acetophenone ⟶ Oxime (11.2) $\xrightarrow[\text{rearrangement}]{\text{Beckmann}}$ Acetanilide (11.2)

96.2 *p*-Aminoazobenzene

Aniline ⟶ Diazoaminobenzene (30.6.3) ⟶ *p*-Aminoazobenzene (30.6.4)

96.3 Acridone

o-chloro-toluene $\xrightarrow[\text{[o]}]{KMnO_4}$ o-Chloro benzoic acid (70.1.2) $\xrightarrow[\text{reaction}]{\text{Ullmann}}$ N-Phenylanthranilic acid (54.14) ⟶ Acridone (54.14)

96.4 Adipic Acid

Cyclohexanol ⟶ Cyclohexene (27.1.1) ⟶ Adipic acid (70.3.3)

96.5 Anthraquinone

Benzene $\xrightarrow[\text{reaction}]{\text{Friedal Crafts}}$ *o* - Benzoyl benzoic acid (46.2.3) ⟶ Anthraquinone (95.6)

96.6 Benzalacetophenone

$$\text{Benzene} \longrightarrow \underset{(46.2.1)}{\text{Acetophenone}} \xrightarrow[\text{reaction}]{\text{Claisen – Schmidt}} \underset{(22.1)}{\text{Benzalacetophenone}}$$

96.7 Benzanilide

$$\text{Nitrobenzene} \longrightarrow \underset{(77.16.1)}{\text{Aniline}} \longrightarrow \text{Banzanilide}$$

or

$$\text{Benzophenone} \longrightarrow \underset{(11.1)}{\text{Oxime}} \xrightarrow[\text{rearrangement}]{\text{Backmann}} \underset{(11.1)}{\text{Benzanilide}}$$

96.8 Benzidine

$$\text{Nitrobenzene} \longrightarrow \underset{(77.13.2)}{\text{Azobenzene}} \longrightarrow \underset{(13.1)}{\text{Hydrazobenzene}} \xrightarrow[\text{rearrangement}]{\text{Benzidine}} \underset{(13.1)}{\text{Benzidine}}$$

96.9 Benzilic Acid

$$\text{Benzaldehyde} \xrightarrow[\text{condensation}]{\text{Benzoin}} \underset{(9.1)}{\text{Benzoin}} \longrightarrow \underset{(14.1)}{\text{Benzil}} \xrightarrow[\text{rearrangement}]{\text{Benzil – Benzillic acid}} \underset{(14.1)}{\text{Benzilic acid}}$$

96.10 *p*-Bromoaniline

$$\text{Aniline} \longrightarrow \underset{(1.8)}{\text{Acetanilide}} \longrightarrow \underset{(51.6)}{p\text{-Bromoacetanilide}} \longrightarrow \underset{(95.2)}{p\text{-Bromoaniline}}$$

96.11 Dibenzyl

$$\text{Benzaldehyde} \xrightarrow[\text{condensation}]{\text{Benzoin}} \underset{(9.1)}{\text{Benzoin}} \longrightarrow \underset{(14.1)}{\text{Benzil}} \xrightarrow[\text{reduction}]{\text{Clemmensen}} \underset{(23.2)}{\text{Dibenzyl}}$$

96.12 2,5-Dihydroxyacetophanone

$$p\text{-Benzoquinone} \longrightarrow \underset{(1.2)}{\text{Hydroquinone diaacetate}} \xrightarrow[\text{migration}]{\text{Fries}} \underset{(47.1)}{\text{2, 5-Dihydroxyacetophenone}}$$

96.13 Diphenylacetylene (Tolan)

Benzaldehyde ⟶ Benzoin (9.1) ⟶ Benzil (14.1) ⟶ Benzildihydrazone (70.7.1) ⟶ Diphenylacetylene (70.7.1)

96.14 *p*-Dinitrobenzene

Aniline ⟶ Acetanilide (1.8) ⟶ *p*-Nitroacetanilide (69.5) ⟶ *p*-Nitroaniline (95.1) ⟶ *p*-Dinitrobenzene (95.3)

96.15 2,4-Dinitrophenylhydrazine

Chlorobenzene ⟶ 2, 4-Dinitrochlorobenzene (69.7) ⟶ 2, 4-Dinitrophenylhydrazine (95.5)

96.16 Ethyl Resorcinol

Resorcinol $\xrightarrow[\text{reaction}]{\text{Nencki}}$ β–Resacetophenone (68.1) $\xrightarrow[\text{reduction}]{\text{Clemmensen}}$ Ethylresorcinol (23.1)

96.17 Flavone

o-Hydroxyacetophenone ⟶ *o*-Benzyloxy acetophenone (10.1) ⟶

o-Hydroxydibenzyl methane (10.1) ⟶ Flavone (10.1)

96.18 Hydantoin

Glycine ⟶ Hydantoic acid (95.11) ⟶ Hydantoin (95.11)

96.19 7-Hydroxyflavone

Resorcinol ⟶ 2, 4-Dihydroxy acetophenone $\xrightarrow[\text{condensation}]{\text{Allan - Robinson}}$ 7-Hydroxyflavone

96.20 Methyl Orange

Aniline ⟶ Sulphanilic acid ⟶ Methyl orange (30.6.5)

96.21 *m*-Nitroaniline

Benzene ⟶ Nitrobenzene ⟶ *m*-Dinitrobenzene ⟶ *m*-Nitroaniline
(69.1) (69.6) (77.10.1)

96.22 *p*-Nitrophenylhydrazine

Aniline ⟶ Acetanilide ⟶ *p*-Nitroacetanilide ⟶ p-Nitroaniline
(1.8) (69.5) (95.3)

⟶ *p*-Nitrophenylhydrazine

96.23 *N*-Phenyl-2,4-dinitroaniline

Chlorobenzene ⟶ 2, 4-Dinitrochlorobenzene ⟶ *N*-Phenyl-2, 4-dinitroaniline
(69.7) (95.10)

96.24 Phthalylglycylglycine

Phthalic anhydride ⟶ Phthaloyl glycine ⟶ Phthaloyl glycine chloride ⟶ Phthalylglycylglycine
(95.16) (95.17) (95.17)

96.25 2,4,5-Trimethoxybenzaldehyde (Asarlaldehyde)

p-Benzoquinone $\xrightarrow[\text{Acetylation}]{\text{Thiele's}}$ Hydroxyhydroquinone triacetate
(1.7)

$\xrightarrow{\text{Methylation}}$ 1, 2, 4-Trimethoxybenzene $\xrightarrow{\text{Vilsmeir-Hack reaction}}$ 2, 4, 5-Trimethoxybenzaldehyde
(65.3) (91.2)

97. QUESTIONS

1. Discuss the role of sulphuric acid in the acetylation of salicylic acid with acetic anhydride.
2. What conditions are used for acetylation of the amino group (not the phenolic group) in *p*-amino phenol?
3. In acetylation of *D*-glucose, with zinc chloride as the catalyst α-pentacetate is the main product whereas with sodium acetate as the catalyst β-anomer in the major product. Explain.
4. What is the Thiele acetylation? Discuss the mechanism also.

5. In the acetylation of aniline, along with the reagent acetic anhydride-acetic acid, zinc dust is also used. Why?
6. Give the mechanism for the hydrolysis of nitriles under acidic as well as basic conditions.
7. What is the carbonation of Grignard reagent?
8. What are the different reagents used for the preparation of acid chlorides from acids?
9. Give the mechanism for the conversion of alcohols into alkyl chlorides with thionyl chloride. What happens when pyridine is also used with thionyl chloride?
10. What is the role of zinc chloride in the conversion of alcohol into alkyl chloride with hydrochloric acid?
11. Discuss the mechanism of the formation of alkyl bromides from alcohols by using phosphorous and bromine.
12. What is the role of diazomethane in the Arndt Eisert synthesis?
13. How can you convert aldehydes or ketones to esters? Give the name and mechanism of the reaction.
14. Which reagent is used in benzoin condensation and why?
15. Give the name of the rearrangement which converts *o*-acyloxyketone to β-diketone.
16. Under which conditions acidic or basic does Beckmann rearrangement take place. Explain.
17. Which is the reagent used for chloromethylation?
18. What happens when hydrazobenzene reacts with hydrochloric acid?
19. Give the different steps involved in the synthesis of 5.5 dimethylhydantoin from acetone.
20. What reagents are used for the generation of carbene? Explain the mechanism involved.
21. Explain why the Cannizzaro reaction is limited to the aromatic aldehydes, tertiary aliphatic aldehydes and formaldehyde.
22. Completely dry conditions are used when sodium metal is used in a reaction. Why?
23. What is zinc amalgam? How can you prepare this?
24. What are Schiff's bases? Give the structure of the Schiff's base formed in the synthesis of quinoline.
25. Discuss the role of ferric chloride in the coupling reaction of β-naphthol.
26. What is the catalyst used in Ullmann coupling?
27. In Dakin reaction, the phenolic group must be present on *ortho* or *para* to the carbonyl substituent. Why?
28. What are the commonly used dehydrating agents?
29. Give the mechanism for the synthesis of 1, 3, 5-tribromo benzene from 2, 4, 6-tribromoaniline.
30. For diazotization, low temperatures (0–5°C) is used. Why?
31. Give two examples of azo dyes along with the structures.
32. What is Aspirin? Give its synthesis and uses.
33. Which synthetic unit is common for the synthesis of sulpha drugs?
34. What is a role of lead peroxide in the synthesis of malachite green?
35. Discuss the stereochemistry of Diels Arder reaction with reference to the starting materials and the products.
36. Give the mechanism and product of Elbs persulphate oxidation of Salicylaldehyde.
37. What are enamines? Discuss the use of enamine obtained from morpholine and cyclohexanone.
38. Give the structure of Azalactone obtained from hippuric acid.
39. What is azeotropic mixture? How can you remove this from the reaction mixture?

40. Give the mechanism of Favorskii rearrangement by using α-halogeno cyclohexanones as a starting compound.
41. Give the synthesis of *o*-benzoyl benzoic acid from benzene.
42. Can ethyl benzene be prepared from a mixture of chlorobenzene and ethane in the presence of $AlCl_3$? Explain.
43. Why does aniline not undergo Friedel craft-reaction even though the NH_2 group is strongly activating?
44. What happens when phenyl acetate is heated with anhydrous aluminium chloride?
45. Give the synthesis of triphenylcarbinol from methyl benzoate using a suitable grignard reagent.
46. Why is the addition of iodine sometimes necessary in the reaction during the preparation of grignard reagent.
47. Explain why a catalyst is required in the halogenation of aromatic compounds, whereas side chain halogenation takes place without a catalyst.
48. For iodination of aromatic compounds, an oxidising agent is also required. Why?
49. In the bromination of benzene, pyridine can also be used as a catalyst. Discuss the role of pyridine.
50. Explain why acetanilide gives *p*-bromo acetanilide whereas aniline gives 2, 4, 6 tribromoaniline on bromination.
51. What are the type of compounds that undergo haloform reaction?
52. What is vat dyeing? Which dye is used for this purpose and how?
53. What is absolute alcohol? How can you obtain it?
54. Which compound is prepared by using Hantzsch synthesis?
55. Give the preparation and use of polyphosphoric acid.
56. Discuss the role of hypobromite in Hofmann reaction.
57. How will you convert phthalimide into anthranilic acid?
58. For introduction of which group is the Houben-Hoesch reaction used. Give the mechansism involved.
59. What happens when a solution of congo red is made acidic? Give the chemistry involved.
60. In the presence of dilute alkali, phenolphthalein produces red colour but into concentrated alkali this red colour disappears. Why?
61. Discuss the Mannich reaction by using the conversion of indole into indole-3-acetic acid.
62. Discuss the role of aluminium alkoxide in MPV reduction.
63. In the methylation of phenols with dimethyl sulphate, alkaline conditions are used. Why?
64. Which is the sugar present in milk? Give its structure.
65. *o*-Nitrophenol is steam volatile while *p*-nitrophenol is not. Explain?
66. Arrange the following compounds according to ease of nitration: benzene, nitrobenzene, acetanilide, phenol, chlorobenzene and bromobenzene.
67. In oxidation reactions with CrO_3 and $K_2Cr_2O_7$ discuss the change in the oxidation state of chromium.
68. Which are the starting materials required for selenium dioxide oxidation.
69. What is pulverised sodium?
70. What are coumarins? How can you synthesize coumarins from (i) resorcinol and (ii) from salicylaldehyde. Give mechanisms.
71. What happens in photochemical reactions?
72. Discuss the mechanism of the addition polymerisation by using benzoyl peroxide as a catalyst.

73. What is Thiokol rubber?
74. Which organometallic compound is formed in Reformatsky reaction?
75. Acid chlorides are reduced to aldehydes by using which reagent? Give the name of the reaction also.
76. What is a difference between Sarett oxidation and Jones oxidation?
77. Discuss the mechanism of Vilsmeier Haack reaction.
78. Discuss the role of morpholine in Willegerodt-Kindler reaction.
79. What are ylides?
80. In Perkin reaction, why do aromatic aldehydes react more readily than aliphatic aldehydes?

Section B: Quantitative Analysis

1. ESTIMATION OF ELEMENTS

1.1 Estimation of Nitrogen

There are two methods for the estimation of nitrogen.

Method 1. Dumas combustion method.

Principle: In this method, a known weight of nitrogen containing organic compound is heated in an atmosphere of carbon dioxide by using cupric oxide as an oxidising agent. Nitrogen oxides thus produced are passed over heated copper. The liberated nitrogen is then collected in a graduated nitrometer containing a 50 per cent potassium hydroxide solution. Potassium hydroxide solution absorbs all other gases except nitrogen. The change in the nitrometer reading gives the volume of liberated nitrogen from which the percentage of nitrogen in an organic compound can be calculated.

Reagents

Nitrogen containing organic compound
Copper oxide
Metallic copper
Potassium hydroxide solution (50%)
Dilute hydrochloric acid
Mercury

Apparatus

The apparatus for this experiment consists of a carbon dioxide generator, combustion tube and nitrometer (Fig. 1).

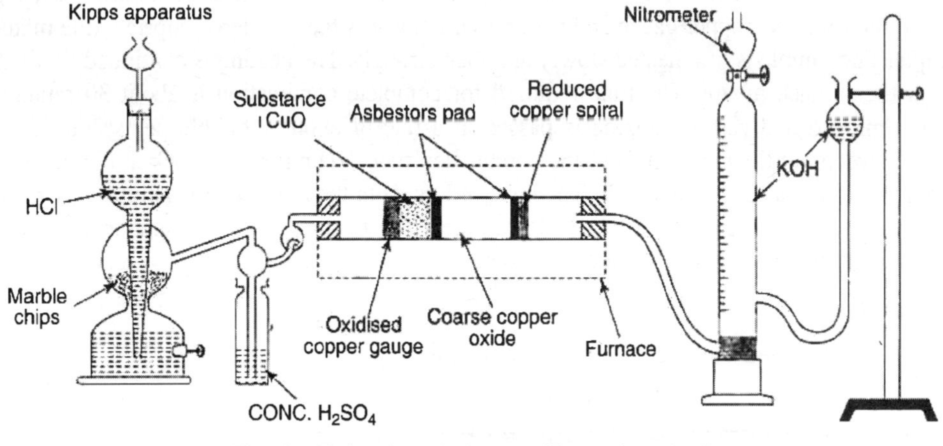

Fig. 1 Estimation of nitrogen (Dumas' method)

Carbon dioxide generator Carbon dioxide is prepared in Kipp's apparatus by using clean and pure marble chips and hydrochloric acid. The important precaution is that the carbon dioxide supplied should be air-free.

Combustion tube—The combustion tube (50–60 cm long, diameter 1 cm) of transparent silica or hard glass is used. The tube is filled with reduced copper spiral, cupric oxide in the form of wire and reduced copper guaze. A small tubular electric furnace is used for heating the tube.

Nitrometer—The one end of the combustion tube is connected to a nitrometer which consists of a graduated tube with a funnel and a tap at its upper end. The graduated tube is connected to a potassium hydroxide solution (50%) reservoir through a pressure tube. The graduated tube is filled with potassium hydroxide solution (50%) with a little mercury at the bottom. The mercury allows the nitrogen gas to enter the nitrometer and prevents the back flow of potassium hydroxide solution.

Procedure

Filling up of combustion tube—The tube is filled from the right hand side leaving 5 cm space. It is then filled with ignited asbestos plug (5 mm), freshly reduced copper spiral (50 mm), asbestos plug (2–3 mm), cupric oxide wire (100 mm) and asbestos plug (2–3 mm). This is a permanent filling of the tube. Before estimation, the tube is temporarily filled with a cupric oxide wire, powdered cupric oxide (25 mm), a mixture of the given sample and cupric oxide powder (50 mm) and oxidized copper gauze (50 mm).

Combustion—The given sample is weighed accurately (20–50 mg) in a sample tube and mixed with powdered cupric oxide. This mixture is then carefully transferred to the combustion tube. The empty sample tube is filled with powdered cupric oxide, shaken and transferred to the combustion tube. This process is repeated 2–3 times to ensure that whole of the solid is being transferred. Finally oxidized copper gauge is placed and the combustion tube is connected to the carbon dioxide supply. A slow stream of carbon dioxide is passed for 10 minutes through the apparatus and the gas is allowed to escape in air. Then, the nitrometer is connected and gas is allowed to pass into the nitrometer till the bubbles are obtained. The carbon dioxide supply is disconnected, the electric furnace is switched on and the temperature is raised to 700–750°C. The level of liquid in the nitrometer is adjusted and the initial reading is noted. The combustion is then started by heating the tube with a Bunsen burner starting from the side of the oxidised copper gauze and then moving towards the powdered cupric oxide mixed with the sample. The sample is first heated slowly and then strongly. The heating is continued till the burner arrives at the furnace mouth. The time required for complete combustion is about 30 minutes. The burner is removed and carbon dioxide is passed at a rate of about 1 bubble per second for 15–20 minutes to sweep out the nitrogen from the combustion tube. Then the stop-cock at the nitrometer, is stopped and the furnace is switched off. The level of liquids in the nitrometer is adjusted by raising the levelling bulb and allowed to stand for 10–15 minutes. The reading of the nitrometer is recorded and barometric pressure and temperature are also noted.

Observations

The observed volume of nitrogen = V_1 ml

Barometric pressure = P

Temperature at nitrometer = t°C

Calculations

Now volume of N_2 at N.T.P. can be calculated as follows:

$$\frac{760 \times 22.415}{273} = \frac{P \times V}{(273 + t)}$$

$$V = \frac{760 \times 22.415 \times (273 + t)}{273 \times P} \text{ litres}$$

Now V of N_2 at N.T.P. $\equiv$ 28.016 g of N_2

$$\frac{V_1}{1000} \text{ of } N_2 \text{ at N.T.P.} \equiv \frac{28.016 \times V_1}{1000 \times V} \text{ g of } N_2$$

Thus, w g of sample produces $\frac{28.016 \times V_1}{1000 \times V}$ g of N_2

100 g will produce $\frac{28.016 \times V_1}{1000 \times V} \times \frac{100}{w}$ g

$$\therefore \quad \% \text{ of Nitrogen} = \frac{28.016 \times V_1 \times 273 \times P}{760 \times 22.415 \times (273 + t) \times 1000} \times \frac{100}{w}$$

Method 2. Kjeldahl method

Principle

A known weight of the given nitrogen containing organic compound is digested with concentrated sulphuric acid in presence of a catalyst. The nitrogen of the organic compound is converted quantitatively into ammonium sulphate. Then the solution is made alkaline with excess of sodium hydroxide solution, and ammonia is liberated which is then distilled in steam. The distilled ammonia is absorbed in saturated boric acid solution. The ammonium borate thus formed, is titrated with standard dilute hydrochloric acid using methyl red as an indicator. From the amount of liberated ammonia one can calculate the percentage of nitrogen in the given organic compound.

$$\text{Organic compound containing nitrogen} \xrightarrow[\Delta]{H_2SO_4} (NH_4)_2SO_4$$

$$(NH_4)_2SO_4 + 2NaOH \longrightarrow Na_2SO_4 + 2H_2O + NH_3\uparrow$$

$$NH_3 + H_3BO_3 \longrightarrow NH_4^+ + H_2BO_3^-$$

$$H_2BO_3^- + \underset{(HCl)}{H^+} \longrightarrow H_3BO_3$$

Reagents

An organic compound containing nitrogen

Conc. sulphuric acid (analytical reagent)

A powdered mixture of potassium sulphate, copper sulphate and selenium (in ratio 20:1:1) as catalyst Standard sodium hydroxide solution (0.02 N)

Saturated boric acid solution

Standard hydrochloric acid solution (0.04 N)

Methyl red indicator

Apparatus

The apparatus consists of two parts, one is the digestion unit and the second one is the steam distillation unit.

The digestion unit consists of 5–6 burners, a support for flasks and a glass fume hood with a series of holes. The neck of the flask fits into these holes and the fumes are removed from the hood by suction with a water pump. The distillation unit consists of a steam generator, distillation flask, splash-head with steam inlet tube, Liebig condenser and receiver (Fig. 2).

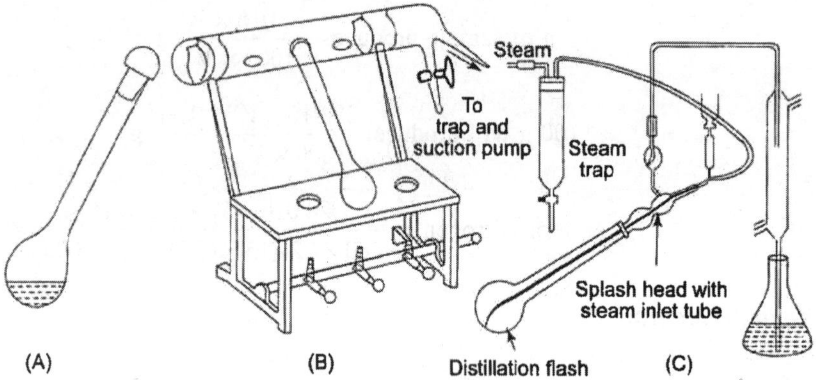

Fig. 2 Kjeldahl apparatus for the estimation of nitrogen; A : Kjeldahl flask; B : digestion stand; C : distillation assembly

Procedure

An accurately weighed given sample (25–50 mg) is taken in a clean and dry Kjeldahl flask and a catalyst mixture (1 g) and concentrated sulphuric acid (5 ml) are added. The flask is stoppered loosely by a glass bulb and heated in an inclined position slowly till the frothing subsides and then vigorously for one hour. The liquid should be colourless at the end of heating. The mixture is cooled and diluted with distilled water (10 ml). The mixture is then transferred to a distillation flask (before use, the distillation apparatus should be steamed out and cooled). The Kjeldahl flask is washed 3–4 times with distilled water and the washings are transferred to the same distillation flask. Saturated boric acid solution or standard hydrochloric acid (0.04 N) or sulphuric acid solution (0.04 N) (25 ml) is taken in a receiving flask and the condenser is adjusted, so that the lower end of the condenser should dip in the solution of the receiving flask. Then sodium hydroxide solution (40%, 20 ml) is added to the mixture in distillation flask through the funnel, leaving some solution in the funnel which serves as a liquid seal. The steam is passed into the distillation flask and the mixture is distilled for 45–50 minutes so that all the ammonia has passed over into the acid solution. Finally, the outside portion of the condenser is washed with distilled water and the washings are collected in the receiving flask. In case of standard acid solution (HCl or H_2SO_4), the mixture of the receiving flask (excess acid) is titrated with standard sodium hydroxide solution (0.02 N) using phenolphthalein as an indicator. If boric acid solution is used then it is titrated with standard hydrochloric acid solution (0.02 N) by using methyl red as an indicator (violet colour indicates the end point).

A blank experiment is also carried out by the using the same quantity of glucose instead of nitrogen containing organic compound.

Observations

Weight of the given sample taken = w g

Volume of hydrochloric acid used in sample = v_1 ml

Volume of hydrochloric acid used in blank = v_2 ml

Normality of hydrochloric acid = Nx

Calculations

Volume of hydrochloric acid used = $v_1 - v_2$ ml

1000 ml of 1N HCl ≡ 1000 ml of 1 N NH_3 ≡ 14 g of nitrogen

$$(v_1 - v_2) \text{ ml of } Nx \text{ HCl} \equiv \frac{14 \times (v_1 - v_2) \times Nx}{1000} \text{ g of nitrogen}$$

Now

$$w \text{ g of sample contains } \frac{14 \times (v_1 - v_2)\, Nx}{1000} \text{ g of nitrogen}$$

$$100 \text{ g of sample will contain } \frac{14 \times (v_1 - v_2)\, Nx}{1000} \times \frac{100}{w} \text{ g}$$

∴

$$\% \text{ of nitrogen} = \frac{14 \times (v_1 - v_2) \times Nx}{1000} \times \frac{100}{w}$$

1.2 Estimation of Sulphur

Following methods are used for the estimation of sulphur.

Method 1 Carius method

Principle

A known weight of sulphur containing organic compound is heated with fuming nitric acid in a sealed Carius tube in a bomb furnace. The sulphur of the compound is converted into sulphuric acid which is estimated as barium sulphate gravimetrically.

Reagents

Sulphur containing organic compound
Fuming nitric acid
Barium chloride solution (0.1 M)

Procedure:

An accurately weighed given sample (0.1–0.2 g) is placed in a small glass tube and sealed at one end. The tube is placed in the Carius tube (50 cm long) containing fuming nitric acid and the upper end of the tube is sealed carefully. The Carius tube is then transferred in an iron tube and heated in the furnace at 250–300°C for 5–6 hours. After this period, the heating is switched off and it is allowed to cool. The sealed end of the tube is softened by heating in a Bunsen flame and cut carefully. The contents of the Carius tube are transferred to the beaker and washed with distilled water. The washings are also collected in the same beaker. The contents of the beaker are heated with dilute hydrochloric acid (1 ml) barium chloride solution (0.1 M) is added slowly to the beaker, till the precipitation is complete. The solution is digested for 10 minutes and allowed to stand overnight. The precipitate is filtered, dried, ignited in a silica crucible and weighed.

Observations

Weight of the sample taken = w g

Weight of barium sulphate = w_1 g

Calculations

$$233.4 \text{ g of } BaSO_4 \equiv 32 \text{ g of sulphur}$$

$$w_1 \text{ g of } BaSO_4 \equiv \frac{32 \times w_1}{233.4} \text{ g of S}$$

Now $$w \text{ g of sample contains } \frac{32 \times w_1}{233.4} \text{ g of S}$$

$$100 \text{ g of sample will contain } \frac{32 \times w_1 \times 100}{233.4 \times w} \text{ g}$$

$$\therefore \quad \% \text{ of S} = \frac{32 \times w_1 \times 100}{233.4 \times w}$$

The limitation of this method is that in many cases an explosion may occur during heating, and many compounds, e.g., aliphatic sulphides cannot be analysed by this method as they are oxidised to highly decomposition resistant sulphones.

Method 2 Na_2CO_3–KNO_3 fusion method

Principle

The sulphur in non-volatile organic compounds is estimated by fusion with a potassium nitrate-sodium carbonate mixture. The sulphur of the compound is converted into sodium sulphate, which is estimated gravimetrically as barium sulphate.

Reagents:

Sulphur containing organic compound

Fusion mixture of Na_2CO_3 and KNO_3 (4:3)

Barium chloride solution (0.1 M)

Procedure

The nickel crucible is filled one-fourth with the fusion mixture. An accurately weighed sample (0.1 – 0.2 g) is placed in the crucible in the form of a thin film and covered with a thick layer of the fusion mixture so that the nickel crucible is three-fourth filled. The crucible is covered with a close-fitting cover and placed in another silica or nickel crucible. The latter is also covered with its lid. The crucible is heated slowly with a low Bunsen flame until the mass melts (15 minutes) and then strongly for half an hour. The crucible is cooled and transferred with its lid into a beaker containing distilled water. The contents of the beaker are boiled to dissolve the melt. The crucible and lid are removed with the help of a clean pair of tongs and washed with distilled water. The washings are also collected in the same beaker. The solution of the beaker is then concentrated to half of its volume after filtration (if necessary) and acidified with dilute hydrochloric acid. The barium chloride solution (0.1 M) is added drop by drop with stirring till the precipitation is complete. The precipitate settles down and is filtered using a gravimetric filter paper. The precipitate is washed with hot distilled water till the filtrate is free from chloride ions and dried. The precipitate along with the filter paper are transferred into a weighed silica crucible. The crucible is heated gently and then strongly until the carbon has burnt out. Finally, the crucible is cooled in a desiccator and weighed. The ignition process is repeated to get a constant weight.
Note: The filtration can also be done in preweighed sintered crucible.

Observations and Calculations

Same as in Carius method.

Method 3 Messenger's method

Principle

A known weight of organic compound is oxidised with alkaline potassium permanganate solution. The sulphur of the organic compound is converted into sulphate. The reaction mixture is neutralised with concentrated hydrochloric acid and sulphate is estimated gravimetrically as barium sulphate.

Reagents

The given sample
Potassium permanganate
Potassium hydroxide
Conc. hydrochloric acid
Barium chloride

Procedure

An accurately weighed quantity of the given sample (0.2 – 0.3 g) is placed in a 250 ml round bottom flask fitted with a reflux condenser. Water (50 ml), potassium hydroxide pellets (0.5–1.0 g) and solid potassium permanganate (2–3 g) are added. The mixture is refluxed on sand bath for 5–6 hours and then cooled. Concentrated hydrochloric acid (20–30 ml) is slowly added through the condenser to decolourize any unreacted permanganate and the mixture is heated again to get a colourless solution. It is filtered (if necessary) and transferred quantitatively into a beaker. The flask is washed 3–4 times and the washings are also collected in the same beaker. Concentrated hydrochloric acid (1–2 ml) is added to the solution, boiled and barium chloride solution (2%) is added dropwise till the precipitation is complete. The precipitate is digested, filtered, washed, dried and weighed.

Observations and calculations

Same as in the Carius method.

1.3 Estimation of Halogens

Method 1 Carius method

Principle

A known weight of the given compound is heated with fuming nitric acid and silver nitrate in a sealed carius tube. The silver halide is formed quantitatively from halogen. By knowing the weight of silver halide, percentage of halogen can be calculated.

Reagents

Given organic compound
Fuming nitric acid
Silver nitrate

Procedure

Powdered silver nitrate (0.2 g) and fuming nitric acid (1–2 ml) are taken in a Carius tube. An accurately weighed given sample (0.02–0.03 g) in a thin glass tube in transferred to the same Carius tube and is heated in the same way as in case of sulphur (1.2). Finally, the contents of the Carius tube are transferred to the pre-weighed sintered crucible, washed with water, ethyl alcohol (5 ml) and acetone (5 ml). The crucible is then dried in an oven at 110°C and weighed.

Observations

Weight of the given sample $= w$ g
Weight of silver chloride $= w_1$ g

Calculations

For chlorine.

$$143.5 \text{ g of AgCl} \equiv 35.5 \text{ g of chlorine}$$

$$w_1 \text{ g of AgCl} \equiv \frac{35.5 \times w_1}{143.5} \text{ g of chlorine}$$

Now, w g of organic compound contains $\frac{35.5 \times w_1}{143.5}$ g of chlorine.

$$100 \text{ g will contain} = \frac{35.5 \times w_1 \times 100}{143.5 \times w}$$

$$\text{Thus, \% of chlorine} = \frac{35.5 \times w_1 \times 100}{143.5 \times w}$$

For bromine

$$\% \text{ of bromine} = \frac{80 \times w_1 \times 100}{188 \times w}$$

For iodine

$$\% \text{ of iodine} = \frac{127 \times w_1 \times 100}{235 \times w}$$

Method 2: Stepnow's metho

Principle

A known weight of the given sample is reduced with sodium and ethanol. The formed sodium halide is acidified and precipitated as silver halide by adding silver nitrate. Then the weight of silver halide is determined and percentage of halogen is calculated.

$$RX + C_2H_5OH + 2Na \longrightarrow RH + C_2H_5ONa + NaX$$

$$NaX + H^+ \longrightarrow HX + Na^+$$

$$HX + AgNO_3 \longrightarrow AgX \downarrow + HNO_3$$

But this method is not good for aryl halides and poly halogen compounds. In a modification of this method, a mixture of monoethanolamine and dioxan is used instead of ethanol. Monoethanolamine has a high boiling point (171°C) and gives satisfactory results.

$$RX + NH_2—CH_2—CH_2—OH + 2Na \longrightarrow NaX + RH + NH_2CH_2CH_2ONa$$

Reagents

The given sample

Monoethanolamine

Sodium metal

Dioxan

Silver nitrate

Procedure

An accurately weighed amount of the given sample (20–30 mg) is placed in a round bottom flask fitted with a reflux condenser. Freshly distilled monoethanolamine (5 ml), halogen free dioxan (5 ml) and sodium metal (0.4 g) are added into the flask and warmed till the vigorous reaction subsides. Then the solution is refluxed gently for an hour with occasional shaking. Additional sodium metal may be added if whole of the sodium has disappeared. The flask is then cooled and excess of sodium is destroyed by adding distilled water (2–3 ml) carefully through the condenser followed by more water (5 ml). The mixture is cooled and acidified to congo red with nitric acid (10%). The contents are then transferred to a beaker, the flask is washed with distilled water 2–3 times and the washings are collected in the same beaker. A dilute solution of silver nitrate is added drop by drop till the precipitation of silver halide is complete. The mixture is warmed on a water bath and allowed to stand overnight. The precipitate is filtered, washed with water, dried and weighed.

Observations and Calculations: *Same as in the Carius method*

Method 3

Fusion method (Na_2CO_3—KNO_3 fusion). Principle and procedure is same as in case of sulphur. The only difference is that acidification is done with dilute nitric acid instead of hydrochloric acid and silver nitrate solution is added in place of barium chloride solution for precipitation of silver halide.

Observations and Calculations: *Same as in the Carius Method*

2. ESTIMATION OF FUNCTIONAL GROUPS

2.1 Estimation of Hydroxyl Group (Alcohols and Phenols)

Acetylation method

Principle

The acetylation method involves the replacement of hydrogen of an hydroxyl group in alcohols or phenols by the acetyl group, by heating with a definite volume of a mixture of acetic anhydride and pyridine. The excess of unchanged acetic anhydride is hydrolyzed by water. The total free acetic acid is then estimated by titration with standardized sodium hydroxide solution, using phenolphthalein as an indicator.

$$R(OH)_n + n(CH_3CO)_2O \longrightarrow R(O.COCH_3)_n + nCH_3COOH$$

$$(CH_3CO)_2O + H_2O \longrightarrow 2CH_3COOH$$

A blank experiment is also performed simultaneously under similar conditions by using the same volume of a mixture of acetic anhydride and pyridine. The difference in the volume of sodium hydroxide solution required in the two experiments is equivalent to the acetic acid derived from the acetic anhydride consumed in the acetylation of the sample. By knowing the molecular weight of the compound (sample), the number of hydroxyl groups in the compound can be calculated.

Reagents

Acetylating mixture: Acetic anhydride and dry pyridine (1:3, v/v) are mixed, the mixture is transferred to a clean dry burette and a calcium chloride guard tube is inserted into the top of the burette.
Sodium hydroxide solution (1N): Standardized with standard oxalic acid solution.
Phenolphthalein indicator.

Procedure

Two round bottom flasks (100 ml) fitted with a reflux water condenser are taken and marked as 'A' and 'B'. An accurately weighed amount of the sample (0.5 g) is placed in flask 'A' and a mixture of acetic anhydride and pyridine (10 ml) is also added to this flask and also to the blank flask 'B'. The flasks are heated on a boiling water bath for about an hour and then cooled. Distilled water (20 ml) is added to each of the flasks, through the condenser and shaken gently for a few minutes. The contents of the flasks are refluxed again for 2–5 minutes, cooled and allowed to stand for 10 minutes. Finally, the contents of the flasks are titrated with sodium hydroxide solution (1 N) using phenolphthalein as an indicator.

Observations

Weight of the sample taken	$= w$ g
Volume of NaOH used in blank experiment	$= v_1$ ml
Volume of NaOH used with sample	$= v_2$ ml
Normality of NaOH solution	$= Nx$

Calculations

Volume of NaOH equivalent to the acetic acid consumed in acetylation = $(v_1 - v_2)$ ml

$$1000 \text{ ml N NaOH} \equiv 1 \text{ g mol of NaOH} \equiv 1 \text{ g mol of } CH_3COOH \equiv 1.OH \text{ group}$$

$$(V_1 - V_2) \text{ ml } Nx \text{ NaOH} \equiv \frac{(v_1 - v_2)\, Nx}{1000} \text{ OH group}$$

$$w \text{ g of the sample contains } \frac{(v_1 - v_2)}{1000} \times Nx \text{ OH group}$$

$$M \text{ g of the sample will have } \frac{(v_1 - v_2)}{1000} \times Nx \times \frac{M}{w} \text{ OH group}$$

(M = molecular weight of the sample)

$$\therefore \quad \text{Number of (OH) groups} = \frac{(v_1 - v_2)}{1000} \times Nx \times \frac{M}{w}$$

For determination of percentage of hydroxyl group

$$1000 \text{ ml 1N NaOH} \equiv 17 \text{ g of OH group}$$

$$(v_1 - v_2) \text{ ml of } Nx \text{ NaOH} = \frac{17 \times (v_1 - v_2)}{1000} \times Nx \text{ g of OH group}$$

$$w \text{ g of the sample contain } \frac{17 \times (v_1 - v_2)}{1000} \times Nx \text{ g of OH group}$$

$$100 \text{ g of the sample will have } \frac{17 \times (v_1 - v_2)}{1000} \times Nx \times \frac{100}{w} \text{ g of OH group}$$

$$\% \text{ of OH group} = \frac{17 \times (v_1 - v_2)}{1000} \times Nx \times \frac{100}{w}$$

Estimation of Hydroxyl Groups (Phenol)—Bromate-Bromide Method

Principle

A known mass of phenol is treated with brominating mixture of potassium bromate and potassium bromide in concentrated hydrochloric acid and then with potassium iodide solution. The unreacted bromine gives an equivalent amount of iodine, the amount of which is determined by titration with hypo solution.

The equations involved are

$$KBrO_3 + 5KBr + 6HCl \longrightarrow 3Br_2 + 6KCl + 3H_2O$$

$$Br_2 + KI \longrightarrow I_2 + 2KBr$$

$$I_2 + 2Na_2S_2O_3 \longrightarrow Na_2S_4O_6 + 2NaI$$

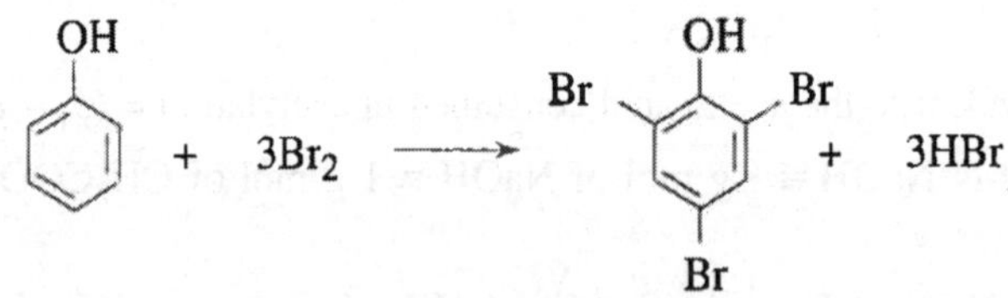

The variables in the brominating procedure include temperature, amount of bromine and time of reaction. The best results are obtained at room temperature or below. Since bromination of the side chain under the conditions is usually very slow, the amount of bromine must be controlled in order to reduce the bromination of the side chain.

The tribromo phenol precipitate formed from phenol may contain in addition to tribromo phenol, some tribromo phenol bromide.

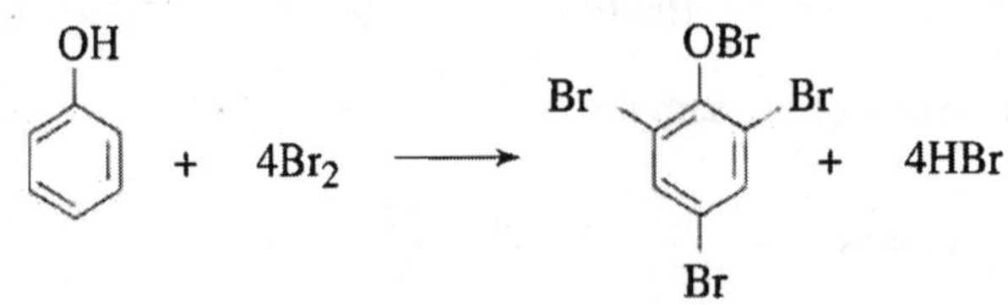

This is of no consequence as it is converted into tribromo phenol when potassium iodide is added to the acid solution.

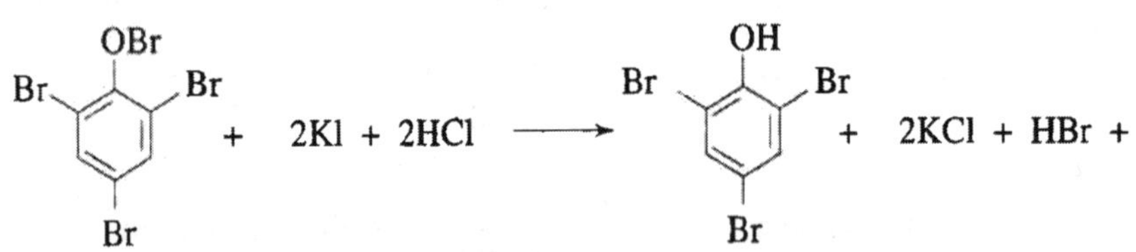

The extra bromine is thus reacting as if it was free bromine. This is the reason why the reaction mixture, after the reaction with potassium iodide, is to be kept for 5–10 minutes to ensure that all the tribromo bromide is decomposed.

The presence of the halogen or nitro group in *ortho* and *para* positions do not hinder the bromination, but the —OH and —NH_2 group in the *o* and *p* positions hinder the entrance of bromine into the ring. Carboxyl and sulphonic acid groups seem to behave in an intermediate manner. A dibromo product is first formed. Upon standing with excess of bromine, the tribromo derivative is produced with the elimination of the carboxyl or sulphonic acid group.

An average bromination period is 30 minutes, using about 100 percent excess of bromine. The number of bromine atoms consumed for some typical phenols is given in parentheses: phenol (6); *p*-chloro phenol (4); *o*-or *p*-nitrophenol (4); *m*-nitrophenol (6); 2, 4-dinitrophenol (2); *m*-cresol (6); resorcinol

(6); β-naphthol (2); salicylic acid (6); *m*-hydroxybenzoic acid (6); methyl salicylate (6); and acetyl salicylic acid (6). The esters of salicyclic acid must be hydrolyzed with dilute alkali before bromination.

Reagents

Brominating mixture (0.2 N) Anhydrous potassium bromate (5.567 g) and potassium bromide (75 g) are dissolved in distilled water and diluted to 1 litre in a volumetric flask (an excess of KBr ensures complete reduction of bromate and keeps the free bromine in the solution).
Sodium thiosulphate solution (0.1 N)
Starch solution
Potassium iodide solution (20%)
Phenol solution
Conc. hydrochloric acid

Procedure

Three iodine flasks of 500 ml capacity are taken. In two flasks phenol solution (25 ml) and bromate-bromide solution (25 ml) are taken from the burette respectively. Distilled water (25 ml) and concentrated hydrochloric acid (5 ml) are added and the flasks are stoppered immediately to prevent any loss of bromine. The flask are allowed to stand for 30 minutes with occasional shaking. These are then cooled in ice and potassium iodide solution (20%, 10 ml) is added. The flasks are shaken for 30 seconds and allowed to stand for 20 minutes. The stopper is removed, washed with distilled water and the washings are collected in the same flask. Finally the liberated iodine is titrated with standard sodium thiosulphate solution using starch as an indicator.

A blank tituration is also performed by taking brominating mixture (25 ml), hydrochloric acid (25 ml) and water (25 ml) in third flask.

Calculations

Volume of thiosulphate used in blank $= v_1$

Volume of thiosulphate used for unknown phenol solution $= v_2$

Normality of sodium thiosulphate solution $= Nx$

Molecular weight of phenol $= M$

Number of bromine atoms substituted in the phenol $= z$

$(v_1 - v_2)$ ml of hypo solution $\equiv$ Volume of Br_2 which has reacted with phenol

$(v_1 - v_2)$ ml of Nx hypo $\equiv (v_1 - v_2)$ ml of Nx $I_2 \equiv (v_1 - v_2)$ ml of Nx Br_2 used $\equiv \dfrac{\text{Mass}}{2 \times z}$ of the phenol taken

Now, 1000 ml of N hypo $\equiv$ 1000 ml of N $I_2 \equiv$ 1000 ml of N $Br_2 \equiv \dfrac{M}{2 \times z}$ of the phenol

$$(v_1 - v_2) \text{ ml of Nx hypo} \equiv \frac{(v_1 - v_2) \times Nx \times M}{2 \times z \times 1000} \text{ of the phenol}$$

$$\therefore \text{ Strength of the phenol solution } = \frac{(v_1 - v_2) \times Nx \times M}{2 \times z \times 1000} \times \frac{1000}{25} \text{ g/l}$$

$$= \frac{(v_1 - v_2) \times Nx \times M}{2 \times 25 \times z} \text{ g/l}$$

If weight of the impure sample dissolved in 1 litre = w g

$$\text{Then \% purity of the sample} = \frac{(v_1 - v_2) \times Nx \times M}{2 \times 25 \times z} \times \frac{100}{w}$$

2.2 Estimation of Acetyl Group (o-acetyl)

Principle

o-Acetyl compounds are estimated by hydrolysing them with ethanolic potassium hydroxide solution followed by dilution of the reaction mixture with magnesium sulphate solution containing sulphuric acid. The acetic acid formed is steam distilled and estimated by titrating with a standard alkali solution.

$$-OCOCH_3 + KOH \longrightarrow -OH + CH_3CO\bar{O}\overset{+}{K}$$

$$2CH_3COOK + H_2SO_4 \longrightarrow 2CH_3COOH + K_2SO_4$$

$$CH_3COOH + NaOH \longrightarrow CH_3COO^-Na^+ + H_2O$$

Reagents

Ethanolic potassium hydroxide solution (1 N): Potassium hydroxide (5.6 g) is dissolved is 100 ml ethanol : water mixture (1:1)

Magnesium sulphate solution: Magnesium sulphate (50 g) is dissolved in concentrated sulphuric acid (0.75 ml) in sufficient water and diluted to 90 ml.

Sodium hydroxide or barium hydroxide solution (0.02 N): Standardised with a standard oxalic acid solution.

Phenolphthalein Indicator

Apparatus

The apparatus consists of a long necked round bottom flask (50 ml), a condenser, and a ground glass stopper having an inlet tube (to pass steam) fitted into the neck of the flask. The flask has a side arm with a ground glass socket and is slightly bent upwards. The condenser is used in the upright position during refluxing and sloping positions during distillation.

Procedure

An accurately weighed (20–30 mg) given sample is placed in the flask and ethanolic potassium hydroxide solution (3 ml) is added. The reflux condenser is inserted in a vertical position and the contents are heated till the sample has dissolved. The mixture is heated for 4 minutes more, cooled, and magnesium sulphate reagent (20 ml) is added. The condenser is rotated in a slanting position for distillation and the mixture is steam distilled in such a manner that the liquid in the flask distills at a fairly rapid rate and is concentrated to about 15 ml during the collection of 50 ml of distillate (use a measuring cylinder as receiver). The distillate is titrated with sodium hydroxide or barium hydroxide solution (0.02 N) using phenolphthalien as an indicator. An additional 5 ml of distillate should not change the end point appreciably. A blank experiment is also performed.

Observations

Weight of the sample taken = w g

Volume of the alkali used for blank titration = v_1 ml

Volume of the alkali used for sample = v_2 ml

Normality of the alkali = Nx

Calculations

Volume of Nx alkali used = $v_2 - v_1$ ml

$$1000 \text{ ml of 1N NaOH} \equiv 1000 \text{ ml of 1N } CH_3COOH \equiv 43.04 \text{ g of } —COCH_3$$

$$(v_2 - v_1) \text{ ml N}x \text{ NaOH} \equiv \frac{43.04 \times (v_2 - v_1) \times Nx}{1000} \text{ g of } —COCH_3$$

$$w \text{ g of the sample contains } \frac{43.04 \times (v_2 - v_1) \times Nx}{1000} \text{ g of } -COCH_3$$

$$100 \text{ g of the sample will contain } \frac{43.04 \times (v_2 - v_1) \times Nx \times 100}{1000 \times w} \text{ g of } -COCH_3$$

$$\therefore \quad \% \text{ of acetyl group} = \frac{43.04 \times (v_2 - v_1) \times Nx}{w \times 10}$$

If molecular weight of the compound = M

$$\text{then number of acetyl group} = \frac{(v_2 - v_1) \times Nx \times M}{w \times 1000}$$

For estimation of *N*-acetyl groups, the principle and method is the same as described above for *O*-acetyl estimation. However, better results are obtained when hydrolysis of *N*-acetyl compounds is performed with potassium hydroxide in *n*-butanol or amyl alcohol.

2.3 Estimation of Methoxyl Group-Zeisel's Method

A known weight of the substance is decomposed by heating with constant boiling hydriodic acid.

$$ROCH_3 + HI \longrightarrow ROH + CH_3I$$

The evolved volatile methyl iodide can be estimated either gravimetrically or volumetrically.

Gravimetric method: The methyl iodide is removed from the boiling reaction mixture by a slow stream of carbon dioxide and transferred to an absorption vessel containing 4 per cent alcoholic solution of silver nitrate. Silver iodide is precipitated, filtered and washed with dilute nitric acid, water, dried and finally weighed.

Volumetric method: In this method, methyl iodide, in a current of carbon dioxide is transferred to an absorption vessel containing a mixture of sodium acetate-acetic acid in bromine solution. Iodine monobromide is first formed, which is oxidized to iodic acid.

$$CH_3I + Br_2 \longrightarrow CH_3Br + IBr$$
$$IBr + 3H_2O + 2Br_2 \longrightarrow HIO_3 + 5HBr$$

The iodic acid is then determined iodometrically by adding sodium acetate solution and formic acid (to destroy the excess of bromine), followed by dilute sulphuric acid and potassium iodide solution. The liberated iodine is titrated with sodium thiosulphate solution.

$$Br_2 + HCOOH \longrightarrow 2HBr + CO_2$$
$$HIO_3 + 5HI + H_2SO_4 \longrightarrow 3I_2$$
$$3I_2 + 6Na_2S_2O_3 \longrightarrow 3Na_2S_4O_6 + 6NaI$$

Apparatus

The apparatus consists of a reaction flask connected to a Liebig condenser and carbon dioxide generator (Kipp's apparatus or gas cylinder) through a sulphuric acid trap. The condenser is attached to a trap which is filled with sodium antimonyl tartrate solution (10%). The trap is connected with an absorption vessel which contains sodium acetate solution in glacial acetic acid (20%, 10 ml) and bromine (6-7 drops). The opening of the trap and tube which goes into the absorption vessel are closed with tight fitting rubber stoppers.

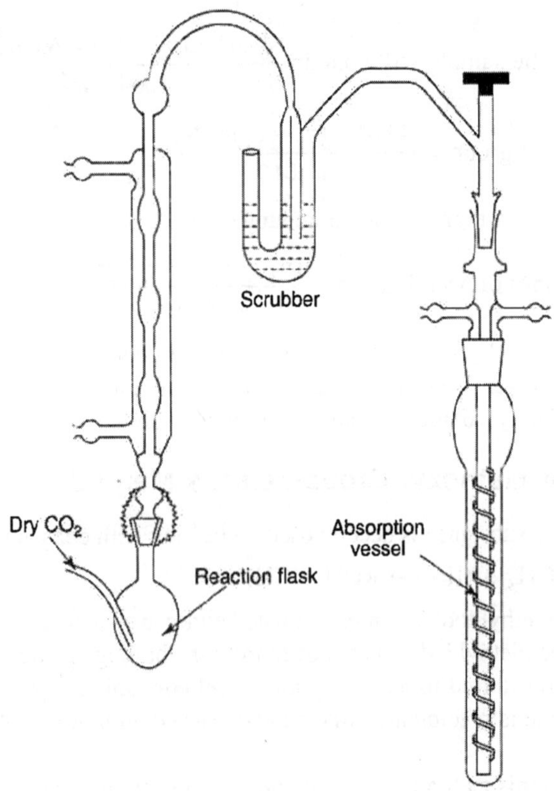

Fig. 3 Apparatus for the estimation of the methoxyl group

Reagants

The given sample
Hydriodic acid (d 1.7) (freshly distilled)
Phenol (freshly distilled)
Propionic or acetic anhydride
Sodium acetate-glacial acetic acid (20%)
Aqueous sodium acetate solution (20%)
Bromine
Aqueous sodium antimonyl tartrate (10%)
Sodium thiosulphate solution (0.05 N)
Formic acid (98%)
Sulphuric acid (10%)
Dil. hydrochloric acid (1:1) }
Marble chips } *For generation of* CO_2
Potassium iodide
Starch solution (indicator)

Procedure

The given sample is weighed (5–10 mg) accurately in a small, tin foil cup and closed tightly by folding it. The sample is introduced into the reaction flask and acetic or propionic anhydride (3–4 drops) is added, followed by phenol (5–6 drops), to dissolve the sample. Freshly distilled hydriodic acid (5 ml) is then added into the reaction flask and carbon dioxide gas is passed through the reaction mixture at the rate of one or two bubbles per second. The flask is gently heated at 125–130°C for an hour. Then the absorption vessel is removed and the tube is washed with distilled water and then with aqueous sodium acetate solution. The washings are collected in a clean iodine flask and the contents of the absorption vessel are also transferred into the same iodine flask. The vessel is washed 3–4 times with sodium acetate solution and mixed with the contents of the iodine flask. The excess bromine is then destroyed by the addition of formic acid drop by drop. The complete removal of bromine is tested by adding methyl red indicator, the colour of the indicator persists when bromine is completely destroyed. The flask is shaken and allowed to stand for 2–3 minutes. Then potassium iodide (1 g) is added followed by sulphuric acid (10%, 10 ml). The flask is stoppered, shaken well and allowed to stand for 5-10 minutes. The liberated iodine is then titrated with sodium thiosulphate solution using starch as an indicator. A blank experiment is also performed simultaneously with all reagents except the sample.

Observations

Weight of the sample taken	= w g
Volume of standard sodium thiosulphate solution used with sample	= v_1 ml
Volume of standard thiosulphate solution used with blank	= v_2 ml
Normality of thiosulphate solution	= Nx

Calculations

$$-OCH_3 \equiv CH_3I \equiv HIO_3 \equiv 3I_2 \equiv 6Na_2S_2O_3$$

Thus, 1000 ml of 1N thiosulphate solution $\equiv \frac{31}{6}$ g of $-OCH_3$ group

$(v_1 - v_2)$ ml of Nx thiosulphate solution $\equiv \frac{31 \times (v_1 - v_2) \times Nx}{1000 \times 6}$ g of —OCH_3 group

Now, w g of compound contains $\frac{31 \times (v_1 - v_2) \times Nx}{6 \times 1000}$ g of —OCH_3 group

100 g will contain $\frac{31 \times (v_1 - v_2) \times Nx}{6 \times 1000} \times \frac{100}{w}$ g

$\therefore$ % of —OCH_3 group is $\frac{31 \times (v_1 - v_2) \times Nx}{6 \times 1000} \times \frac{100}{w}$

2.4 Estimation of C-Methyl Group

Principle

The compounds containing C—CH_3 group on treatment with an oxidising agent, CrO_3—H_2SO_4 mixture, produce acetic acid which can be determined quantitatively by titration with alkali, after separation from the reaction mixture by steam distillation.

$$-\overset{|}{\underset{|}{C}}-CH_3 \xrightarrow{CrO_3-H_2SO_4} CH_3COOH$$

But the limitation of this process is that the methyl groups present on the aromatic ring, *gem*-dimethyl and *tert.* butyl groups are not oxidised to acetic acid quantitatively. Only the straight chain compounds and alicyclic compounds containing methyl group give the quantitative results.

Reagents

Oxidising mixture Chromic anhydride (16.8 g) is dissolved in water (100 ml), and conc. sulphuric acid (20 ml) is added. The solution is then cooled.

Sodium hydroxide solution (0.02 N)

or *barium hydroxide solution (0.02 N)*

Magnesium sulphate

Procedure

An accurately weighed amount (25 mg) of the sample is taken in the round bottom flask and oxidising mixture (5 ml) is added. 2–3 pieces of porcelain are added to prevent bumping and the reaction mixture is refluxed at 165–170°C for 90 minutes on a sand bath. The contents are cooled and the condenser is washed with distilled water and the washings are collected in the same reaction flask. Then magnesium sulphate (7 g) is added and the mixture is steam distilled slowly. The distillate (50 ml) is collected and titrated with standard alkali (NaOH or $Ba(OH)_2$) using phenolphthalein as an indicator. An additional 5 ml of the distillate should not change the end point appreciably. A blank experiment is also carried out in the similar manner, i.e., without using sample.

Observations

Weight of the sample taken	$= w$ g
Volume of alkali used in titration with the sample	$= v_1$ ml
Volume of alkali used in blank	$= v_2$ ml
Normality of alkali	$= Nx$

Calculations

$$1000 \text{ ml of 1N NaOH} \equiv 1000 \text{ ml of 1N } CH_3COOH \equiv 1\text{—}CH_3 \text{ group}$$

$$\equiv 15.04 \text{ g of } CH_3 \text{ group}$$

$$(v_1 - v_2) \text{ ml of } Nx \text{ NaOH} \equiv \frac{15.04 \times (v_1 - v_2) \times Nx}{1000} \text{ g of } CH_3 \text{ group}$$

$$\text{Now } w \text{ g of sample contains } \frac{15.04 \times (v_1 - v_2) \times Nx}{1000} \text{ g of } CH_3 \text{ group}$$

$$100 \text{ g of sample will contain } \frac{15.04 \times (v_1 - v_2) \times Nx}{1000} \times \frac{100}{w} \text{ g of } CH_3 \text{ group}$$

$$\% \text{ of —}CH_3 \text{ group} = \frac{15.04 \times (v_1 - v_2) \times Nx}{1000} \times \frac{100}{w}$$

If the molecular weight (M) of the compound is given then one can calculate the number of C—CH_3 groups present in the compound.

$$\text{Number of C—}CH_3 \text{ groups} = \frac{(v_1 - v_2) \times Nx}{1000} \times \frac{M}{w}$$

2.5 Estimation of Carbonyl Group

There are various methods for the estimation of carbonyl groups.

Method 1 Hydroxylamine hydrochloride-pyridine method

Principle

The method is based on the reaction of the carbonyl compound (aldehyde or ketone) with hydroxyl amine hydrochloride to form oxime.

$$>C{=}O + H_2NOH \cdot HCl \rightleftharpoons\ >C{=}NOH + H_2O + HCl$$

The reaction is reversible and the equilibrium is shifted towards the right hand side on addition of pyridine. Pyridine forms pyridinium hydrochloride with hydrochloric acid.

Pyridinium hydrochloride, thus formed is acidic and is titrated with standard sodium hydroxide using bromophenol blue as an indicator. The reagent hydroxylamine hydrochloride is neutral to this indicator.

The rate of reaction depends on the carbonyl compound taken. Some compounds react at room temperature (aldehydes) while others require heating (e.g. ketones and certain hindered aldehydes).

Reagents

The given carbonyl compound

Hydroxylamine hydrochloride reagent (0.5 N): Pure hydroxylamine hydrochloride (17.5 g) is dissolved in distilled water (80 ml) and diluted to 500 ml with 95 per cent ethanol.

Methanolic sodium hydroxide solution (0.5 N): Sodium hydroxide pellets (A.R., 19 g) are dissolved in distilled water (50 ml) and diluted to 500 ml with absolute methanol. The solution is standardized with hydrochloric acid (0.5 N).

Indicator solution: A solution of bromophenol blue in alcohol (4%, 0.2 ml) is mixed with pyridine (10 ml) and diluted to 1 litre with 95 per cent ethanol.

Procedure

Hydroxylamine hydrochloride solution (30 ml) is placed in a conical flask (250 ml) fitted with a glass stopper. An accurately weighed quantity (~ 1 g) of carbonyl compound is added followed by bromophenol blue indicator solution (100 ml). The flask is stoppered and allowed to stand for 30 minutes at room temperature (or reflux the mixture on a boiling water bath for 2 hours in case of ketones, particularly aromatic ketones). Then the mixture is titrated with standard methanolic sodium hydroxide solution to a blue-green colour end point.

A blank experiment is also performed. The colour of the end point of the sample containing solution should match with that of the blank experiment.

Observations

Weight of the carbonyl compound taken = w g

Volume of the standard sodium hydroxide solution used with sample = v_1 ml

Volume of the sodium hydroxide solution used for blank = v_2 ml

Normality of the sodium hydroxide solution = Nx

Molecular weight of the given carbonyl compound = M g

Calculations

1000 ml of 1N NaOH ≡ 1000 ml of 1N pyridinium hydrochloride ≡ M g of compound

$$v_1 - v_2 \text{ ml of } Nx \text{ NaOH} \equiv \frac{M \times (v_1 - v_2)}{1000} \times Nx \text{ g of carbonyl compound}$$

$$\% \text{ Purity of carbonyl compound} = \frac{M\,(v_1 - v_2)}{1000} \times Nx \times \frac{100}{w}$$

Method 2. Gravimetric method

Principle

2, 4-Dinitrophenylhydrazine reacts quantitatively with carbonyl compounds and forms a precipitate of 2, 4-dinitrophenylhydrazone which are filtered, dried and weighed.

$$>C{=}O + NH_2NHC_6H_3(NO_2)_2 \longrightarrow >C{=}NNHC_6H_3(NO_2)_2 + H_2O$$

Reagents

2, 4-DNP reagent. A saturated solution of 2, 4–DNP is prepared in hydrochloric acid (2N) at 0°C (approx. 0.4 g of 2, 4-DNP is present in 100 ml of solution).

Carbonyl compound

Procedure

2, 4-DNP reagent (50 ml) is placed in a glass stoppered conical flask (250 ml) and an accurately weighed quantity (~ 0.03 milli mole) of carbonyl compound is added. The flask is shaken well for 5 minutes and allowed to stand in an ice bath for one hour with occasional shaking. The precipitate is filtered in a pre-weighed sintered crucible and washed with hydrochloric acid (2 N), followed by water. The crucible is dried in a vacuum desicator over sulphuric acid or in an oven at 100°C and weighed. The heating and weighing process is repeated to a get a constant weight.

Observations

Weight of the carbonyl compound taken $= w$ g

Constant weight of empty crucible $= y_1$ g

Constant weight of crucible with precipitate $= y_2$ g

Weight of precipitate of hydrazone $= y_2 - y_1 = y$ g

Calculations

$$\%\text{ purity of the carbonyl compound} = \frac{y \times F \times 100}{w}$$

where F is the gravimetric factor.

2.6 Estimation of carboxyl (–COOH) group

There are three methods for the estimation of the carboxyl group.

Method 1

By direct titration with alkali in aqueous, aqueous-alcoholic, alcoholic or non-aqueous medium. For estimation of water soluble carboxylic acids, the titration with standard aqueous alkali (potassium, sodium or barium hydroxide) is performed by using phenolphthalein as an indicator.

$$RCOOH + {}^{-}OH \longrightarrow RCOO^{-} + H_2O$$

Other indicators which can also be used in the titration are thymol blue (pH range 8.0–9.6, colour change is from yellow to blue) and thymolphthalein (pH range 9.4–10.6; colour change is from colourless to blue).

Neutralization equivalent—The number of grams of acids required to neutralize one litre of normal alkali. If the basicity of an acid is known then its molecule weight can also be determined.

The water insoluble acids are dissolved in aqueous alcoholic or alcoholic medium and then titrated with standard alkali.

Reagents

Carboxylic acid solution. An accurately weighed given acid (0.6 g) is dissolved in distilled water in a 100 ml measuring flask.

Standard sodium hydroxide solution (0.05 N)

Phenolphthalein indicator

Procedure

The acid solution (20 ml) is pipetted out into a 100 ml conical flask and 1 – 2 drops of phenolphthalein indicator are added. It is titrated with standard sodium hydroxide solution until a faint pink colour is obtained. The titration is repeated to get three concordent readings.

For water insoluble acids, the solution is prepared in aqueous alcohol or absolute methanol or ethanol and then the titration is performed in the usual manner with alkali. In these cases where alcohol is used, a blank experiment is also performed simultaneously.

Observations

Weight of the acid dissolved in 100 ml = w g

Volume of the alkali used = v_1 ml

Volume of acid solution taken = 20 ml

Normality of alkali used = Nx

Calculations: Equivalent weight determination

$$v_1 \text{ ml of } Nx \text{ NaOH} \equiv 20 \text{ ml of } Nx \text{ acid} \equiv \frac{w}{5} \text{ g of acid}$$

$$1000 \text{ ml of 1N NaOH} \equiv \frac{w}{5} \times \frac{1000}{v_1 \times Nx} \text{ g of acid}$$

$$\text{Equivalent weight of the acid} = \frac{w \times 1000}{5 \times v_1 \times Nx}$$

If basicity of the acid is given, then

Molecular weight = Equivalent weight × basicity

% Purity of acid – If molecular weight is given = M, strength of acid solution = y g/l

$$1000 \text{ ml of 1 N NaOH} \equiv 1000 \text{ ml of 1 N acid} \equiv \frac{M}{n} \text{ g of acid } (n = \text{basicity})$$

$$v_1 \text{ ml of } Nx \text{ NaOH} \equiv 20 \text{ ml of } Nx \text{ acid}$$

$$\equiv \frac{M}{n} \times \frac{v_1 \times Nx}{1000} \text{ g of acid}$$

$$20 \text{ ml of acid solution contain } \frac{M}{n} \times \frac{v_1 \times Nx}{1000} \text{ g of acid}$$

$$1000 \text{ ml of acid solution contain } \frac{M}{n} \times \frac{v_1 \times Nx}{20} \text{ g of acid}$$

$$\% \text{ purity} = \frac{M \times v_1 \times Nx}{n \times 20 \times y} \times 100$$

Method 2 Silver salt method

Principle

The silver salts of carboxylic acids can be easily prepared by treatment of ammonium salt of carboxylic acid with silver nitrate solution. The silver salts of most of the carboxylic acids are insoluble or sparingly soluble in cold water. The silver salts on ignition, decompose and give a quantitative yield of metallic silver which can be determined.

$$RCOOH + NH_4OH \longrightarrow RCO\bar{O}\overset{+}{N}H_4$$

$$RCOONH_4 + AgNO_3 \longrightarrow RCOOAg + NH_4NO_3$$

$$RCOOAg \xrightarrow{\Delta} Ag + \text{Oxides of carbon, hydrogen etc.}$$

Many silver salts are sensitive to light; therefore drying of salt should be done under minimum exposure to light. Further, this method is not used for acids containing halogens and sulphur due to the formation of silver halides and silver sulphide containing silver sulphate on ignition.

Reagents:

Carboxylic acid sample
Silver nitrate solution (5%)
Ammonium hydroxide solution

Procedure

A weighed quantity (0.5 g) of the given sample is dissolved in distilled water (25–30 ml) in a 250 ml beaker and ammonium hydroxide solution is added dropwise with continuous stirring till the solution gives a smell of ammonia. The solution is boiled gently to remove excess of ammonia (check the vapours with red litmus paper). It is cooled and silver nitrate solution (5%) is added with stirring, until the precipitation of the silver salt is complete. The precipitate is filtered, washed thoroughly with water to remove excess of silver nitrate and drained well by pressing the solid with a glass stopper. The precipitates are then transferred on a pours plate, dried in a desiccator and covered with brown paper to protect the precipitate from exposure to light.

The dried silver salt (0.3–0.4 g) is placed in a dried and pre-weighed silica crucible with a lid and weighed accurately. The crucible with lid is placed on a clay-pipe triangle and heated gently with a Bunsen burner from its upper portion. The heating is extended to the base of the crucible slowly. After 5 minutes of slow heating, it is heated strongly for 15–20 minutes, cooled in a desiccator and weighed. The heating process and weighing is repeated to attain a constant weight.

Observations and calculations

Weight of silver salt of carboxylic acid taken = w g

Weight of silver produced = w_1 g

Equivalent weight of an acid = E g

Equivalent weight of silver salt of an acid = E + 106.88 (RCOOAg)

Now w_1 g of silver is present in w g of silver salt of acid

107.88 g of silver is present in $\dfrac{w \times 107.88}{w_1}$ g of silver salt

Then
$$\frac{w \times 107.88}{w_1} = E + 106.88$$

$$E = \frac{w \times 107.88}{w_1} - 106.88$$

If basicity of carboxylic acid is known, then,

$$\text{Molecular weight of acid} = \left[\left(\frac{w}{w_1} \times 107.88\right) - 106.88\right] \times \text{basicity}$$

Method 3 By iodometric titration

Principle

The solution of strong carboxylic acids can be estimated via iodometric titration. The method involves the addition of excess of a neutral solution of potassium iodate and potassium iodide to the dilute acid solution. The hydrogen ions produce an equivalent amount of iodine.

$$IO_3^- + 5I^- + 6H^+ \longrightarrow 3I_2 + 3H_2O$$

The liberated iodine is titrated with a standard sodium thiosulphate solution.

For weak organic acids (dissociation constant < 10^{-6}) this method is not very good because of the formation of weak acid and its salt buffer.

$$6RCOOH + KIO_3 + 5KI \longrightarrow 6RCOOK + 3I_2 + 3H_2O$$

For better results in such cases, an excess of standard sodium thiosulphate solution is added, which removes the iodine and increases the rate of the reaction. Finally, the excess of thiosulphate is titrated with standard iodine solution, from which the amount of iodine liberated can be calculated and then the carboxylic acid is estimated.

Reagents

Potassium iodate-potassium iodide reagent Potassium iodate (A.R., 1.5 g) and potassium iodide (A.R.. 12.5 g) are dissolved in distilled water (100 ml).

Sodium thiosulphate solution (0.1 N)

Iodine solution (0.1 N) Potassium iodide (A.R., 10 g) is dissolved in distilled water (20 – 30 ml) in a 500 ml measuring flask. Roughly weighed iodine (6.3 g) is then added into the potassium iodide solution and the flask is shaken to dissolve the iodine. The solution is diluted with distilled water upto the mark. It is then standarized with sodium thiosulphate solution (0.1 N).

Starch solution (indicator)

Procedure

A accurately weighed sample (0.2 – 0.3) of the given acid is taken in an iodine flask and dissolved in water. Potassium iodide-potassium iodate reagent (20 ml) is added, followed by standard sodium thiosulphate solution (0.1 N, 40 ml). The mixture is allowed to stand for 28–30 minutes with occasional shaking and excess of thiosulphate is titrated with standard iodine solution (0.1 N) by using starch as an indicator. A blank experiment is also carried out.

Observations

Volume of iodine solution used with sample	$= v_1$ ml
Volume of iodine solution used with blank	$= v_2$ ml
Normality of iodine solution	$= Nx$
Weight of sample taken	$= w$ g
Molecular weight of acid	$= M$
Basicity of acid	$= n$

Calculations

$$1000 \text{ ml of 1N } I_2 \equiv 1000 \text{ ml of 1 N thiosulphate} \equiv \frac{M}{n} \text{ g of acid}$$

$$(v_2 - v_1) \text{ ml of } Nx\ I_2 \equiv \frac{M}{n} \times \frac{(v_2 - v_1) \times Nx}{1000} \text{ g of acid}$$

$$\therefore \quad \% \text{ purity of acid sample} = \frac{M \times (v_2 - v_1) \times Nx \times 100}{n \times 1000 \times w}$$

2.7 Estimation of Esters

Principle

The esters are quantitatively estimated by hydrolysis with a known volume of standardized alkali (saponification). The unreacted alkali is titrated with standard acid.

$$RCOOR' + OH^- \longrightarrow RCOO^- + R'OH$$

The water soluble esters are hydrolysed with aqueous alkali, while water insoluble esters are hydrolysed with alcoholic alkali. Alcohol used can be ethanol, isopropanol, *n*-propanol or *n*-butanol. The choice of alcohol depends on the temperature required for hydrolysis of esters. The high boiling alcohol like diethylene glycol is generally used for the esters which are difficult to hydrolyse, e.g., di-*n*-butyl phthalate.

Saponification equivalent. The saponification equivalent of an ester is defined as the weight of the ester, in grams, which reacts with one gram equivalent of a strong base.

Reagents

Alcoholic sodium hydroxide solution (0.5 N). Sodium (small pieces) (0.6 g) is added slowly with caution into absolute alcohol (100 ml) in a flask (500 ml). When the sodium has reacted, water (10 ml) is added and mixed wall. It is then standarized with hydrochloric acid.

Alcoholic potassium hydroxide solution (0.5 N). Potassium hydroxide pellets (1.2 g) are dissolved in 95% ethanol (100 ml) and allowed to stand overnight. The solution is filtered and standarized with hydrochloric acid.

Hydrochloric acid (0.5 N)

Phenolphthalein indicator

Procedure

An accurately weighed quantity (0.5 g) of ester is placed in a round bottom flask (100 ml) fitted with reflux condenser, and aqueous sodium hydroxide solution (25 ml) for water soluble ester, or alcoholic sodium hydroxide solution (25 ml) for water insoluble ester is added. The contents are refluxed gently for one hour. In case of alcoholic solution, water bath is used for refluxing. The reaction mixture is cooled and the condenser is washed with water (25 ml) and the washings are collected in the same flask. Finally, the excess of alkali is titrated with standard hydrochloric acid using phenolphthalein as an indicator. The blank experiment is also performed simultaneously.

Observations

Weight of the given sample taken $= w$ g

Volume of standard hydrochloric acid used in blank $= v_1$ ml

Volume of standard hydrochloric acid used with sample $= v_2$ ml

Normality of hydrochloric acid $= Nx$

Calculations

$$(v_1 - v_2) \text{ ml of } Nx \text{ HCl} \equiv (v_1 - v_2) \text{ ml of } Nx \text{ NaOH}$$

Volume of alkali used in hydrolysis of ester $= v_1 - v_2$ ml

Now 1000 ml of 1N NaOH contains 1 g equivalent of NaOH

$(v_1 - v_2)$ ml of Nx NaOH contains $\dfrac{(v_1 - v_2) \times Nx}{1000}$ g equivalent of NaOH

$\dfrac{(v_1 - v_2) \times Nx}{1000}$ g equivalent of NaOH is required for w g of ester

1 g equivalent of NaOH is required for $\dfrac{w \times 1000}{(v_1 - v_2) \times Nx}$ g of ester

$$\text{Thus, saponification equivalent} = \frac{w \times 1000}{(v_1 - v_2) \times Nx}$$

Purity of the ester:

If molecular weight of an ester is M g

$$1000 \text{ ml of 1 N HCl} \equiv 1000 \text{ ml of 1 N NaOH} \equiv M \text{ g of ester}$$

$$(v_1 - v_2) \text{ ml of } Nx \text{ HCl} \equiv \frac{M \times (v_1 - v_2) \times Nx}{1000} \text{ g of an ester}$$

$$\% \text{ Ester} = \frac{M \times (v_1 - v_2) \times Nx}{1000 \times w} \times 100$$

2.8 Estimation of Amines

There are various methods for the estimation of amines.

Method 1 Acetylation method

Principle

The amino group of organic compounds (aliphatic or aromatic) can be estimated by acetylation with acetic anhydride in presence of pyridine at room temperature.

$$RNH_2 + (CH_3CO)_2O \longrightarrow RNHCOCH_3 + CH_3COOH$$

$$RR'NH + (CH_3CO)_2O \longrightarrow RR'N{\cdot}COCH_3 + CH_3COOH$$

The excess of anhydride is also decomposed with water to give acetic acid and total free acid is then titrated with standard sodium hydroxide solution.

The acetylation reaction involves the following steps.

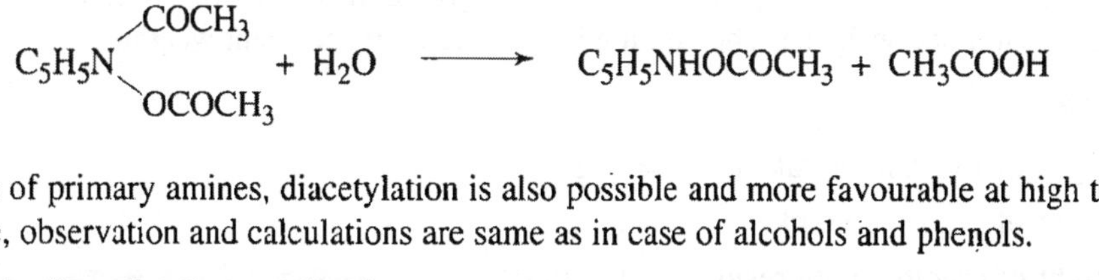

In case of primary amines, diacetylation is also possible and more favourable at high temperatures. Procedure, observation and calculations are same as in case of alcohols and phenols.

Method 2 Bromination method

Principle

Aromatic amines can also be estimated by the bromination method. The sample is treated with excess of standard bromate-bromide solution in presence of hydrochloric acid. The excess of bromine is then determined iodometrically.

$$KBrO_3 + 5KBr + 6HCl \longrightarrow 3Br_2 + 6KCl + 3H_2O$$

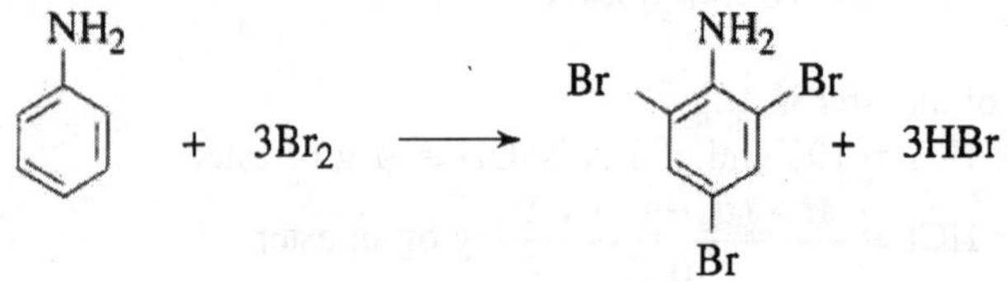

$$\underset{\text{(excess)}}{Br_2 + 2KI} \longrightarrow I_2 + 2KBr$$

(The solution of aniline is prepared by taking this required amount of amine in a volumetric flask, dissolved in minimum amount of dilute hydrochloric acid and diluted to the mark with distilled water).

Method 3 Chloroplatinate method (Gravimetric method)

Principle

Amines (base) or their halide salts on reaction with chloroplatinic acid (H_2PtCl_6) produce insoluble chloroplatinates of the amines $(B_2H_2)PtCl_6$ (B is one equivalent of the base) in cold water. The above salt on ignition gives pure platinum. By knowing the weight of the platinum produced, equivalent weight of the base (amine) can be calculated.

Reagents

Sample of base (amine)

Dilute hydrochloric acid

Chloroplatinic acid

Procedure

The given sample is weighed accurately (~ 0.2 g) and dissolved in dilutle hydrochloric acid (5 m) by warming (if necessary). A slight excess of chloroplatinic acid solution (5 – 6 ml of solution containing 1.0 g of chloroplatinic acid) is added. The precipitated salt is then filtered, washed with ice-cold water and dried in an oven at 100°C.

The dried chloroplatinate salt is weighed accurately (~ 0.2 g) in a pre-weighed silica crucible with lid. The crucible is then heated, first gently and then strongly till the crucible is red hot. It is cooled and placed in a desiccator and weighed. The ignition is repeated to get constant weight.

Observations

Weight of the sample taken $= w$ g

Weight of empty crucible $= w_1$ g

Weight of crucible with residue $= w_2$ g

Atomic weight of platinum $= 195.2$ g

Calculations

$$\text{Weight of platinum produced} = w_2 - w_1 \text{ g}$$

$$\text{Equivalent weight of base (amine)} = \frac{w \times 195.2}{(w_2 - w_1)} \text{ g}$$

If acidity of base is known, then the molecular weight of base can be calculated.
Molecular weight = Equivalent weight × acidity

2.9 Estimation of Amide Group

Principle

The organic compounds containing the amide group are estimated by their hydrolysis with known volume of alkali. The excess alkali is then titrated with a standard acid.

$$RCONH_2 + NaOH \longrightarrow RCOONa + NH_3\uparrow$$

Reagents

Sodium hydroxide solution (2N)
Hydrochloric acid (1 N)
Phenolphthalein indicator

Procedure

An accurately weighed (~ 0.5 g) given sample is placed in a round bottom flask (100 ml) fitted with a reflux condenser. The sample is dissolved in minimum amount of aqueous ethanol and sodium hydroxide solution (2N, 20 ml) is added. The mixture is refluxed on a sand bath till the hydrolysis is complete (2–3 hours). The completion of the reaction is checked by testing the evolution of ammonia. The flask is cooled, the condenser is washed with water and the washings are collected in the same flask. Finally, the excess alkali is titrated with standard hydrochloric acid using phenolphthalein as an indicator. A blank experiment is also performed.

Observations

Weight of the sample taken	= w g
Volume of hydrochloric acid used in titration with sample	= v_1 ml
Volume of hydrochloric acid used in titration with blank	= v_2 ml
Normality of hydrochloric acid	= Nx
Molecular weight of sample	= M g
Number of amide groups present in a given compound	= n

Calculations

$$1000 \text{ ml of 1 N HCl} \equiv 1000 \text{ ml of 1 N NaOH} \equiv \frac{M}{n} \text{ g of sample}$$

$$(v_2 - v_1) \text{ ml of } Nx \text{ HCl} \equiv \frac{M}{n} \times \frac{(v_2 - v_1)}{1000} \times Nx \text{ g of sample}$$

$$\% \text{ purity} = \frac{M}{n} \times \frac{(v_2 - v_1)}{1000} \times Nx \times \frac{100}{w}$$

2.10 Estimation of Urea

Method 1 Gasometric or hypobromite method

Principle

Urea gets decomposed with sodium hypobromite solution in alkaline medium and nitrogen is liberated.

$$H_2NCONH_2 + 3NaOBr + 2NaOH \longrightarrow N_2 \uparrow + Na_2CO_3 + 3NaBr + 3H_2O$$

The evolved nitrogen is then measured. But the reaction is not perfectly quantitative due to the presence of small amount of ammonium cyanate.

$$NH_2CONH_2 \rightleftharpoons NH_4CNO$$

Thus, practically 1 g of urea gives 357.0 ml of nitrogen at N.T.P. instead of 378.2 ml (22415/60.06).

Reagents

Urea sample

Sodium hypobromite solution. Sodium hydroxide (100 g) is dissolved in water (500 ml), cooled to 0–5°C (ice bath) and bromine (25.0 ml) is added slowly with stirring.

Apparatus The apparatus for the experiment consists of a conical flask with a rubber stopper which is connected to a graduated gas burette and a levelling tube (Fig. 4).

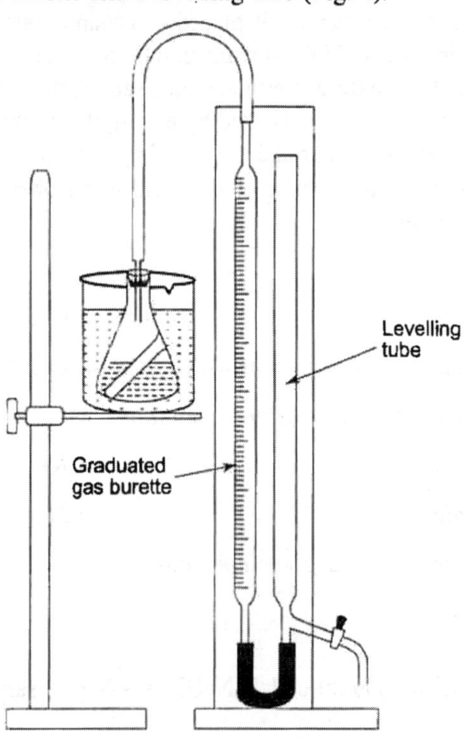

Fig. 4 Apparatus for the estimation of urea

Procedure: The given urea sample is weighed (0.1 g) accurately in a small tube and placed in a conical flask containing sodium hypobromite solution (60 ml), in such a way that the two reactants do not come in contact with each other. The flask is stoppered, connected to the graduated gas burette through a rubber tube and immersed in a water bath. When the thermal equilibrium is attained, the levels in the gas burette and levelling tube are equialized by using a stopcock and the reading of the gas burette is noted. Then the flask is tilted so that the two reactants get mixed with each other and the liberated nitrogen is collected in the gas burette. The same water level is maintained in both the tubes by removing excess water from the tap. When all the nitrogen gas has evolved, the reaction mixture is cooled to the temperature of the bath and the levels in both the tube are equalised. The final reading of the gas buratte is recorded.

Observations

Weight of the urea sample taken = w g

Initial gas burette reading = v_1

Final gas burette reading = v_2

Temperature of the bath = t°C

Barometric pressure = P mm Hg

Vapour pressure of water at t°C = p mm Hg

Calculations

Volume of N_2 evolved = $v_2 - v_1$ ml

Now at N.T.P., the volume of N_2 is V

then
$$\frac{V \times 760}{273} = \frac{(v_2 - v_1)(P - p)}{(273 + t)}$$

$$V = \frac{(v_2 - v_1)(P - p) \times 273}{(273 + t) \times 760} \text{ ml}$$

We know that 357.0 ml of N_2 is evolved from 1 g of urea.

Then $\dfrac{(v_2 - v_1)(P - p) \times 273}{(273 + t) \times 760}$ ml of N_2 will be evolved from $\dfrac{(v_2 - v_1)(P - p) \times 273}{(273 + t) \times 760 \times 357}$ g of urea.

$$\% \text{ Purity of urea} = \frac{(v_2 - v_1)(P - p) \times 273}{(273 + t) \times 760 \times 357} \times \frac{100}{w}$$

Method 2 Gravimetric or xanthhydrol method

Principle

Urea reacts with a solution of xanthhydrol quantitatively and gives a precipitate of di-xanthhydryl urea. The precipitate is filtered, dried and weighed.

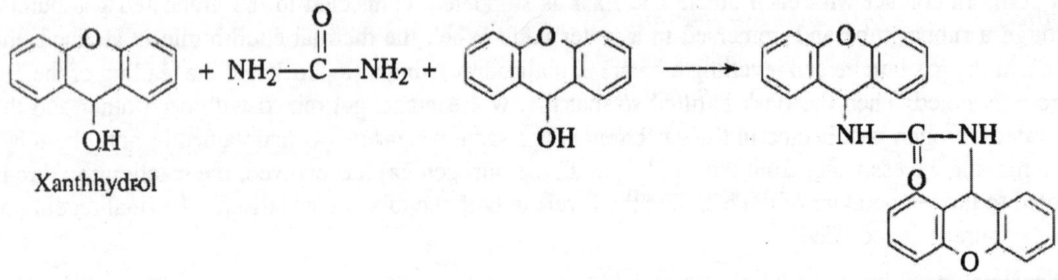

Xanthhydrol   Di-xanthhydryl urea

Reagents

Urea sample
Methanolic xanthhydrol solution (10%, w/v)
Glacial acetic acid

Procedure

An accurately weighed (0.1 g) urea sample is placed into a conical flask and distilled water (40 ml), glacial acetic acid (150 ml), xanthhydrol solution (20 ml) are added. The contents are mixed and allowed to stand for one hour. The precipitate is filtered in a pre-weighed sintered glass crucible or a procelain filter crucible, washed with absolute methanol (30–40 ml), dried at 120°C and weighed. The process of heating and weighing is repeated to get a constant weight.

Observations

Weight of urea sample taken = w g

Constant weight of empty crucible = w_1 g

Constant weight of crucible with precipitate = w_2 g

Calculations

Weight of dixanthdrylurea formed = $w_2 - w_1$ g
Now 420.45 g of dixanthydrylurea is formed from 60.06 g of urea.

$\therefore$ $w_2 - w_1$ g of dixanthydrylurea will from $\dfrac{(w_2 - w_1) \times 60.06}{420.45}$ g of urea.

$$\% \text{ Purity of urea sample} = \frac{(w_2 - w_1) \times 60.06}{420.45} \times \frac{100}{w}$$

2.11 Estimation of Nitro Group

Principle

The nitro groups in organic compounds are estimated by their reduction to amino group. The different reducing agents used are stannous chloride, titanous chloride and titanous sulphate.

Method 1. Stannous chloride method

A known weight of the given sample is treated with a known volume of stannous chloride solution. The excess of stannous chloride is determined by titration with standard iodine solution.

$$RNO_2 + 3SnCl_2 + 6HCl \longrightarrow RNH_2 + 3SnCl_4 + 2H_2O$$

$$\underset{\text{(excess)}}{Sn^{2+}} + I_2 \longrightarrow Sn^{4+} + 2I^-$$

Thus, six equivalents of stannous salt are required for reduction of one nitro group.

Reagents

Stannous chloride solution. Pure tin powder (25 g) is placed in a round bottom flask (500 ml) fitted with an air condenser. Concentrated hydrochloric acid (250 ml) is added and the flask is heated till the reaction commences. When all the tin has dissolved, the clear solution is decanted and diluted to 1 litre. The stannous chloride solution is standardized with standard iodine solution.

Iodine solution

Given sample

Procedure

An accurately weighed amount (0.2–0.4 g) of the given sample is placed into a flask (250 ml) fitted with a two-holed cork, connected to a carbon dioxide generator and a reflux condenses. The compound is dissolved in alcohol (30 ml) by gentle warming. The air of the flask is displaced by carbon dioxide and a solution of stannous chloride (50 ml) is added after removing the cork which is then replaced quickly. The flask is heated over a wire gauze for 2 hours with continuous bubbling of carbon dioxide. The contents are cooled and titrated with standard iodine solution in carbon dioxide atmosphere by using starch as an indicator.

A blank titration is also performed.

Observations

Volume of iodine solution used in blank = v_1 ml

Volume of iodine solution used with sample = v_2 ml

Normality of iodine solution = Nx

Weight of the sample taken = w g

Calculations:

$$1000 \text{ ml of } 1N\,I_2 \equiv 1000 \text{ ml of } 1\text{ N } SnCl_2 \equiv \frac{46}{6} \text{ g of } NO_2 \text{ group}$$

$$(v_1 - v_2) \text{ ml of } Nx\ I_2 \equiv \frac{46}{6} \times \frac{(v_1 - v_2)}{1000} \times Nx \text{ g of } NO_2 \text{ group}$$

Now $\quad w$ g of given sample contains $\dfrac{46}{6} \times \dfrac{(v_1 - v_2)}{1000} \times Nx$ g of NO_2 group

$$100 \text{ g will contain } \frac{46}{6} \times \frac{(v_1 - v_2)}{1000} \times Nx \times \frac{100}{w}$$

Thus, $$\% \text{ of } NO_2 \text{ group} = \frac{46 \times (v_1 - v_2) \times Nx \times 100}{6 \times 1000 \times w}$$

If molecular weight of given sample is M then number of nitro groups can be calculated.

∴ $$\text{Number of nitro groups} = \frac{(v_1 - v_2) \times Nx}{6 \times 1000} \times \frac{M}{w}$$

Method 2. Titanous sulphate method

The given amino compound is reduced with titanous sulphate and excess of titanous salt is titrated with standard ferric ammonium sulphate solution using ammonium thiocyanate as an indicator.

$$R—NO_2 + 3Ti_2(SO_4)_3 + 3H_2SO_4 \longrightarrow RNH_2 + 6Ti(SO_4)_2 + 2H_2O$$

$$Ti^{3+} + Fe^{3+} \longrightarrow Ti^{4+} + Fe^{2+}$$

Thus, six equivalents of the titanous salt are required for reduction of one nitro group.

Reagents

Standard nitro compound solution. An accurately weighed nitro compound (0.2 – 0.4 g) is dissolved in water or ethanol or glacial acetic acid (choice of solvent depends upon the solubility of nitro compound) and diluted with the same solvent to 250 ml in a volumetric flask.

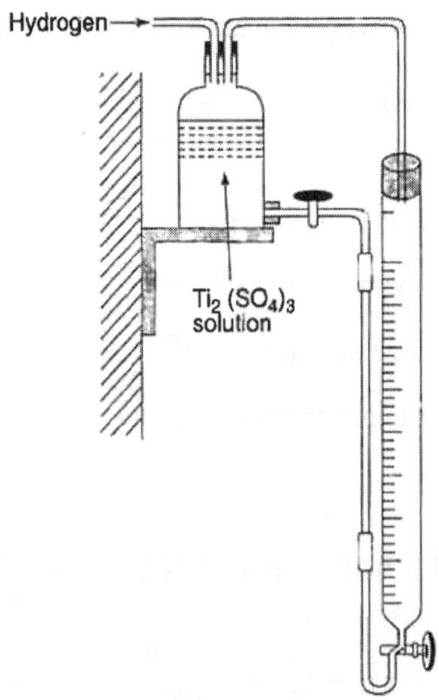

Fig. 5 Assembly for the storage of titanous sulphate solution

Standard titanous sulphate solution (0.1 N). Pure titanium metal (6 g) is dissolved in dilute sulphuric acid (1:3 v/v, 600 ml) by warming on a water bath (7 hours). It is cooled in a stream of hydrogen and diluted to 1 litre with freshly boiled and cooled distilled water. The solution is stored in a special type of bottle (1 litre capacity) which is connected to a burette and a hydrogen gas generator (to keep the solution out of contact with air). It is standardized with standard ferric ammonium sulphate solution by using ammonium thiocyanate as an indicator.

Ferric ammonium sulphate solution (0.1 N). Ferric ammonium sulphate (4.822 g) is dissolved in boiled and cooled distilled water, and sulphuric acid (50%, 1 ml) is added. The solution is diluted to 100 ml in a volumetric flask (100 ml) with boiled and cooled distilled water.

Ammonium thiocyanate solution (10%). This is used as the indicator.

Procedure

A nitro compound solution (20 ml) is pipetted out into a conical flask (250 ml) fitted with a two-hole rubber stopper and dilute sulphuric acid (2:5 v/v, 20 ml) is added. Carbon dioxide gas is passed through one hole for 5 minutes to remove air and then titanous sulphate solution (0.1 N, 40 ml) is added through the burette from the second hole. The mixture is refluxed for 10 minutes and cooled at 35–40°C while maintaining the current of carbon dioxide gas. Ammonium thiocyanate solution (10%, 10 ml) is added, which gives red colour to the mixture and it is titrated with standard ferric ammonium sulphate solution (0.1 N) to a pale red and point. The process is repeated to get concordant readings.

A blank experiment is also performed.

Observations

Weight of the nitro compound dissolved in 100 ml of water = w g

Volume of sample solution (nitro compound solution) taken in each titration = 20 ml

Volume of ferric ammonium sulphate solution used for sample = v_1 ml

Volume of ferric ammonium sulphate solution used for blank = v_2 ml

Normality of ferric ammonium sulphate solution = Nx

Calculations

Now we know that $NO_2 \equiv 6Ti^{3+} \equiv 6Fe^{3+}$

Then 1000 ml 1N of ferric ammonium sulphate ≡ 1000 ml of 1N titanous sulphate solution

$\equiv \frac{46}{6}$ g of NO_2 group.

$(v_2 - v_1)$ ml of Nx of ferric ammonium sulphate $\equiv \frac{46}{6} \times \frac{(v_2 - v_1) \times Nx}{1000}$ g of NO_2 group

Now 20 ml of sample solution contains $\frac{46}{6} \times \frac{(v_2 - v_1) \times Nx}{1000}$ g of NO_2 group

Then 100 ml of sample solution will contain

$$\frac{46}{6} \times \frac{(v_2 - v_1) \times Nx}{1000} \times \frac{100}{20} \text{ g of } NO_2 \text{ group}$$

$$w \text{ g of sample contains } \frac{46}{6} \times \frac{(v_2 - v_1) \times Nx}{1000} \times \frac{100}{20} \text{ g of } NO_2 \text{ group}$$

$$100 \text{ g of sample contains } \frac{46}{6} \times \frac{(v_2 - v_1) \times Nx}{1000} \times \frac{100}{20} \times \frac{100}{w}$$

Thus, $$\% \text{ of } NO_2 \text{ groups is } \frac{46}{6} \times \frac{(v_2 - v_1) \times Nx}{1000} \times \frac{100}{20} \times \frac{100}{w}$$

It molecular weight of nitro compound is M, then number of NO_2 groups can be calculated as follows:

$$w \text{ g of sample contains } \frac{(v_2 - v_1) \times Nx}{6 \times 1000} \times \frac{100}{20} NO_2 \text{ group}$$

$$M \text{ g of sample will contain } \frac{(v_2 - v_1) \times Nx}{6 \times 1000} \times \frac{100}{20} \times \frac{M}{w} NO_2 \text{ group}$$

% purity of the sample:

If number of nitro groups in the given sample is n then % purity of the sample can be calculated:

$$1000 \text{ ml of 1 N ferric ammonium sulphate solution} \equiv \frac{M}{6n} \text{ g of sample}$$

$$(v_2 - v_1) \text{ ml of } Nx \text{ ferric ammonium sulphate solution} \equiv \frac{M}{6n} \times \frac{(v_2 - v_1)}{1000} \times Nx$$

$$\text{Now, 20 ml of sample solution contains } \frac{M}{6n} \times \frac{(v_2 - v_1)}{1000} \times Nx \text{ g of sample}$$

100 ml of sample solution contains

$$\frac{M}{6n} \times \frac{(v_2 - v_1)}{1000} \times Nx \times \frac{100}{20} \text{ g of sample}$$

$$\% \text{ purity} = \frac{M}{6n} \times \frac{(v_2 - v_1)}{1000} \times Nx \times \frac{100}{20} \times \frac{100}{w}$$

3. ESTIMATION OF SUGARS

Method 1 Fehling solution method

Principle

A known volume of the Fehling solution (blue coloured solution of cupric ions) is reduced to cuprous ions (precipitated as red cuprous oxide) by glucose (reducing sugar) solution. The strength of glucose is then determined from the volume of glucose solution consumed in the reaction.

$$\underset{\text{Glucose}}{C_6H_{12}O_6} + \underset{\text{Fehling solution}}{2Cu^{2+} \text{ (tartarate complex)}} \longrightarrow \underset{\text{red ppt.}}{Cu_2O} + \underset{\text{Gluconic acid}}{C_6H_{12}O_7}$$

Reagents

Fehling solution A. Copper sulphate ($CuSO_4.5H_2O$, 17.32 g) is dissolved in distilled water (250 ml).

Fehling solution B. Sodium potassium tartrate (Rochelle salt) (86.5 g) and sodium hydroxide (25 g) are dissolved in water and diluted to 250 ml with distilled water.

Methylene blue indicator. Methylene blue (0.5 g) is dissolved in distilled water (250 ml).

Standard glucose solution. An accurately weighed amount (1.25 g) of anhydrous glucose is dissolved in water in a 250 ml measuring flask and diluted with distilled water upto the mark.

Procedure

Fehling solution is prepared by mixing solutions A and B in equal volumes. The solution (20 ml) is pipetted out into a conical flask (250 ml) and diluted with distilled water (20 ml). The contents are boiled and standard glucose solution is added in 1 ml portions from the burette till the solution acquires a faint blue colour and then drop by drop until the blue colour just disappears (the solution is boiled after every addition of glucose solution). The titration is repeated to get three concordant readings. The same experiment is performed with unknown glucose solution.

In the titration, the supernatant liquid is blue and the precipitate formed is red in colour, thus there may be difficulty in determination of the end point. In such a situation, methylene blue indicator may be added at the point when the addition of just 0.5–1.0 ml of the glucose solution is left to complete the reduction, and titrated until the blue colour just disappears.

Observations

Weight of glucose in 250 ml of standard solution = w g

Volume of Fehling solution taken in each titration = 20 ml

Volume of standard glucose solution used = v_1 ml

Volume of unknown glucose solution used = v_2 ml

Calculations

20 ml of Fehling solution ≡ v_1 ml of standard glucose solution ≡ $\frac{w \times v_1}{250}$ g of glucose.

Now v_2 ml of unknown glucose solution ≡ 20 ml of Fehling solution

$$\equiv \frac{w \times v_1}{250} \text{ g of glucose}$$

$$1000 \text{ ml of unknown glucose solution} \equiv \frac{w \times v_1 \times 1000}{250 \times v_2}$$

$$= \frac{w \times v_1 \times 4}{v_2} \text{ g}$$

Thus strength of unknown solution is $\frac{w \times v_1 \times 4}{v_2}$ g/l

The % purity of glucose solution can be determined if the strength of impure sample (unknown) is given.

If strength of impure sample is Y g/l

$$\text{Then \% purity} = \frac{w \times v_1 \times 4}{v_2} \times \frac{100}{y}$$

Method 2 Benedict's solution method

Principle

A known volume of Benedict's solution is reduced to cuprous ions by glucose solution. The strength of the given glucose is then determined from the volume of solution consumed.

$$C_6H_{12}O_6 + Cu^{2+} \text{ (citrate complex)} \longrightarrow \underset{\substack{\text{(as CuSCN} \\ \text{white ppt)}}}{Cu^+} + C_6H_{12}O_7$$

Benedict reagent contains potassium thiocyanate, potassium ferrocyanide and cupric citrate. The reduced copper is precipitated as white cuprous thiocyanate instead of red cuprous oxide. The advantage of this reagent is that the end point can be easily detected and is more stable. Ferrocyanide helps to prevent the deposition of cuprous oxide.

Reagents

Standard glucose solution

Benedict's solution. Crystalline sodium carbonate (50 g), crystalline sodium citrate (50 g) and potassium thiocyanate (31.25 g) are dissolved in 200 ml of hot water. It is filtered (if necessary) and cooled. Then a solution of crystalline copper sulphate (4.5 g) in 25 ml of water is added, followed by 5 per cent of ferrocyanide solution (1–5 ml). Finally it is diluted to 250 ml with distilled water (the solution is commercially available).

Anhydrous sodium carbonate

Procedure

The Benedict's solution (20 ml) is pipetted out into a conical flask and anhydrous sodium carbonate (3 g) and distilled water (20 ml) are added. The function of sodium carbonate is to liberate carbon dioxide which prevents the atmospheric oxidation of glucose solution. The contents of the flask are boiled over a wire gauze and the titration is carried out as in case of Fehling solution.

The titration with unknown solution is also performed similarly.

Observation and Calculations: Same as in case of Fehling solution method.

4. ESTIMATION OF UNSATURATION

The carbon-carbon double bond in organic compound is estimated by a number of methods. By the catalytic hydrogenation of double bond, by knowing the volume of hydrogen gas absorbed, one can calculate the number of double bonds present. Another method utilizes the bromination of the double bond. The bromine required for the reaction is generated from potassium bromate and potassium bromide in presence of mercuric sulphate as catalyst.

$$KBrO_3 + 5KBr + 3H_2SO_4 \longrightarrow 3Br_2 + 3K_2SO_4 + 3H_2O$$

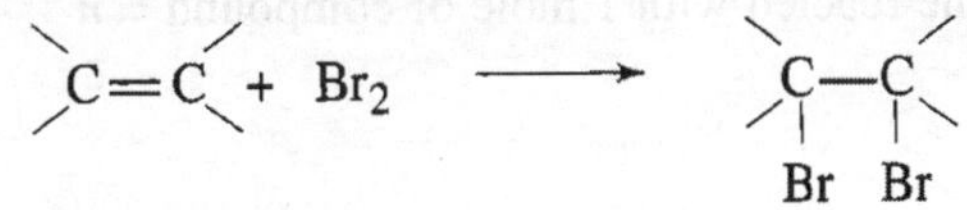

The excess of bromine is titrated iodometrically.

$$Br_2 \text{ (excess)} + 2KI \longrightarrow 2\ KBr + I_2$$

$$I_2 + 2Na_2S_2O_3 \longrightarrow 2NaI + Na_2S_4O_6$$

In Wij's method of estimation of unsaturation, the reagent used is iodine monochloride in glacial acetic acid. Hanus method uses the reagent iodine monobromide in acetic acid. The use of the latter is similar to that of Wij's method.

Bromination method

Reagents

Olefinic compound
Potassium bromate-potassium bromide solution. Anhydrous potassium bromate (2.784 g) and potassium bromide (30 g) are dissolved in distilled water and diluted to 1 litre.
Mercuric sulphate solution. Mercuric sulphate (12 g) is dissolved in a solution of conc. sulphuric acid (11.2 ml) and distilled water (380 ml).
Standard sodium thiosulphate solution (0.05 N)
Potassium iodide
Sulphuric acid (10%)
Sodium chloride solution (2 N)
Starch solution

Procedure

An accurately weighed quantity of given compound (0.1–0.2 g) is dissolved in water (25 ml) (for water soluble compounds) or carbon tetrachloride (for water insoluble compounds) in an iodine flask. Bromate-bromide solution (25 ml) and sulphuric acid (10%, 20 ml) are added. The contents are shaken for 5 minutes and mercuric sulphate solution (10 ml) is added. The mixture is allowed to stand for 7 – 10 minutes with occasional shaking. Then sodium chloride solution (2N, 15 ml) is added to liberate free bromine from the mercury sulphate complex followed by addition of potassium iodide (2 g). The flask is shaken for 1 minute and liberated iodine is titrated with sodium thiosulphate solution (0.05 N) using starch as an indicator. Blank experiment is also carried out simultaneously.

If the given sample is cinnamic acid or benzal acetone ($C_6H_5CH{=}CHCOCH_3$) then there is no need for a catalyst. The substitution reaction will also take place in such cases in presence of catalyst. For malic acid the reaction time required is longer, i.e., ~ 30 minutes.

Observations

Weight of the compound taken = w g

Volume of thiosulphate used in experiment = v_1 ml

Volume of thiosulphate used in blank = v_2 ml

Normality of thiosulphate = Nx

Number of moles of bromine reacted with 1 mole of compound = n

Calculations

$$2000 \text{ ml of 1 N thiosulphate} \equiv 1000 \text{ ml of 1 N } I_2$$

$$\equiv 1000 \text{ ml of 1 N } Br_2 \equiv \frac{M}{n} \text{ of given compound}$$

$$(v_2 - v_1) \text{ ml of } N_x \text{ thiosulphate } \frac{v_2 - v_1 \times Nx \times M}{2000 \times w} \equiv g$$

$$\text{No. of double bonds present} = n = \frac{v_2 - v_1 \times Nx \times M}{2000 \times w}$$

$$\% \text{ purity} = \frac{M \times (v_2 - v_1) \times Nx \times 100}{n \times 2000 \times w}$$

5. ESTIMATION OF OILS AND FATS

Oils and fats are glyceryl esters of higher fatty acids. Animal fat is made up largely of triglycerides containing fully saturated fatty acids with straight chains and an even number of carbon atoms. Palmitic acid (*n*-hexadecanoic) and stearic acid (*n*-octadecanoic) are the commonest fatty acids.

Vegetable oils contain unsaturated fatty acids. Oleic acid is the most common.

Glycerides containing unsaturated fatty acids have lower melting points than the corresponding compounds with fully saturated fatty acids.

5.1 Saponification Value of the given Oil or Fat Sample

On refluxing with alkali, glyceryl esters are hydrodyzad to give glycerol and the potassium salts of the fatty acids (soaps).

$$\begin{array}{l} CH_2OCOR \\ | \\ CHOCOR \\ | \\ CH_2OCOR \\ \text{Trigyceride} \end{array} + 3KOH \longrightarrow \begin{array}{l} CH_2OH \\ | \\ CHOH \\ | \\ CH_2OH \\ \text{Glycerol} \end{array} + 3RCOOK$$

The saponification value is the number of miligrams of KOH required to neutralise the fatty acids resulting from the complete hydrolysis of 1 g of fat. The saponification value gives an indication of the nature of the fatty acids in the fat, since the longer the carbon chain, the less acid is liberated per gram of fat hydrolyzed.

Reagents

Oil or fat sample

Fat solvent. Equal volume of 95 per cent ethanol and ether or neutral alcohol (neutralized exactly to phenolphthalein by alcoholic potassium hydroxide solution).

Alcoholic potassium hydroxide solutions (0.5 N) Potassium hydroxide pellets (24 g) are dissolved in a small amount of distilled water in a 1 litre measuring flask and filled up to the mark with rectified spirit.
Standard hydrochloric acid (0.5 N)
Phenolphthalein indicator

Procedure

An accurately weighed (1 g) given fat sample is transferred in a round bottom flask fitted with a reflux condenser with the help of fat solvent (3 – 5 ml). An alcoholic potassium hydroxide solution (25 ml) is added through a pipette or burette and the contents are refluxed on a boiling water bath for 30 – 40 minutes with occasional shaking. The flask is cooled to room temperature and the contents are titrated with standard hydrochloric acid (0.05 N) using phenolphthalein as an indicator. A blank titration is also carried out, i.e., without taking oil or fat.

Observations

Weight of the sample taken = w g

Volume of standard hydrochloric acid used in blank = v_1 ml

Volume of standard hydrochloric acid used with sample = v_2 ml

Normality of hydrochloric acid used = Nx

Calculations

$$1000 \text{ ml of } 1 \text{ N HCl} \equiv 1000 \text{ ml of } 1 \text{ N KOH} \equiv 56 \text{ g of KOH}$$

$$(v_1 - v_2) \text{ ml of } Nx\text{HCl} \equiv \frac{56 \times (v_1 - v_2) \times Nx}{1000} \text{ g of KOH}$$

$$= 56 \times (v_1 - v_2)) \times Nx \text{ mg of KOH}$$

$$\text{Then saponification value} = \frac{56 \times (v_1 - v_2) \times Nx}{w} \text{ mg/g of sample}$$

If we know the saponification value then we can calculate the average molecular weight of a fat as follows:

$$\text{Saponification value } (S) = \frac{3 \times 56 \times 1000}{\text{Av. mol. wt. of fat}}$$

$$\therefore \quad \text{Av. mol. wt. of fat} = \frac{3 \times 56 \times 1000}{S}$$

5.2 Iodine Value of an Oil or Fat

Principle

Halogens add across the double bonds of unsaturated fatty acids to form dihalogen compounds. Thus, iodine monochloride in acetic acid (Wij's solution) is allowed to react with the fat in the dark*. The amount of iodine consumed is then determined by titrating the iodine (released from the unused iodine mono chloride) with sodium thiosulphate, which is compared with a blank experiment

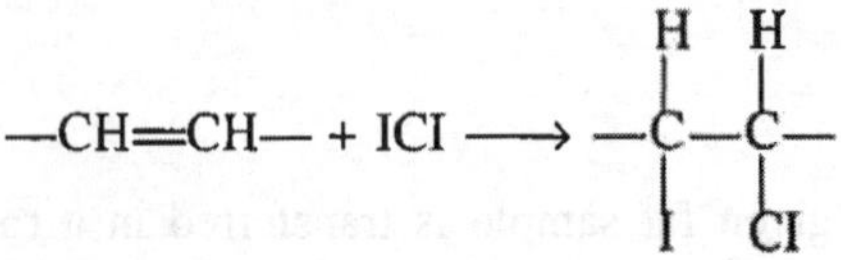

$$ICl + KI \longrightarrow KCl + I_2$$

$$I_2 + 2Na_2S_2O_3 \longrightarrow 2NaI + Na_2S_4O_6$$

The iodine number is the number of grams of iodine taken up by 100 g of fat or oil.

Reagents

Fat or oil sample (2% solution in chloroform or carbon tetrachloride).

Wij's solution Pure iodine (13 g) is dissolved in 1 litre of glacial acetic acid and dry chlorine gas is passed into the solution until a change in colour from dark brown to orange-yellow takes place (or determine the strength of solution before and after passing chlorine gas. In the latter case double the original amount of thiosulphate should be used).

Standard sodium thiosulphate solution (0.1 N)

Potassium iodide (10%)

Starch indicator

Procedure

The fat solution (10 ml) is pipetted out into a stoppered flask (500 ml) and iodine monochloride solution (25 ml) is added. The flask is stoppered and allowed to stand in the dark for 1 hour with occasional shakings. The stopper and neck of the flask are rinsed with about 50 ml of water and potassium iodide solution (10 ml) is added. The liberated iodine is titrated with standard sodium thiosulphate solution using starch as an indicator. The flask must be shaken thoroughly throughout the titration to ensure that all the iodine is removed from the organic layer. A blank experiment is also carried out without adding the fat or oil simultaneously.

Observations

Amount of oil or fat present in 10 ml of solution = w g

Volume of thiosulphate solution used in the blank = v_1 ml

Volume of thiosulphate solution used in sample solution = v_2 ml

Normality of thiosulphate solution used = Nx

Calculations

1000 ml of 1 N hypo solution ≡ 1000 ml of 1 N I_2 solution ≡ 127 g of I_2

---

* The reaction mixture is kept in the dark and the titration is carried out as quickly as possible since the halogens are oxidized in the light.

$(v_1 - v_2)$ ml of Nx hypo solution $\equiv (v_1 - v_2)$ ml of Nx I_2 solution

$$\equiv \frac{127 \times (v_1 - v_2) \times Nx}{1000} \text{ g of } I_2$$

$$\therefore \quad \text{Iodine number} = \frac{127 \times (v_1 - v_2) \times Nx}{1000} \times \frac{100}{w}$$

6. ESTIMATION OF AMINO ACIDS

6.1 Sorensen Formol Titration Method

Principle:

Amino acids reacts with formalin to form methylene amino acids.

$$H_2NCH_2CO_2H + HCHO \longrightarrow CH_2{=}NCH_2CO_2H + H_2O$$

But, it was found that the reaction is more complex and the main product is monomethylol and dimethylol amino acids.

$$—NH_2 + HCHO \longrightarrow —NH(CH_2OH)$$

$$—NH(CH_2OH) + HCHO \longrightarrow —N(CH_2OH)_2$$

Amino acids exist in zwitter ionic forms and cannot be titrated directly with alkali. Thus, amino groups of amino acids are blocked by reaction with formaldehyde (or it reduces the basic character of amino group). But formaldehyde does not react with the charged amino groups ($—NH_3^+$), thus, first the amino acid reacts with sodium hydroxide solution to give $NH_2CHRCOO^-$, which condenses with formaldehyde to give a stable anion.

$$^-OH + H_3NCHRCOO^- \rightleftharpoons H_2NCHRCOO^- + H_2O$$

$$CH_2O + H_2NCHRCOO^- \longrightarrow CH_2{=}N\ CHRCOO^- + H_2O$$
$$+ CH_2OHNHCHRCOO^-$$
$$+ (CH_2OH)_2N\ CH_2COO^-$$

Further, formaldehyde solutions contain formic acid and also amino acids are not exactly neutral species, thus it is necessary that both the formaldehyde and amino acid solution should have the same pH before mixing and for this purpose each solution is first made just alkaline to phenolphthalein by means of dilute sodium hydroxide solution.

Reagents

Neutral formalin solution. It is prepared by neutralization of formalin solution with alkali to phenolphthalein end point.

Standard glycine solution. An accurately weighed (0.8 g) glycine is dissolved in distilled water in a measuring flask (100 ml).

Unknown glycine solution
Sodium hydroxide solution (0.1 N)
Phenolphthalein indicator

Procedure

A standard glycine solution (20 ml) is pipetted out into a conical flask and phenolphthalein (2–3 drops) is added. It is titrated with sodium hydroxide solution until a faint pink colour appears. Then the neutralized formaldehyde solution (10 ml, excess) is added. The pink colour of the solution in the titration flask disappears immediately and the solution becomes acidic. Then the contents of the flasks are titrated with sodium hydroxide solution till the pink colour is obtained. The titration is repeated to get three concordant readings. The same experiment is repeated for the unknown glycine solution.

Observations

Weight of glycine dissolved in 100 ml = w g

Volume of sodium hydroxide solution used with standard glycine solution = v_1 ml

Volume of sodium hydroxide solution used with the unknown glycine solution = v_2 ml

Volume of glycine solution taken = 20 ml

Calculations

$$v_1 \text{ ml of NaOH} \equiv 20 \text{ ml of standard glycine solution} \equiv \frac{w}{5} \text{g of glycine}$$

$$v_2 \text{ ml of NaOH} \equiv 20 \text{ ml of unknown glycine solution} \equiv \frac{w}{5} \times \frac{v_2}{v_1} \text{ g of glycine}$$

$$\text{Thus strength} = \frac{w}{5} \times \frac{v_2}{v_1} \times \frac{1000}{20} \text{ g/l}$$

$$= \frac{w \times v_2 \times 10}{v_1} \text{ g/l}$$

6.2 Percentage Purity of Amino Acid

Procedure

A solution of the given amino acid is prepared by dissolving amino acid (0.8 g) in distilled water in a 100 ml measuring flask and the titration is performed in the same fashion as in the previous experiment.

In this case a blank titration is also performed, i.e., without taking formaldehyde solution.

Observations

Strength of the prepared amino acid solution = w g/l

Volume of the standard sodium hydroxide solution used in titration = v_1 ml

Volume of standard sodium hydroxide solution used in blank titration = v_2 ml

Normality of sodium hydroxide solution = Nx

Molecular weight of amino acid = Mg

Calculations

$$1000 \text{ ml of 1N NaOH solution} \equiv 1000 \text{ ml of 1 N amino acid solution} \equiv \text{M g of amino acid}$$

$$(v_1 - v_2) \text{ ml of } Nx \text{ NaOH solution} \equiv \frac{\text{M} \times (v_1 - v_2) \times Nx}{1000} \text{ g of amino acid}$$

Now, 20 ml of amino acid solution contains $\frac{\text{M} \times (v_1 - v_2) \times Nx}{1000}$ g of amino acid

Then 1000 ml of amino acid solution will contain

$$\frac{\text{M} \times (v_1 - v_2) \times Nx}{1000} \times \frac{1000}{20} \text{ g of amino acid}$$

$$\text{Thus \% purity} = \frac{M \times (v_1 - v_2) \times Nx}{20} \times \frac{100}{w}$$

7. ESTIMATION OF ASCORBIC ACID (VITAMIN C)

Vitamin C is L-ascorbic acid. It can exist either in the oxidized or reduced form, both of which have vitamin activity.

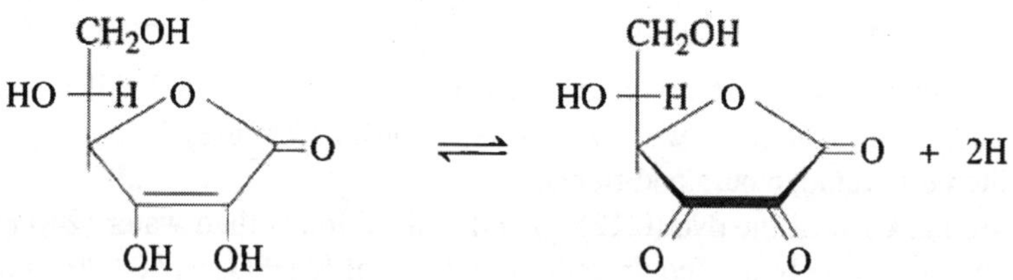

L-Ascorbic acid   Dehydroascorbic acid

Vitamin C is needed to maintain the normal health and function of intracellular materials, bone cartilage and dentine. It is probably a cofactor for the synthesis of collagen. Lack of vitamin C causes scurvey. The vitamin C is widely distributed, but occurs in high concentration in green plants and citrus fruits.

The daily intake of ascorbic acid is helpful in the prevention and treatment of common cold, prevention of heart attack and certain types of cancer, such as bladder cancer.

7.1 Determination of Ascorbic Acid Concentration in Fruit Juice

Principle

Ascorbic acid is oxidized by the coloured dye 2, 6-dichlorophenolindophenol to dehydro ascorbic acid. At the same time, the dye is reduced to a colourless compound so that the end point of the reaction can be easily determined.

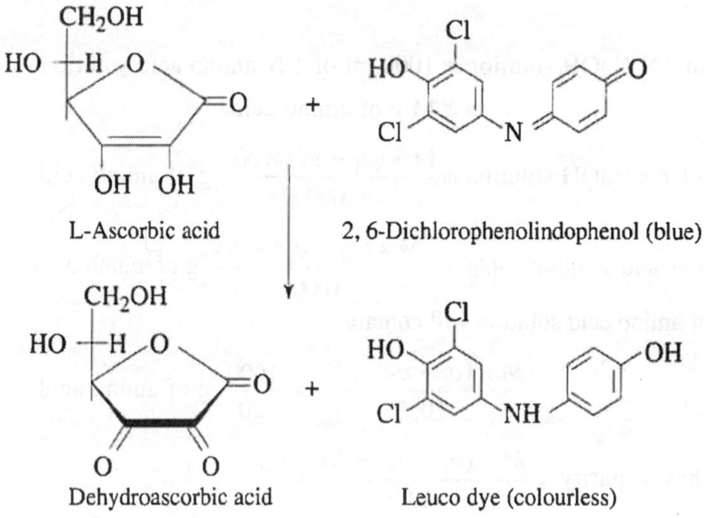

The reaction is carried out in an acidic medium as the dye is decolorized by other compounds, and also in an acid solution the interfering substances react only slowly.

Thus, a known volume of juice solution and standard ascorbic acid solution are titrated separately with a solution of 2, 6-dichlorophenolindophenol in an acid medium until a faint pink colour (end point) appears. From the volumes of dye solution consumed, the concentration of ascorbic acid in the juice solution is calculated.

Reagents

Ascorbic acid standard solution. Pure ascorbic acid (10 mg) is dissolved in distilled water in a 100 ml measuring flask and filled to the mark with water (prepare just before use).

Fruit juice. Filtered to remove pulp and fibres.

Dye solution. Sodium salt of the dye (0.125 g) is dissolved in distilled water (240 ml) in a 250 ml measuring flask and phosphate buffer (equivalent to 0.0165 g of K_2HPO_4 and 0.0202 g of KH_2PO_4/250 ml of the solution) is added and filled to the mark.

Glacial acetic acid

Chloroform

Procedure

Diluted fruit juice (5 ml) (1 ml should contain nearly 0.1 mg of ascorbic acid) is pipetted out in a conical flask (100 ml). Glacial acetic acid (1 ml) and 1 ml of chloroform (in case the solution is coloured) is added. The contents are titrated with the dye solution to a faint permanent pink colour. In case of coloured solution, the end point in the chloroform layer is observed. The titration is repeated to get concordant readings.

Similarly, titration with the standard ascorbic acid solution and with distilled water (blank) is also carried out.

Observations

Weight of ascorbic acid dissolved in 100 ml of solution = w g

Volume of dye solution consumed with fruit juice = v_1 ml

Volume of dye solution consumed with standard ascorbic acid solution = v_2 ml

Volume of dye solution consumed in the blank = v_3 ml

Calculations

$$\text{Concentration of ascorbic acid in the fruit juice} = \frac{v_1 - v_3}{v_2 - v_3} \times w \times \text{dilution of fruit juice}$$

(If 50 ml of fruit juice is diluted to 100 ml, its dilution will be 2).

8. ESTIMATION OF ACTIVE HYDROGEN-ZEREVITINOFF'S METHOD

Principle

The organic compounds containing active hydrogen, i.e., easily replaceable hydrogen, react with grignard reagents, viz., methyl magnesium iodide and give methane quantitatively. Every reactive hydrogen atom gives one molecule of methane

$$RH + CH_3MgI \longrightarrow RMgI + CH_4$$

This method of determination of active hydrogen is known as Zerevitinoff's method. The reaction is performed in an inert atmosphere and in an anhydrous organic solvent. The volume of liberated methane is measured with the help of a gas burette.

Thus, compounds like carboxylic acid, phenols, alcohols, thiols, amines, sulphonic acid, acetylenes, amides and sulphonamides are estimated by this method.

The solvents used in the reaction may include anhydrous anisole, anethole, di-*n*-amyl ether, di-iso-amyl ether, di-*n*-butyl ether, xylene or pyridine.

Reagents

The given sample
Magnesium turnings
Methyl iodide
Dry anisole

Procedure

Preparation of grignard reagent (methyl magnesium iodide). Dried magnesium turnings (2.4 g) are placed in a 250 ml round bottom flask fitted with a condenser and dropping funnel. A solution of methyl iodide (9.8 ml) and anisole (60 ml) is placed in a dropping funnel and this solution (15 ml) is added to the magnesium turnings. If the reaction does not start after a few minutes, a small crystal of iodine is added and the mixture is warmed on a water bath until the reaction commences. The remaining methyl iodide solution is added during 30–35 minutes and heated on a boiling water bath for one hour. The unreacted methyl iodide is distilled off and the solution is decanted through a funnel with a loose glass wool plug into a bottle from which air has been displaced by oxygen-free nitrogen.

Reaction of grignard reagent with the given sample. An accurately weighed (0.1–0.2 g) given sample is placed in a 100 ml conical flask and dissolved in dry anisole (15–20 ml) by warming gently. A small tube containing methyl magnesium iodine (10–15 ml) is placed inside the same conical flask by taking care that the two reactants do not react with each other (fig. 6). The flask is stoppered and connected through a gas burette and a levelling tube. The reaction flask is placed in a beaker of water at room temperature. After 15–20 minutes, when the system has attained a constant temperature, the pressure is released by opening the tap *S* and the levels of mercury (or di-*n*-butyl phthalate) are adjusted in the gas burette and levelling tube. The reading of the burette is noted. Then the beaker of water is removed and the flask is tilted so that the two reactants get mixed with each other. The contents of the flask are shaken gently until gas evolution ceases. To ensure the completion of the reaction the flask is heated at 90°C in a hot water bath for 15 minutes. Finally, the hot water bath is replaced with a water bath at room temperature and allowed to stand for 10–15 minutes. The final burette reading is recorded. A blank experiment is also performed with anisole and grignard reagent.

Observations

Weight of the given sample taken = w g

Initial burette reading with sample = x_1

Final burette reading with sample = x_2

Initial burette reading with blank = y_1

Final burette reading with blank = y_2

Temperature of water bath = t°C

Atmospheric pressure = P

Vapour pressure of mercury at t°C = p

Molecular weight of given sample = M

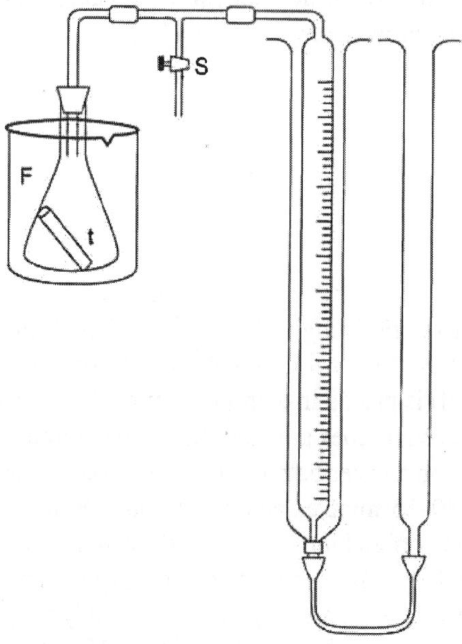

Fig. 6 Apparatus for the estimation of active hydrogen.

Calculations

$$\text{Volume of gas evolved with sample} = x = x_2 - x_1$$
$$\text{Volume of gas evolved with blank} = y = y_2 - y_1$$
$$\text{Volume of methane gas produced} = x\text{–}y \text{ ml}$$
$$= V_{obs}$$

We know that 1 mole of methane or 22.415 litres at N.T.P. is evolved from 1 mole of active hydrogen.

Now by using $\frac{P_1 v_1}{T_1} = \frac{P_2 v_2}{T_2}$ relationship we can calculate the volume of gas evolved.

$$\frac{760 \times 22415}{273} = \frac{(P - p) \times V_2}{(273 + t)}$$

$$V_2 = \frac{760 \times 22415}{(P - p)} \times \frac{(273 + t)}{273}$$

Now for M g of sample the volume of gas liberated is

$$\frac{760 \times 22415}{(P - p)} \times \frac{(273 + t)}{273}$$

Then for w g of sample liberated gas should be

$$\frac{760 \times 22415}{(P - p)} \times \frac{(273 + t)}{273} \times \frac{w}{M} \text{ ml} = V_{cal}$$

$$\text{The number of active hydrogen} = \frac{V_{obs}}{V_{cal}} = \frac{(x - y)}{V_{cal}}$$

9. MOLECULAR WEIGHT DETERMINATION

9.1 Rast's method

Principle

The depression in melting (or freezing) point of pure substance is observed when an impurity is mixed. The lowering in melting point can be expressed in quantitative form.

Thus a known weight of the given compound is mixed with a known weight of comphor. A homogeneous solution is obtained by heating and then the melting point of the melt (solid) is determined. The molecular weight is then determined by the following relationship:

$$M = \frac{K \times w_1 \times 1000}{\Delta t \times w_2}$$

where K = molecular depression constant

w_1 = weight of the compound

w_2 = weight of the comphor

Δt = lowering in melting point

Reagents

The given compound
Camphor
Naphthalene

Procedure

An accurately weighed amount of the given sample (~ 0.2 g) is placed in a test tube and a weighed quantity of powdered camphor (2 g) is added. The tube is corked loosely and heated gently on a flame to obtain a clear homogeneous solution. The tube is shaken and allowed to cool. A small portion of the solidified mixture is taken, powdered and its melting point determined. The determination of melting point is repeated to get consistent results. The melting point of pure camphor is also determined so that the depression in the melting point can be calculated.

The same experiment is repeated with a known compound, i.e., naphthalene and the value of K calculated for camphor.

Observations

Weight of the sample taken = w_1 g

Weight of camphor taken = w_2 g

Melting point of pure camphor = t°C (176°C)

Melting point of the mixture (camphor + sample) = t°_m C

Weight of naphthalene taken = w'_1

Weight of camphor taken = w'_2

Melting point of the mixture (camphor + naphthalene) = t'_m°C

Molecular weight of naphthalene = 128 g = M

Calculation

$\Delta t'$= Lowering in melting point of camphor on mixing naphthalene = $(176° - t'_m)$°C

$$\text{K for camphor} = \frac{M \times \Delta t' \times w_2'}{w_1' \times 1000} = \frac{128 \times w_2' \,(176 - t_m')}{w_1' \times 1000}$$

$$\text{Molecular weight of the sample} = \frac{K \times w_1 \times 1000}{\Delta t \times w_2}$$

10 QUESTIONS

1. Why is it necessary to place a silver gauze in the combustion tube while estimating carbon and hydrogen in compounds containing halogens?
2. How can you avoid a flash-back during the estimation of carbon and hydrogen elements in an organic compound?
3. Why is the Kjeldahl method of the estimation of nitrogen not suitable in the cases of compounds having nitrogen covalently linked to an oxygen or another nitrogen atom? How is this difficulty overcome?
4. In the estimation of phenol and aniline, potassium bromate-bromide reagent is used and not other brominating reagent like bromine in acetic acid. Why?
5. Give the method for the estimation of the methoxyl group.
6. Which oxidising agent is used in the estimation of C-methyl group?
7. Discuss the role of pyridine in the estimation of carbonyl group.
8. What is saponification equivalent?
9. Discuss the estimation of urea sample?
10. In which form are the nitro groups estimated?
11. In the estimation of sugars which reagent is better—Fehling or Benedict? Why?
12. What is iodine number?
13. Discuss the role of formaldehyde in amino acid estimation.
14. How will you estimate the ascorbic acid concentration in fruit juice?

Section C: Chromatography

Chlorosulphonic acid (25 ml) is taken in a three necked round bottom flask, fitted with a mechanical stirrer with a mercury seal (a magnetic stirrer can also be used) and a dropping funnel. The flask is cooled by the use of an ice bath and dry acetanilide (10 g, 0.074 mole) is added through the third neck (removing the cork when necessary) in small lots while the mixture is kept stirred. The top of the condenser is connected to a gas trap to absorb the liberated hydrogen chloride. After the addition is over the flask is warmed on a water bath at 60°C to complete the reaction (10–15 minutes). The reaction mixture is poured into ice water (300 ml). The separated product is filtered, washed with cold water and pressed to drain off excess water. The crude product ([illegible] g, [illegible]%) is immediately used as such for the next stage. However, if required the crude product can be crystallised from chloroform or benzene-acetone mixture.

Note: Acetanilide is obtained as described on page 6[illegible].

Step 2 p-Acetamidobenzenesulphonamide

p-Acetamidobenzene sulphonyl chloride — p-Acetamidobenzene sulphonamide

The crude p-acetamidobenzenesulphonyl chloride (from step 1) is taken in a conical flask (250 ml). Ammonia solution (35 ml concentrated ammonia solution ([illegible]) and 35 ml water) is added and the contents mixed thoroughly. The mixture is heated (with occasional shaking) to 80–90°C (boiling water bath) for 15 minutes. The formed p-acetamidobenzene sulphonamide is cooled [illegible] with [illegible] sulphuric acid [illegible] next step through [illegible] quantity can be crystallised from water. The [illegible]

Step 3 p-Aminobenzene sulphonamide (sulphanilamide)

p-Acetamidobenzene sulphonamide — p-Aminobenzene sulphonamide (sulphanilamide)

The [illegible] (obtained above in step 2) is refluxed gently for [illegible] with dilute hydrochloric acid ([illegible] ml concentrated hydrochloric acid and [illegible] ml water). The solution is cooled (ice water). It is to be ascertained that no solid product separates on cooling; in case it separates (indicating that the hydrolysis is incomplete) the mixture is further heated [illegible] to boiling and filtered through a fluted filter paper. To the [illegible]

Chromatography is an important technique employed to isolate, separate and identify the compounds or mixture of compounds even when they are present in small amounts.

This phenomenon was first introduced by Iswett (1906). He used it for the separation of coloured substances from plants (In Greek language chromon means colour). Since the scope of this technique has been widely extended to colourless mixtures, the idea of colour is no longer connected with the concept.

Principle of chromatography

Chromatography is a separation process which depends on the differential distribution of the components of a mixture between a mobile phase and a stationary phase.

The stationary phase is usually in the form of a thin film and it may be either liquid or solid. The moving phase is either liquid or gas.

When the moving phase is a solid, the basis of separation is adsorption, i.e., distribution of the components of a mixture depends on the selective adsorption on the surface of the solid. When the stationary phase is liquid, the basis is partition, i.e., partitioning of components between two liquid phases.

Depending upon the various combinations of the moving phase and stationary phase there are different techniques of chromatography and these are:

(i) Liquid-solid chromatography (thin layer chromatography and column chromatography)
(ii) Gas-liquid chromatography.
(iii) Liquid-liquid chromatography (paper chromatography).

1. COLUMN CHROMATOGRAPHY

This technique of chromatography is based on the difference in the adsorbing power of the components on an adsorbent. It is a solid-liquid partitioning technique. The stationary phase is a solid, which separates the components of a liquid passing through it by selective adsorption on its surface. These type of interaction, causing adsorption, are the same as those that cause attractions between any molecules, i.e., electrostatic attraction, complexation, hydrogen bonding, Van der Waals forces etc.

Procedure

A glass tube with a stopcock is used as a column (Fig. 7). The adsorbent is packed into the column as a slurry. A slurry is a mixture of a solvent and an undissolved solid and is prepared by adding the solid adsorbent (a little at a time) to a quantity of the solvent. Enough adsorbent is added to the solvent with swirling to form a thick but flowing slurry. The slurry should be homogeneous and free of entrapped air bubbles. Then the column is half filled with solvent and the stopcock is opened to allow the solvent to drain slowly into a beaker and the slurry is poured in portions onto the top of the draining column. The column is tapped constantly and gently on the side. The tapping promotes even settling and gives an evenly packed column free of air bubbles. Tapping is continued until all the material has settled to give a well defined level at the top of the column.

The solvent used to pack the column is normally the least polar solvent. The compounds chromatographed should not be infinitely soluble in the solvent.

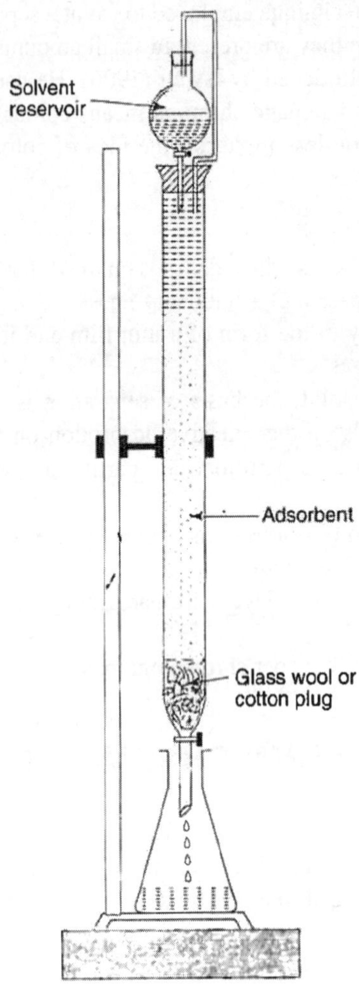

Fig. 7 Adsorption column chromatography apparatus

The solvent level is lowered to the top of the adsorbent column by draining. The liquid mixture or dissolved mixture or slurry of mixture with adsorbent is added to form a small layer on the top of the adsorbent. Then the solvent is added carefully so as to not disturb the surface. As the solvent passes through the top layer, it dissolves the mixture adsorbed in this layer, and the solution so obtained flows down. As the solution run through the adsorbent, the components of the mixture are subjected to a competition between the two phases. The stationary phase (adsorbent) tries to adsorb the components on its surface while the moving phase (solvent or eluent) tries to take away the components along with it downwards.

Different constituents of the mixture give different response to these competing phases. The result is that the different components are adsorbed at different heights. The layers of adsorbent with different constituents adsorbed in them are called *zones* or *bands*. Zones generally possess different colours in case of mixtures of coloured pigments. The most strongly adsorbed constituent is present in the band near the top while the least readily adsorbed constituent is found in the band near the bottom.

These bands or chromatograms are then separated and the individual components are recovered. This process is known as elution. There are two ways of elution:

(i) *Physically*: The solvent is drained off, the bands are physically extruded and are cut by a spatula and the adsorbed component is recovered by means of a suitable solvent.

(ii) *By flow of solvent*: In this method, the flow of the solvent is continued till the bands are washed out one by one and the components are recovered by distillation of the solvent. The elution normally begins with a non-polar solvent like hexane or petroleum ether and the polarity of the elution solvent can be increased gradually by adding successively greater percentages of either ether or benzene or some other solvent of greater solvent polarity. The transition from one solvent to another solvent should not be too rapid in solvent changes.

Elution is successful when the components of the mixture are coloured. But when the components are not coloured, special devices are to be adopted to locate the bands on the column. Some of these are listed below.

(i) When components are colourless, small fractions of the eluent are collected in different flasks and analysed by thin layer chromatography. The fractions which give the same spot in thin layer chromatography are mixed, concentrated and the component is crystallised out.

(ii) Inspection of colourless chromatogram under U.V. light: This was particularly applied by Winterstein (1934) and Karrer independently. They observed that certain compounds give fluorescence in U.V. light and hence can be separated, e.g., carbazole gives deep blue fluorescence, anthracene gives blue violet fluorescence and naphthalene gives yellow colour (non fluorescent).

(iii) Formation of coloured areas by washing with an indicator or other substances which react with colourless bands to form coloured compounds, e.g., in the separation of amines by this method, the aqueous solution of anthranilic acid containing sodium nitrite and hydrochloric acid is sprayed on the chromatogram. The formation of the coloured dye indicates the presence and location of the individual amines.

(iv) Preparation of coloured derivatives of the colourless compounds, e.g., dinitrophenylhydrazones of carbonyl compounds give coloured zones.

Adsorbent

A few of the solid adsorbents commonly used include alumina, silica gel, florisil, charcoal, magnesia, calcium carbonate, starch and sugar. Alumina (Al_2O_3) is a highly active, strongly adsorbing polar compound and is available in three forms: neutral, basic and acidic. Basic and acidic alumina offer good separating power for acids and bases respectively. For compounds that are sensitive to chemical reactions under acidic or basic conditions, neutral alumina should be used. Being highly polar, alumina adsorbs polar compounds quite tenaciously, so that they may be difficult to elute from the column. The activity of alumina may be reduced by addition of small amounts of water. Silica gel and florisil are also polar but less than alumina.

Solvent

The nature of the liquid phase (solvent) to be used is an important factor. The solvent may also be adsorbed on the solid, thereby competing with the solute for the adsorptive sites on the surface. If the solvent is more polar and more strongly adsorbed than the components of the mixture, these components will remain almost entirely in the mobile liquid phase and little separation will occur during the experiment. For effective separation, the eluting solvent must be significantly less polar than the components of the mixture. Further, the components must be soluble in the solvent, for if they are not they will remain permanently adsorbed on the stationary phase of the column. The eluting powers of various solvents, i.e., their ability to move a given component down a column is given below:

Solvent	
Hexane	↓
Carbon tetrachloride	
Toluene	
benzene	
Dichloromethane	
chloroform	Increasing
Ethyl ether	eluting
Ethyl acetate	power
Acetone	
Propanol	
Ethanol	
Methanol	↓
Water	

Exercise 1

Separation of a mixture of *o*-and *p*-nitro anilines. The mixture of *o*- and *p*-nitro anilines can be separated by using a column packed with alumina in benzene. Elution with benzene will give first *o*-nitro aniline and then *p*-nitro aniline.

Exercise 2

Separation of syn and anti azobenzene. The mixture of syn and anti azobenzene is separated by using a column packed with alumina in petroleum ether. On elution with petroleum ether anti-isomer will come first and then elution with 5 per cent methanol in petroleum ether will give syn isomer.

2. THIN LAYER CHROMATOGRAPHY

Thin layer chromatography (t.l.c) like column chromatography is a solid-liquid chromatography and is based on the principle of adsorption.

In this technique of chromatography, the moving liquid phase is allowed to ascend a thin layer of adsorbent coated onto a backing support. The most typical backing support is a glass plate. Thus, a 1:1 slurry of the adsorbent and methanol is commonly coated (0.25 mm) on glass plates (5 × 20 cm or 20 × 20 cm) and the adsorbent activated by drying the plates at 110°C for 30 minutes. The most common adsorbent in use are alumina G (aluminium oxide) and silica G (silica acid). The G designation stands for gypsum (calcium sulphate) which works as a binder to bind the adsorbent together and to the glass plates. The other adsorbing materials are similar to those used in column chromatography.

The mixture to be separated is dissolved in a suitable solvent and spotted on to the bottom of the t.l.c. plate, 1.0 – 1.5 cm from the end. A suitable dropping tube may be made by drawing out the middle of a melting point capillary tube with a burner and breaking the tube in the middle (Fig. 8). The dropper is filled by capillary action and is discharged when the liquid at the tip drops on to the untouched adsorbent surface. The spot should be 2 – 5 mm in diameter.

When the solvent around the spot has evaporated, the plate is placed vertically in a jar containing a small quantity of the solvent (moving phase) and lined with a piece of filter paper thoroughly moistened with the developing solvent. The solvent saturated lines help to keep the jar saturated with solvent vapours, thereby speeding up the development process. The level of the solvent in a jar is adjusted, so that the lower edge of the adsorbent layer is under the solvent but the spot should be above this level and the jar is then covered (Fig. 9). The solvent rises through the adsorbent layer and the components of the mixture ascend at different rates depending on their affinities for the adsorbent. When the solvent has nearly reached the top of the adsorbent layer, the plate is removed from the jar. The position of the solvent front is marked and the solvent is allowed to evaporate from the plate.

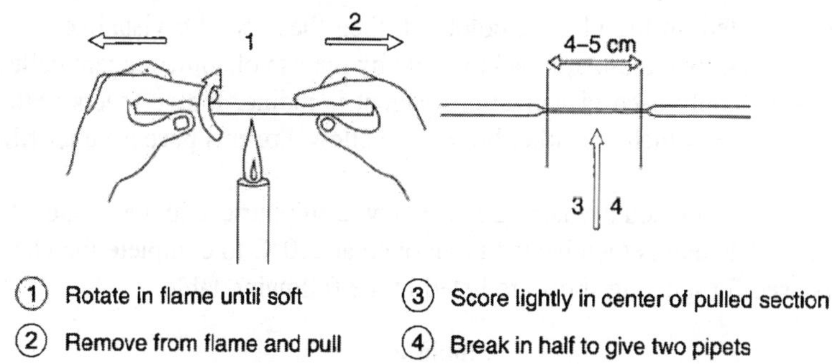

Fig. 8 Construction of two capillary micropipets

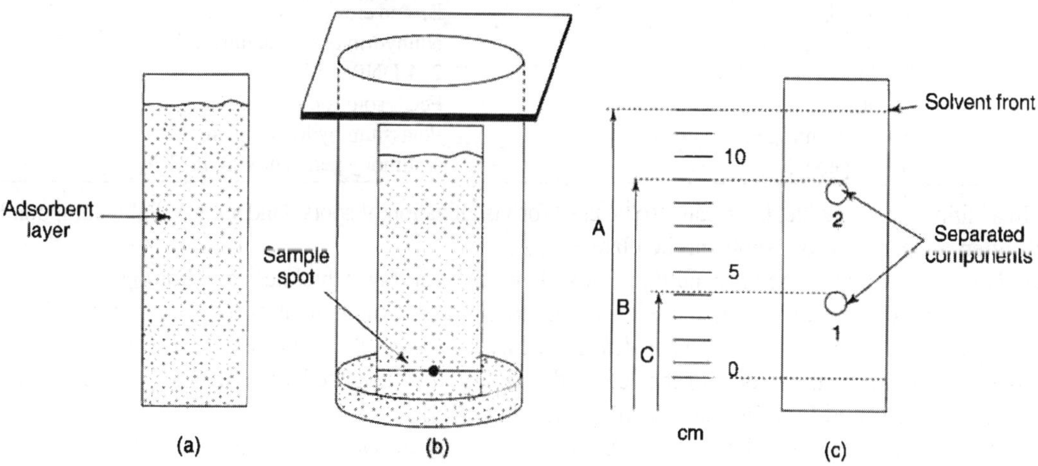

$$R_f = \frac{\text{distance travelled by substance}}{\text{distance travelled by solvent}}$$

$$R_f \text{ (compound 1)} = \frac{C}{A}$$

$$R_f \text{ (compound 2)} = \frac{B}{A}$$

Fig. 9 Thin layer chromatography (a) TLC plate (b) Development (c) developed chromatogram

The choice of solvent (moving phase) depends on the components to be separated. A solvent which causes all the spotted material to move with the solvent front is too polar. One which will not cause any of the material in the spot to move is not polar enough.

Solvents with their relative polarity are listed in a column Chromatography. Chloroform and benzene are solvents of intermediate polarity and are a good choice for a wide variety of functional groups to be separated. Hexane or petroleum ether with varying proportions of benzene or ether give solvent mixtures of moderate polarity which are useful for many common functional groups. For polar components ethyl acetate, acetone or methanol can be used.

Location of the spot

If the components separated on the t.l.c are coloured, then these can be visualized easily. But if the components are colourless, then these are made visible by the use of some reagent called visualization reagents. The most commonly used visualization reagent is iodine. Iodine reacts with many organic materials to form complexes which are either brown or yellow. For this purpose a jar filled with iodine vapours is used.

Another commonly used method is charring of spots with sulphuric acid, i.e., concentrated sulphuric acid is sprayed on the plate and is then heated in an oven at 110°C to complete the charring.

Certain reagents used for locating spots are listed in the following table.

Table 2

Substance	*Reagents*
1. Acids	Bromo cresol green
2. Amino acids	Ninhydrin or cupric nitrate
3. Aldehydes or ketones	2, 4-DNP
4. Phenols	$FeCl_3$ solution
5. Basic nitrogen or sulphur	Potassium hydrogen platinate
6. Terpenoids	Antimony pentachloride in CCl_4

In addition, an ultraviolet lamp can also be used for visualization of spots. Under a UV light, compounds will often appear as bright spots on the plate.

R_f value Under a given set of conditions (adsorbent, solvent, layer thickness and homogencity), R_f is the rate of movement of a compound with respect to the rate of movement of the solvent front. R_f is a property of that compound. The value is determined by measuring the distance travelled by a substance from a starting line to the middle of the spot and dividing, this distance by the distance the solvent has travelled, as measured from the same starting line [Fig. 9(c)].

A distinct advantage of t.l.c. is that very small quantity of sample is required. A lower limit of detection of 10^{-9} grams is possible in some cases.

Preparative t.l.c.

If larger plates (20 × 20 cm) are used, components may be separated and the separated components can be individually recovered from the plates. Plates used in their way are called preparative plates. For preparative plates, a thicker layer of adsorbent is generally used. A mixture is applied as a line about 1 cm from the bottom of the plate. As the plate is developed, the separated components form the bands and are outlined with pin. Then these bands are scraped from the plate and extracted with solvent to remove the adsorbend material. Filtration removes the adsorbent and evaporation of the solvent gives the components of the mixture.

Applications of t.l.c.

1. Establishing that two compounds are identical.
2. Determining the number of components in a mixture.
3. Determining the appropriate solvent for a column chromatographic separation.
4. Monitoring a column chromatographic separation.
5. Monitoring the progress of a reaction.

Exercise 1

Separation of mixture of methyl orange and methylene blue by t.l.c. A spot of the mixture of dyes is placed on the t.l.c plate made up of alumina and two more spots of pure methyl orange and methylene blue are also placed on the same t.l.c. The t.l.c. plate is developed in a jar containing bezene. After developing, the spots are visible as the components are coloured. R_f value of the individual spot is calculated.

Exercise 2

Separation of a mixture of 2, 4-dinitrophenylhydrazones by t.l.c. A spot of the mixture of 2, 4-dinitrophenylhydrazones of acetaldehyde, benzaldehyde and vanilline in chloroform is placed in the middle of the t.l.c. plate. The spots of pure hydrazones of individual aldehydes are also put on the same t.l.c. The t.l.c plate is developed in a jar containing benzene: petroleum ether (3:1) mixture. After developing, the spots are visible and the R_f value is calculated.

3. GAS CHROMATOGRAPHY

In gas chromatography, the compounds are separated by a partitioning process between a moving gas phase (inert gas such as nitrogen or helium) and a stationary liquid phase. Gas chromatography is also known as vapour phase chromatography (vpc) and as gas-liquid partition chromatography (glpc).

The apparatus used to carry out a gas-liquid chromatographic separation is generally called a gas chromatograph (Fig. 10). The sample to be separated is injected into the chromatograph and is vapourized immediately. The vapourized sample is then swept into the column by the carrier gas. The column is packed with a solid, finely divided substance, the surface of which is coated with a liquid (stationary phase) of relatively low volatility. The column is contained within an oven, the temperature within the oven is controlled by a thermostat and heating elements. As the sample passes through the column, its components separate into individual bands in the carrier gas, which then pass through the detector. The detector produces an electronic signal, the voltage of which is proportional to the amount of material present in the gas stream. The recorder plots this voltage as a function of time to give the gas chromatogram.

The time required for a given component to pass from the injection part to the detector is known as retention time. The different factors, which affect the retention time of a compound are (1) the nature of the stationary liquid phase, (2) the length of the column, (3) the temperature of the column and (4) the rate of flow of inert gas. The retention time is independent of the presence of other components in the mixture. This value when obtained under controlled conditions, may be used to identify a compound by a direct comparison of it to the values for known compounds determined under the same conditions.

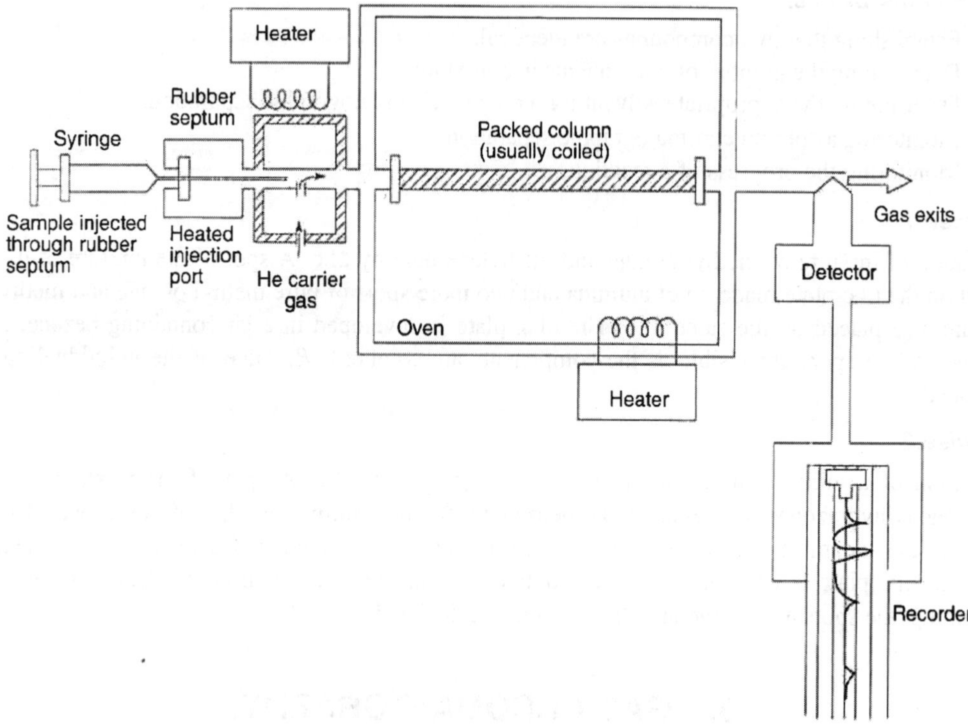

Fig. 10 Schematic diagram of a gas chromatograph

The column used in gas chromatography is usually made of copper or stainless steel tubing and sometimes glass is also used. The tubing is packed with the stationary phase. The most common means of supporting the liquid phase within the column is a thin film coating on an inert solid support. The support should be of small, evenly meshed granules, so that a large surface area is available. This provides a large film area in contact with the vapour phase, which is necessary for efficient separation. The liquid is coated on the solid support by dissolving the liquid in a suitable low boiling solvent and mixing this solution with the solid. The low boiling solvent is then evaporated (rotary evaporator), leaving the solid granules evenly coated. The common solid supports in use are chromosorb "P", chromosorb "W" and crushed firebrick chromosorb "T". An alternative method of supporting the liquid is by making use of capillary columns. These are very long (~ 300 ft) and very small diameter (0.1–0.2 mm) columns. In these columns, the liquid is coated directly on the inner walls of the tubing. These type of columns are highly efficient but are expensive.

The efficiency of the column increases with increase in path length (the separation between bands increases and difference in retention time will be large) and decreases with increase in diameter of the column (a small diameter column will give rise to narrower bands and thus greater efficiency).

The best choice for the liquid phase is the liquid giving rise to the largest differences in the partition coefficients of the components to be separated, which in turn depends on the solubility of components in liquid phase. The solubility of a gas in a liquid depends on its vapour pressure. In general, lower boiling and highly volatile components will move through the column faster than those of lower volatility because the higher the vapour pressure of a gas, the lower its solubility in a liquid. The solubility of a component in a liquid also depends on the polar interactions, e.g., hydrogen bonding and other electrostatic interactions. For polar samples, it is best to use a polar liquid phase and for non-polar samples, a non-polar liquid phase is used. Different liquid phases in use are squalane (hydrocarbon grease, non-polar), apiezon-L (hydrocarbon grease, non-polar), SE–30 (silicon gum rubber, non-polar), carbowax 20 M (hydrocarbon wax, polar), diethyleneglycol succinate (DEGS, polyester, polar), DC–550 (silicon oil, intermediate polarity) and QF–1 [silicone (fluoro), intermediate polarity].

The temperature at which the column is maintained, is an important factor in gas chromatography. Retention time decreases by using high temperature. This is because the solubility of gases in liquids decreases with increasing temperature, due to which band separation will decrease, resulting in poor resolution. For each liquid phase there is a maximum temperature limit, which is determined by the stability and volatility of the liquid being used. The temperature should not be too high to cause volatilization of liquid phase. Further, there is a minimum temperature limit which is governed by the melting point and viscosity of the liquid. The liquid should not be quite viscous otherwise, it will be inefficient in dissolving the components of the gaseous mixture.

Another affecting factor is the rate of flow of carrier gas. The carrier gas must not pass over the liquid phase so rapidly that molecules dissolved in the gas phase cannot equilibrate with the liquid coated particles.

The gas chromatography experiment may be used both as qualitative or quantitative analysis of the sample. As under a given set of conditions, the retention time of a given compound is a property of that compound and may be used to identify it. Further, if the voltage output of the detector is proportional to the mole fraction of the components being detected in the vapour, the relative areas under two peaks of the chromatogram will equal their percentage ratio in the mixture. (Figs 11 and 12).

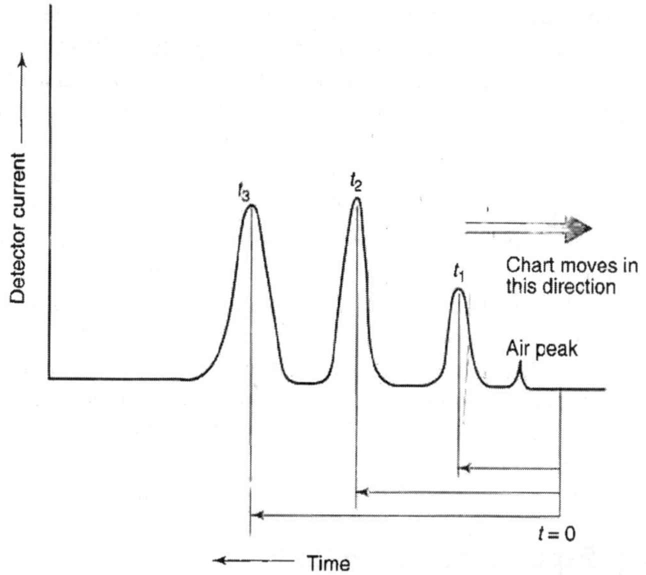

Fig. 11 Gas chromatogram

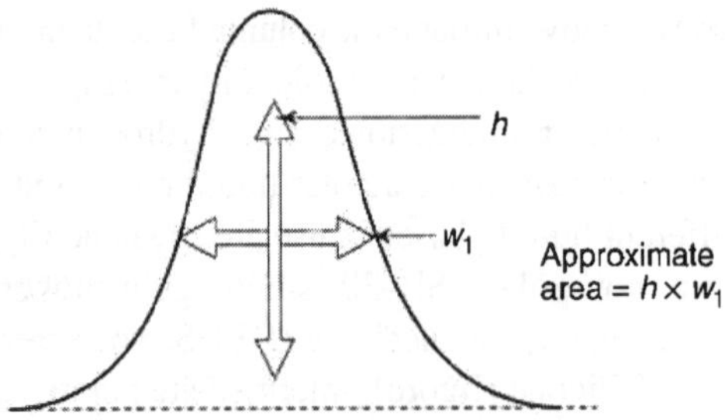

Fig. 12 Triangulation of a peak

4. PAPER CHROMATOGRAPHY

This technique of chromatography is based on the principle of partition. Here, the stationary phase is a filter paper. Filter paper is made up of cellulose which is a polyhydroxy compound and it has water adsorbed in it. It is not the filter paper but the water adsorbed in it which acts as a stationary phase. Thus there is a partition of components between the stationary liquid phase and moving liquid phase.

There are three types of paper chromatography: (i) Ascending paper chromatography; (ii) descending paper chromatography; and (iii) radial chromatography.

4.1 Ascending Paper Chromatography

In this chromatography a strip of filter paper is used and moving phase (solvent) is taken in a jar. The filter paper used should be of a high quality. A drop of a solution of the mixture to be separated is placed near the bottom of the paper strip and allowed to evaporate in the air. The strip is then placed in the closed cylinder or jar with precaution that the bottom of the strip is just immersed in the solvent and the spot of mixture remains, above the surface of the solvent [Fig. 13(a)]. The solvent will rise steadily up the strip via capillary action of the paper, and during this process, the mixture of components also moves along with the moving phase and is continuously extracted by the retained water molecules in the paper. A highly hydrophobic (water repellent) component will move up closely behind the solvent front whereas a highly hydrophilic solute will remain at the original point where the drop of a solution of mixture is placed. In an intermediate case, the extent to which a particular component has moved relative to the extent to which the solvent has moved is known as R_f value of the component and is a function of the partition coefficient of the component between the solvent and film of retained water.

When the solvent front has reached a 3/4th height of the strip, it is removed and the position of the solvent front is marked and dried. The position of the various components, if they are coloured, appear as clear separate spots. In case, the solution contains colourless compounds, the chromatogram is made visible by spraying the paper strip with a suitable reagent which will form coloured compounds with the component on the filter paper, e.g., amino acids and sugars are colourless. The spraying reagent for amino acid is ninhydrin which forms violet colour spot and for sugars is aniline hydrogen phthalate which gives brown coloured spots.

A chromatogram obtained in this way is known as one dimensional chromatogram.

4.2 Descending Paper Chromatography

In this type, the solvent reservoir is placed at the top and the paper strip is bent over so as to dip into the solvent [Fig. 12(b)].

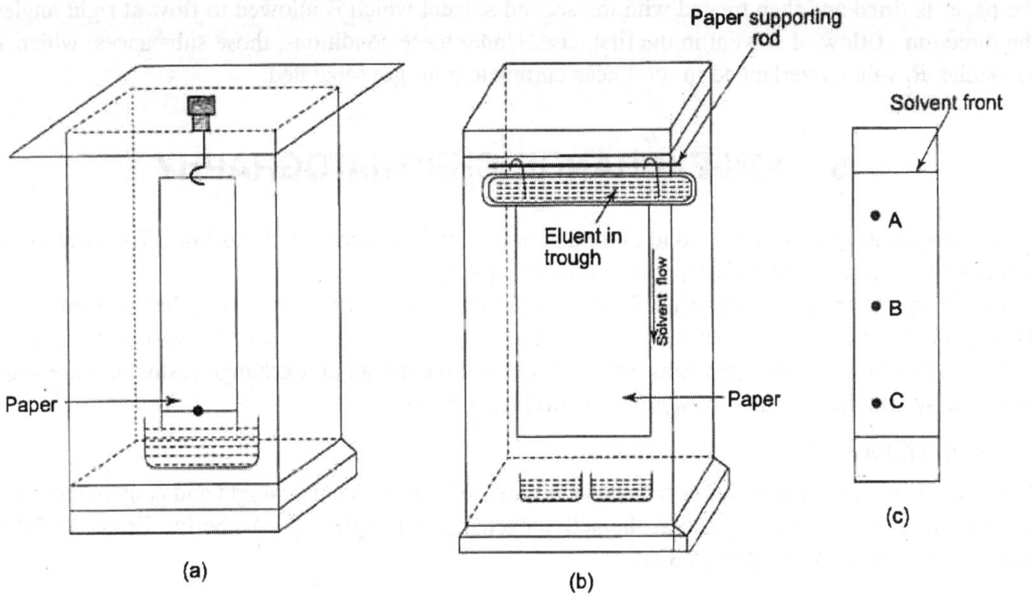

Fig. 13 Paper chromatography apparatus: (a) Ascending, (b) Descending. (c) Developed chromatogram shows resolved components A, B, C

4.3 Radial Chromatography

In this type of paper chromatography, a circular paper is used instead of a paper strip. The moving phase, i.e., solvent is taken in a petridish. A wick or flap is made in the filter paper. A drop of mixture of components to be separated is put in the form of a solution in the centre of the filter paper near the wick and dried. Then the wick or flap is dipped in the solvent and the solvent rises due to capillary action. The petridish is covered with a lid to avoid evaporation of solvent. In this case, different circular bands are obtained.

4.4 Representation of Adsorbing Power

The adsorbing power of compounds is represented by R_f, i.e., ratio of fronts or retenation factor.

$$R_f = \frac{\text{Distance travelled by the compound}}{\text{Distance travelled by the solvent}}$$

$$R_f < 1$$

$$\text{Standard retentation factor } R_{st} = \frac{R_f \text{ of the compound}}{R_f \text{ of the standard compound}}$$

If the substances have same R_f value under same circustances (one set of conditions), the separation is not possible. In such cases, two-dimensional chromatography offers a better means of separation than the one dimensional method. In this method, a drop of solution is placed near one corner of a large square filter paper strip, allowed to dry and then immersed in the solvent. After running one chromatogram, the paper is dried and then treated with the second solvent which is allowed to flow at right angles to the direction of flow of solvent in the first case. Under these conditions, those substances, which, due to similar R_f value, overlapped in the linear chromatogram get separated.

5. ION-EXCHANGE CHROMATOGRAPHY

This is a special type of liquid-solid chromatography and applicable only to ions. The separation of components depends on their attraction to an ion-exchange resin.

In this type of chromatography, a column is packed with the slurry of the ion exchange resin in water. Then a mixture of ions is introduced and the column is run as in case of column chromatography.

Two types of ion exchange resins are used, i.e., cation and anion exchange resins and are usually prepared by copolymerization of styrene and divinyl benzene.

Anion exchange resins

These are strong insoluble bases or salts of such bases, of high molecular weight and contain quarternary ammonium group or amino group as the active exchange principle, e.g., Amberlite IRA-400, Duolite A-42, Amberlite IRA-45, Dowex 3 etc.

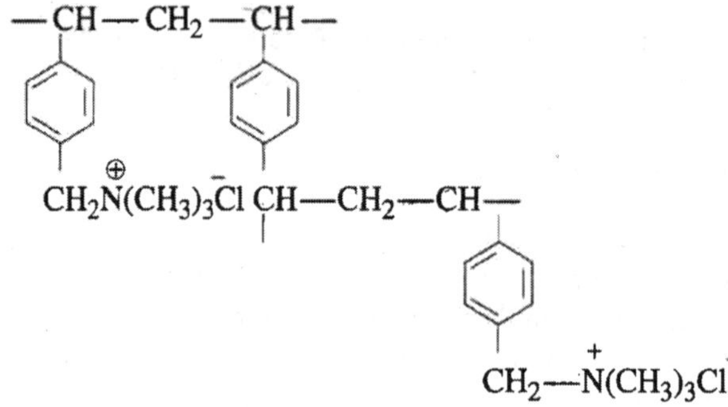

Anion exchange resin

Cation exchange resins

These are insoluble acids of extremely high molecular weight and contain sulphonic acid group or carboxylic acid group as the active exchange principle, e.g., Amberlite IR-112, Dowex 50, Duolite C-25, Amberlite IR-65, Duolite C-225 etc.

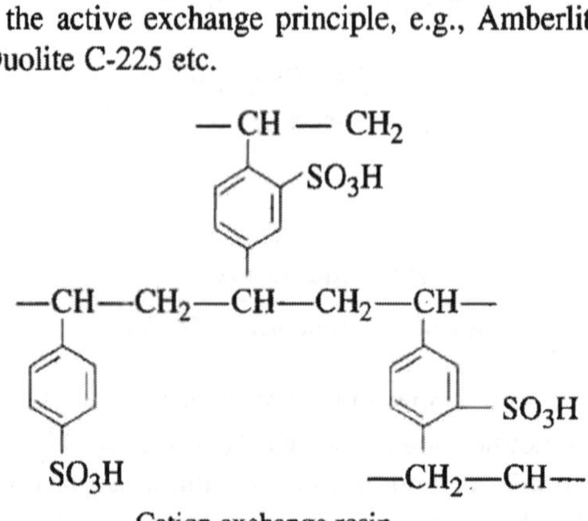

Cation exchange resin

By ion exchange chromatography, a mixture of amino acids, obtained by the hydrolysis of protein with dilute hydrochloric acid, may be separated. The amino acids in the mixture are present as their hydrochloride.

$$R—CH—\overset{\oplus}{N}H_3\ COOH\}\ Cl^-$$

If this solution is passed through a column containing a cation exchange resin, then exchange occurs, whereby the amino acids remain attached to the resin. A suitable resin for this purpose is one that contains benzene sulphonic acid residue incorporated into macromolecules of the resin. This resin may be represented as Res-SO_3H. When the amino acid solution is passed through the column, the following exchange occurs.

$$Res—SO_3^-\}\ H^+ + RCH\overset{\oplus}{N}H_3\ CO_2H\}\ Cl^-$$

$$\rightleftharpoons$$

$$Res—SO_3^-\}\ RCH\overset{\oplus}{N}H_3\ CO_2H + HCl$$

Thus the amino acids are held on the column. Elution of the column with buffer solutions of different pH values causes removal of amino acids. The more weakly basic the amino acid is, the more readily it will be removed and by increasing the pH of the eluant, the more basic amino acid will be removed. Hence the separation is effected by stepwise elution.

Since amino acids contain carboxylic groups also, they may be separated by passing a solution of their sodium salt down a column containing an anion exchange resin.

6. HIGH PERFORMANCE COLUMN CHROMATOGRAPHY

In the classical procedure of column chromatography, the liquid phase passes through the stationary phase under the influence of gravity and the development of the chromatogram and subsequent separation of the components is usually a rather slow process. If we use the stationary phase packing material of uniform, smaller diameter and high surface area, and pump the liquid phase through the column at high pressure, the speed and versatility of column chromatography increases tremendously. In recent years this high performance or high pressure liquid chromatography (HPLC) is very efficient and yields excellent separation in a short time.

There are several advantages of HPLC over other methods of chromatography. In column chromatography the separation that would require several hours may be made in few minutes in HPLC. Efficiency is much higher than in thin layer chromatography. Gas chromatography is also a high performance method but important difference between the two is that GC can only cope with substances that are volatile or can be evaporated intact at elevated temperatures, whereas HPLC can be applied to high molecular weight compounds efficiently. Further, HPLC is normally carried out at room temperature, so that there is no problem of decomposition or molecular rearrangements.

For HPLC, a stainless steel tube of length 20 cm and internal diameter of 5 mm is used as a column. The column packing material is of uniform, smaller diameter and high surface area. The pressure is maintained by a suitable pumping system up to a value of 6000 p.s.i. The sample is injected in the column through a loop type of system which is connected to the main solvent flow tube. The liquid flow through the column may be from 1 μl/min to 10 μl/min according to the conditions required for the analysis. Finally, detectors are there which monitor and record the eluate continuously by means of differential refractometry and/or ultraviolet spectrometry. The former is required for those compounds which do not absorb in the U.V. Other detectors includes a fluorescence detector and a conductivity detector.

HPLC may be used both analytically and preparatively. For preparative use, a column of length up to 50 cm and internal diameter up to 20 mm is used. A quantitative analysis is indicated by the area under the peaks which are displayed by detector. Since the detectors are located at the exit from the column, the various components in the eluate may be identified with the peaks on the detector recording and collected separately.

Index